ALTERNATIVE
FUELS
and the
ENVIRONMENT

Edited by
Frances S. Sterrett

LEWIS PUBLISHERS
Boca Raton   Ann Arbor   London   Tokyo

Cover: SeaWest San Gorgonio Wind Park near Palm Springs, California. Photograph courtesy of Neil Kelley and James Tangler (NREL)

Library of Congress Cataloging-in-Publication Data

Alternative fuels and the environment / edited by Frances Sterrett.
    p. cm.
    Includes bibliographical references and index.
    ISBN 0-87371-978-6
    1. Energy development—Environmental aspects. I. Sterrett, Frances S.
TD195.E49A44   1994
621.042—dc20   94-25419
                                                                 CIP

© 1995 by CRC Press, Inc.
Lewis Publishers is an imprint of CRC Press

No claim to original U.S. Government works
International Standard Book Number 0-87371-978-6
Library of Congress Card Number 94-25419
Printed in the United States of America 1 2 3 4 5 6 7 8 9 0
Printed on acid-free paper

PREFACE

This book is based on the symposium "Alternative Fuels and the Environment" presented before the Division of Environmental Chemistry at the 205th National Meeting of the American Chemical Society in Denver, Colorado on March 29–30, 1993. The symposium consisted of 19 papers presented in three sessions. The chapters in this volume are predominantly presentations from this symposium. The organizer of this symposium was Frances S. Sterrett, Ph.D., Professor of Chemistry Emerita at Hofstra University and an active member of the American Chemical Society.

This volume will be of interest to a wide audience — those who anticipate an approaching energy crisis and recognize that fuels other than fossil fuels and nuclear fuels, namely renewable resources as fuels, will have to play an important role in the next decade. Some scientists believe that we will run out of fossil fuel reserves in 50 years. These alternative fuels will not only have the advantage of being renewable, but will be economically attractive as well and will have a minimal environmental impact (no emissions of waste materials such as SO_2, NO_x, or greenhouse gases). Since 1975, I have been heating the hot water in my house with solar energy and found this method very efficient, profitable, and environmentally safe.

This volume examines and evaluates resources such as the suns energy to be harnessed by photochemical reactions, artificial photosynthesis, and photovoltaic electric power generation. Solar energy can be used for destruction of hazardous chemicals, or to detoxify wastewater streams. Hydrogen as fuel can be produced by solar photoelectrolysis of water. Wind turbines may generate electricity, and oceanthermal energy could produce electric power and fresh water. Geothermal energy and hydroelectric power are renewable forms of energy that can generate electricity. Biofuels and biomass can be useful as energy sources. Surplus land can be used for woody, herbaceous, high-yield energy crop production. Reformulated gasoline with oxygenated fuels such as ethanol or methanol is also investigated.

Recently, it was possible to sustain nuclear fusion for about 4 seconds, the longest time yet, but it will take many years until fusion power can materialize; therefore, nuclear fusion power will not, in the near future, be able to replace fossil fuels or nuclear fission power as developed so far.

FRANCES S. STERRETT, Ph.D. (University of Vienna, Austria) is Professor Emerita of Chemistry at Hofstra University, where she has taught and conducted research for over 30 years. She has numerous publications on various topics including environmental issues and laboratory instruction manuals. She is presently teaching Environmental Chemistry and Environmenal Science. Dr. Sterrett is a Fellow of the New York Academy of Sciences, and has chaired the sections of Public Policy and Environmental Sciences. She is an active member of the American Chemical Society. At the national level, she is a member of the Committee of Environmental Improvement and the Divisions of Environmental Chemistry, Chemical Education, Agricultural and Food Chemistry, and Professional Relations. At the New York Section, she has been a member of the Board of Directors since 1977 and currently serves as a councilor. At the Long Island Subsection, she has been a member of the Board of Directors since 1970. She has chaired the Subsection and several committees. Presently, Dr. Sterrett is chair of the Environmental Chemistry Committee.

ACKNOWLEDGMENTS

I wish to express my appreciation to the following people:

Thomas D. Bath, Ph.D., Director, Analytic Studies Division, National Renewable Energy Laboratory, 1617 Cole Boulevard, Golden, Colorado, who was very helpful in suggesting many speakers for this symposium and in agreeing to be a session chair as well as a reviewer for many of the manuscripts. Without his help the symposium would not have been as successful as it was. I am extremely grateful to him.

John S. Connolly, Ph.D., Principal Scientist, Basic Sciences Division, National Renewable Energy Laboratory, 1617 Cole Boulevard, Golden, Colorado, who also agreed to be a session chair and reviewer for many manuscripts. I am grateful to him for his help.

De Anna C. Schaeffer, Senior Administrative Assistant, National Renewable Energy Laboratory, 1617 Cole Boulevard, Golden, Colorado, who so capably helped gather many of the manuscripts.

Kathy Walters, Associate Editor of Lewis Publishers for her valuable advice.

The following reviewers are thanked for their valuable and useful suggestions:

Larry G. Anderson, Ph.D., Associate Professor, Department of Chemistry, University of Colorado at Denver, Denver, Colorado.

Brandon R. Brygider, Assistant Professor of Environmental Technology, New York Institute of Technology, Old Westbury, New York.

In addition, I'd like to extend my gratitude to the following scientists who are affiliated with the National Renewable Energy Laboratory, Golden, Colorado:

Desikan Bharathan	Senior Engineer
Lynn Coles	Principal Policy Advisor
Brian A. Gregg	Senior Scientist
Carol J. Hammel	Staff Environmental Analyst
Susan M. Hock	Manager, Wind Energy Program
Roland L. Hulstrom	Manager, PV Engineering & Applications Branch
Gary L. Nakarado	Technical Director, Utility Programs
James M. Ohi	Senior Environmental Analyst
Brian K. Parsons	Senior Engineering Analyst
David S. Renne	Program Leader/Senior Scientist
Walter W. Short	Manager, Market Analysis Branch

Last, but not least, my sincere thanks to all the authors of this volume, who so generously gave of their time and knowledge, and without whom this book could not have materialized.

LIST OF CONTRIBUTORS

Larry G. Anderson, Department of Chemistry, Box 194, Center for Environmental Sciences, Box 136, University of Colorado at Denver, P.O. Box 173364, Denver, CO 80217-3364

Regina A. Barrell, Center for Environmental Sciences, Box 136, University of Colorado at Denver, P.O. Box 173364, Denver, CO 80217-3364

Thomas D. Bath, Analytic Studies Division, National Renewable Energy Laboratory, 1617 Cole Boulevard, Golden, CO 80401

John Benner, National Renewable Energy Laboratory, 1617 Cole Boulevard, Golden, CO 80401

Daniel M. Blake, National Renewable Energy Laboratory, 1617 Cole Boulevard, Golden, CO 80401

James R. Bolton, Solarchem Environmental Systems, 130 Royal Crest Court, Markham, Ontario L3R 0A1, CANADA — on leave from the Photochemistry Unit, Department of Chemistry, The University of Western Ontario, London, Ontario N6A 5B7 CANADA

Glenn F. Čada, Environmental Sciences Division, Oak Ridge National Laboratory, P.O. Box 2008, Oak Ridge, TN 37831-6036

Stephen R. Cater, Solarchem Environmental Systems, 130 Royal Crest Court, Markham, Ontario L3R 0A1, CANADA

John S. Connolly, Basic Sciences Division, National Renewable Energy Laboratory, 1617 Cole Boulevard, Golden, CO 80401

J. H. Cushman, Biofuels Feedstock Development Program, Environmental Sciences Division, Oak Ridge National Laboratory, Oak Ridge, TN 37831-6352

D. L. Elliott, Atmospheric Sciences Department, Pacific Northwest Laboratory, Battelle Boulevard, Richland, WA 99352

James E. Francfort, Idaho National Engineering Laboratory, P.O. Box 1625-3875, Idaho Falls, ID 83415-3875

Jeffrey S. Gaffney, Environmental Research Division, Argonne National Laboratory, Argonne, IL 60439

David Ginley, National Renewable Energy Laboratory, 1617 Cole Boulevard, Golden, CO 80401

Devens Gust, Center for the Study of Early Events in Photosynthesis, Department of Chemistry and Biochemistry, Arizona State University, Tempe, AZ 85287-1604

John A. Lanning, Department of Chemistry, Box 194, Center for Environmental Sciences, Box 136, University of Colorado at Denver, P.O. Box 173364, Denver, CO 80217-3364

S. A. Martin, Biofuels Feedstock Development Program, Environmental Sciences Division, Oak Ridge National Laboratory, Oak Ridge, TN 37831-6352

Robert McLaren, Institute for Environmental Chemistry, National Research Council Canada, Ottawa, Ontario K1A 0R6 CANADA

Ana L. Moore, Center for the Study of Early Events in Photosynthesis, Department of Chemistry and Biochemistry, Arizona State University, Tempe, AZ 85287-1604

Thomas A. Moore, Center for the Study of Early Events in Photosynthesis, Department of Chemistry and Biochemistry, Arizona State University, Tempe, AZ 85287-1604

R. Gerald Nix, National Renewable Energy Laboratory, 1617 Cole Boulevard, Golden, CO 80401

A. J. Nozik, National Renewable Energy Laboratory, Basic Sciences Division, 1617 Cole Boulevard, Golden, CO 80401

George Paraskevopoulos, Institute for Environmental Chemistry, National Research Council Canada, Ottawa, Ontario K1A 0R6 CANADA

Carl J. Popp, Department of Chemistry, New Mexico Institute of Mining and Technology, Socorro, NM 87801

Marshall J. Reed, Geothermal Division, U.S. Department of Energy, Washington, DC 20585

David S. Renné, National Renewable Energy Laboratory, 1617 Cole Boulevard, Golden, CO 80401

Joel L. Renner, Idaho National Engineering Laboratory, Idaho Falls, ID 83415-3526

Cynthia J. Riley, National Renewable Energy Laboratory, 1617 Cole Boulevard, Golden, CO 80401

Ali Safarzadeh-Amiri, Solarchem Environmental Systems, 130 Royal Crest Court, Markham, Ontario L3R 0A1, CANADA

M. N. Schwartz, Atmospheric Sciences Department, Pacific Northwest Laboratory, Battelle Boulevard, Richland, WA 99352

Donald L. Singleton, Institute for Environmental Chemistry, National Research Council Canada, Ottawa, Ontario K1A 0R6 CANADA

Andrew R. Trenka, Pacific International Center for High Technology Research (PICHTR), 2800 Woodlawn Drive, Suite 180, Honolulu, HI 96822

K. Shaine Tyson, National Renewable Energy Laboratory, 1617 Cole Boulevard, Golden, CO 80401

Pamela Wolfe, Center for Environmental Sciences, Box 136, University of Colorado at Denver, P.O. Box 173364, Denver, CO 80217-3364

L. L. Wright, Biofuels Feedstock Development Program, Environmental Sciences Division, Oak Ridge National Laboratory, Oak Ridge, TN 37831-6352

Lin Zhang, Department of Chemistry, New Mexico Institute of Mining and Technology, Socorro, NM 87801

Ken Zweibel, National Renewable Energy Laboratory, 1617 Cole Boulevard, Golden, CO 80401

CONTENTS

Chapter 1

Ocean Thermal Energy Conversion (OTEC): A Status Report on the Challenges

Andrew R. Trenka

CONTENTS

Since the concept of ocean thermal energy conversion was first conceived by D'Arsenoval in 1889 and implemented by Claude in 1930, OTEC (ocean thermal energy conversion) technology has run a full gamut of interest. During the early 1970s and 1980s, the U.S. Department of Energy expressed interest in OTEC and some $240,000,000 were expended on developing the technology. Today, interest in OTEC continues among steadfast pockets of believers despite the efforts of recent years to slash OTEC research budgets to zero.

The technology has significantly advanced, problems have been resolved, interest is stirring, and future implementation of the technology looks bright. OTEC technologies for heat exchangers, evaporators, and turbines have advanced significantly and status system designs have been refined. Technological advances have been made in developing direct contact condensers and aluminum surface condensers. Biofouling and corrosion are controllable. Research has culminated in the construction and installation of an open-cycle OTEC Experimental Apparatus on the Big Island of Hawaii.

Interest in OTEC has been rekindled worldwide and the future for the implementation of the technology looks bright once again. By-products associated with OTEC such as fresh water, utilization of cooling, and air conditioning, mariculture and agriculture, make OTEC an attractive technology justifiably described as a *life support system*. OTEC has the potential to be a major contributor, a giant among the renewable energy sources as soon as the reduction of dependence on fossil fuels is seriously undertaken.

HISTORICAL OVERVIEW

Hindsight of OTEC's history is necessary, simply because if one does not understand history, as one famous man said, "we are bound to repeat it." One certainly does not wish to do that with OTEC. After D'Arsenoval and Claude, over 20 years would lapse without any notable activity in the field. And then in 1956, the French revived interest in OTEC

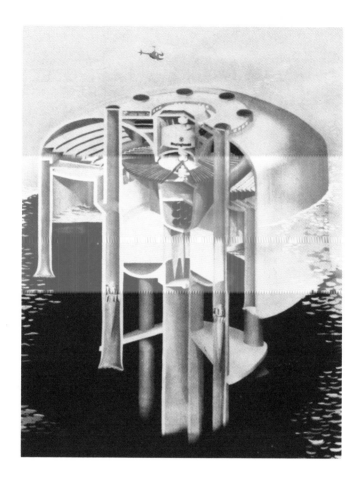

Figure 1 100-MWe OTEC (Westinghouse) floating plant. (Courtesy of U.S. Department of Energy, reprinted from *100 MWe OTEC Alternate Power Systems*, Final Report, March 5, 1979 under DOE contract no. EG-77-C-05-1473 with Westinghouse Electric Corporation.)

by designing a 3-MWe plant for application off the West Coast of Africa. However, the plant was never built.

In 1972, the U.S. Government turned its attention to renewable energy sources to thwart the growing U.S. dependency on foreign oil. One of the alternatives was ocean energy. In 1972, the National Science Foundation initiated an Ocean Energy Program. In 1973, the Ocean Energy Program was bolstered by the oil embargo. The program eventually passed to the U.S. Department of Energy (DOE) where it currently is managed. The focus of the early program was on large closed-cycle OTEC (CC-OTEC) systems.

The early program spanned a decade from about 1971 until about 1981. At that time, designs of large (50 to 100 MWe) floating plants (Figures 1 and 2) were developed, culminating in the OTEC-1 (Figure 3) project in 1981. Running parallel with the federal interest was international interest, industrial interest, and specific U.S. states' interest. The French and Japanese had significant programs. The State of Hawaii in conjunction with a consortium of industry, Lockheed Aircraft Company, and several Hawaii-based

3

Figure 2 40-MWe shore-based OTEC (Kahe Point, Hawaii) plant. (Drawing courtesy of Parsons Hawaii and the State of Hawaii Department of Business and Economic Development.)

Figure 3 OTEC–1. (Photograph courtesy of National Renewable Energy Laboratory.)

Figure 4   Mini-OTEC. (Photograph courtesy of Hawaiian Dredging & Construction Co.)

companies developed the mini-OTEC project (Figure 4). In 1979, the small plant mounted on a barge off Hawaii was the first closed-cycle plant to produce net power.

The passage of office to a new president in 1981 shifted the fundamental approach of the DOE from demonstration programs to high-risk, high-return investigations. In addition, the focus of the DOE program shifted from closed-cycle to open-cycle since it was believed the closed-cycle concept was ready for commercialization except for a few *minor* problems. These *minor problems* were associated with the deployment and development of cost-effective seawater supply systems, system scale-up engineering, and problems of heat exchanger biofouling and corrosion.

The federal program shifted to the development of open-cycle, near-shore and onshore plants in the 2- to 15-MWe size range because studies indicated these systems to be the first market penetration scenarios. A carefully structured program of code and subsystem development, followed by freshwater and then seawater testing, and verification was to lead to a small proof of concept experiment, the OC-OTEC Net Power Producing Experiment (NPPE).

In 1986, the DOE established an OTEC system design team with support from Argonne National Laboratory, the Solar Energy Research Institute (now known as the National Renewable Energy Laboratory [NREL]), and PICHTR. From 1990, PICHTR had sole technical development responsibility. After successfully demonstrating a CC-OTEC system on the island of Nauru, the government of Japan funded a parallel program at PICHTR that aimed at the development of small OTEC plants for applications in the Pacific Islands.

There continued to be some industrial and State of Hawaii interest in OTEC by-products, namely the deep seawater applications for mariculture and aquaculture. The Natural Energy Laboratory of Hawaii (NELH) facilities at Kona on the Big Island of Hawaii represent the world's leading industrial applications of the deep cold water for the production of a variety of mariculture and aquaculture products. It is also the location of the OC-OTEC NPPE.

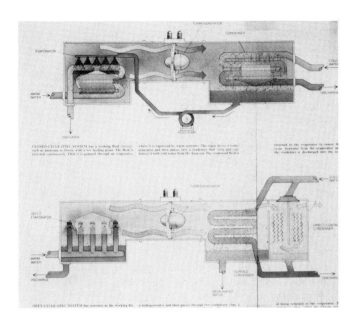

Figure 5 Closed-cycle/open-cycle OTEC system. (Reprinted by permission from T. R. Penney and Desikan Bharathan, "Power from the Sea," *Sci. Am.*, January 1987, p. 88–89.)

OTEC TECHNOLOGY

The resource is the warm and cold water of the ocean. At the surface, throughout the equatorial belt, at latitudes from 0° to 20°, warm water (25 °C) flows year round. About 1000 m beneath the ocean's surface, a river of cold seawater (about 4 °C) flows from the Antarctic and Arctic ice caps. Thus, the available temperature differential is about 20 °C, which is the minimum required for OTEC to work. At these values, the thermal efficiency of OTEC is only about 3 to 4%. This low thermodynamic efficiency requires the pumping of a significant amount of water in order to extract a useful amount of energy. That is one of the drawbacks of the OTEC technology. These values are small (approximately 1/10) compared to efficiencies obtained for conventional power plants; however, OTEC uses a free and renewable resource.

For OTEC to be feasible, it must be a system with very accurate prediction codes that is highly efficient with access to enormous amounts of the resource. A 1-MWe plant requires some 128,000 gallons per minute (gpm) warm and cold water flow. To obtain the 4- to 6- °C cold water needed, the cold water pipe must descend to 800 to 1000 m. The cold ocean water is nutrient rich with an abundance of phosphates and carbonates and is pathogen free. It is a key to OTEC's attractiveness and justifies its descriptor as a *life support system*.

The major OTEC cycles are the closed, open, and hybrid cycles (Figure 5). There are also other OTEC systems such as mist-lift and direct thermal-electric that have been proposed but have not yet been extensively investigated.

CLOSED-CYCLE OTEC

Closed-cycle refers to the fact that the working fluid is in a closed loop that does not come into direct contact with the warm or cold seawater. The working fluids of choice are ammonia and Freon; others have been investigated. Liquid ammonia and warm seawater are brought into a surface heat exchanger where the heat from the warm seawater evaporates the ammonia. The ammonia vapor passes through a turbine generator, and then to a second surface heat exchanger where cold seawater recondenses the vapor to a fluid and the cycle is repeated.

The closed-cycle OTEC power cycle is state-of-the-art. Conventional Rankine cycle systems are available, literally off the shelf, up to 100 MWe. Operating pressures in the range of 90 to 145 psi pose no problem. The heat exchangers were once a major problem area because large volumes of heat exchangers are required. A 100-MWe plant requires approximately 5,600,000 ft^2 heat exchangers.

Biofouling was a major concern and corrosion problems required the use of expensive materials such as titanium, copper-nickel cladding, and stainless steels. Titanium in the early 1980s was about U.S.$7/lb. Thus, concerns about corrosion, biofouling, and large volumes of expensive heat exchangers were perceived to be serious limiting factors affecting the development of CC-OTEC. As will be discussed later, these problems have largely been overcome.

The seawater supply system for the large floating plants being developed consisted of a cold water pipe of 40 ft (12 m) diameter for a 100-MWe plant. The concerns were pipe dynamics, platform control, and platform stability; these problems were also resolved.

Two demonstration projects, the Mini-OTEC in 1979 and OTEC 1 in 1980, clearly showed the technical viability of closed-cycle OTEC. Assembled on a floating platform off the Oahu coast of Hawaii, the Mini-OTEC succeeded in generating 50 kW electrical power. Approximately 35 kW of this power was consumed by seawater pumps and auxiliary equipment, resulting in a net production of 15 kW. The Mini-OTEC success-fully demonstrated pipe deployment technology and production of net power. The project was conducted by a consortium of the state of Hawaii, Lockheed, Dillingham, Makai Ocean Engineering, and other Hawaii-based firms.

In parallel, the DOE program that focused on 100-MWe plants, had developed OTEC-1 for demonstrating large heat exchangers, testing and deploying large pipes, and address-ing the issues of station keeping and grazing. It had a thermal power of 1 MW and was to have a turbine installed to produce net power. Power was not produced because of the change in the U.S. government administration, which refocused the OTEC program on open-cycle plants.

Economic analyses indicated the early market for OTEC to be island and near-shore communities requiring 15 MWe or less. Cost analyses showed that closed-cycle OTEC would be cost effective for only very large-sized (greater than 40 MWe) plants. The simpler and cheaper heat exchangers of the open-cycle system, coupled with the fact that open-cycle systems produced larger volumes of fresh water with less power penalty, helped to shift the emphasis to the development of OC-OTEC.

The successful demonstration by the Japanese of a CC-OTEC system on the island of Nauru in about the same time frame bolstered the DOE decision to declare CC-OTEC ready for commercialization and to focus on OC-OTEC. In recognition of the unresolved problems of biofouling and corrosion in heat exchangers, research continued in these areas.

A number of objectives could only be met by further testing. Two freshwater labora-tories focused on OTEC research, one at Argonne National Laboratory in (Chicago, Illinois) and the other at Solar Energy Research Institute (now NREL in Golden, Colo-rado).

Experimental work was conducted at the Seacoast Test Facility (STF) in Hawaii to obtain needed seawater data. A biofouling and corrosion laboratory was developed and a heat mass transfer scoping test apparatus (HMTSTA) was conceived (Figure 6). Experimental work was conducted using the available cold and warm seawater supplies at NELH.

Figure 6 Heat and mass transfer scoping test apparatus. (Photograph courtesy of National Renewable Energy Laboratory.)

The STF has access to two seawater supply systems in the NELH. The first system installed included a 1600-gpm warm water pipe and a 1000-gpm cold water pipe which fed the biofouling, corrosion, and HMTSTA Facilities. This system also provided cold water for mariculture/agriculture experiments ongoing at NELH. An increase in the demand for water (particularly cold water) led to the installation of a second seawater supply system. A joint effort by the state of Hawaii and the federal government installed a 1-m (40 in.) diameter polyethylene cold water pipe capable of pumping 13,500 gpm cold seawater, the largest cold water pipe in the world. Approximately 6000 ft polyethylene pipe was deployed off the coast of Kona and represents the *state-of-the-art* cold in seawater supply systems today. Also installed as part of the second system was a 28-in. diameter warm water pipe capable of supplying some 9600 gpm. These facilities supported the further development of both open- and closed-cycle OTEC.

Thus, for CC-OTEC, the technology and design can be summarized as follows:

1. There exist a number of OTEC plant designs: e.g., 40-MWe Kahe Point (Hawaii) shore-based plant; 3-MWe (French) plant.
2. Net power has been demonstrated for CC-OTEC at the 50-kW (gross) level on at least two separate systems.
3. Hanging pipe fabrication, deployment, and control has been demonstrated for sizes up to 8 ft [Vega, L. A., and G. C. Nihous, At-Sea Test of the Structural Repsonse of a Large-Diameter Pipe Attached to a Surface Vessel, *Proc. Offshore Technology Conf.*, Houston, Texas, May 1988].
4. Aluminum heat exchangers (AL 5052 and 3006) can be used for OTEC applications. At corrosion rates of 0.0067 μm/d, a 0.035-in. thick aluminum tube would take 36 years to lose 10% of its thickness (see Figures 7 and 8).
5. Biofouling is controllable using chlorination at 70 ppb for 1h/d (Figure 9).

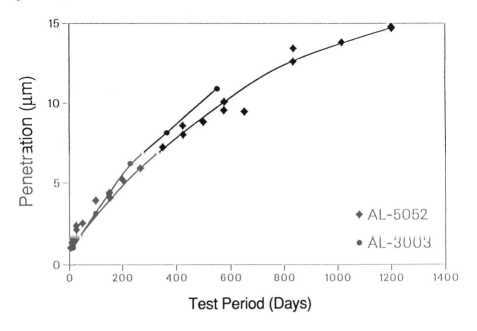

Figure 7 Aluminum corrosion rate in cold water. (Argonne National Laboratory, *OTEC Biofouling and Corrosion Study*, 1983–1987 under U.S. DOE contract no. W-31-109-Eng-38.)

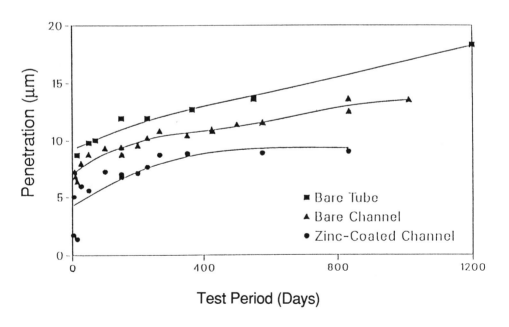

Figure 8 Warm-water aluminum-3003 corrosion rate compatible with long component life. (Figures courtesy of Argonne National Laboratory, *OTEC Biofouling and Corrosion Study,* 1983–1987 under U.S. DOE contract no. W-31-109-Eng-38.)

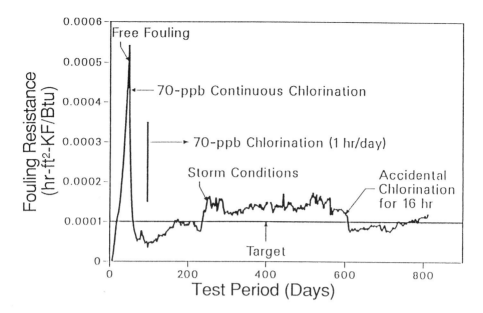

Figure 9 Biofouling management is similar for spirally fluted and smooth tubes. (Argonne National Laboratory, *OTEC Biofouling and Corrosion Study*, 1983–1987 under U.S. DOE contract no. W-31-109-Eng-38.)

6. Thermal and hydraulic performances of CC-OTEC subsystems have been demonstrated at the 1-MWt level; (Argonne National Laboratory, test program of state-of-the-art industrial heat exchangers, 1983).

7. Closed-cycle ammonia power subsystems are established technology.

OPEN CYCLE OTEC

In the open-cycle system, the working fluid is the seawater itself. Warm seawater is drawn into a vacuum vessel at approximately 0.2 psi. The seawater is flashed into steam and the steam runs through a turbine. The steam, after passing through the turbine generator, enters a condenser where cold seawater is utilized to recondense the steam into water. If a direct contact condenser is used, the steam is brought into direct contact with the cold seawater; it is the most efficient method for condensation. If the steam is passed over a surface condenser, the resulting product is fresh water.

In the early days of the U.S. program, as for the CC-OTEC systems, the open-cycle system studies focused on very large 100-MWe-size floating plants. The program focus shifted to near-shore and onshore plants in the 2- to 15-MWe-size range and the necessary technology base for the subsystems for these plants was developed.

Deep seawater systems were investigated in 1988 at a workshop held to address the *state-of-the-art* in terms of deployment, materials, and cost. It was found that deploying polyethylene pipes of up to 1.6 m was feasible, which is sufficient to sustain a 2- to 3-MWe plant. Deployment strategy and deployment bending stress were key problems to be dealt with. The cost goal of $2650/kW in 5 years is achievable only for very large pipes using the current state-of-the-art. A clear message emerged from this workshop: much more work is needed for the seawater supply systems feeding shore-based plants.

Research conducted on evaporator technologies included a number of different types of evaporators for the open-cycle technology. Open-channel flow, falling films, falling

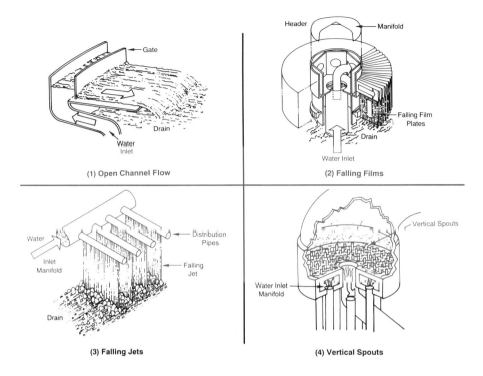

Figure 10 Four types of evaporators tested. (Drawings courtesy of U.S. Department of Energy.)

jets, and vertical spout evaporators were tested (Figure 10). Results led to a focus on the spout evaporator. The development of spout evaporators was successful and high efficiencies were obtained. For liquid loading between 20 and 50 units, one can obtain thermal effectiveness in excess of 90% (Figure 11).

Surface and direct contact condensers were also investigated. Surface condensers cost more than direct contact condensers for the same efficiency but they can produce low-cost freshwater. The cost differences between surface and direct contact condensers is narrowing since the introduction of new fabrication technologies. Investigations led to structured packing and a geometric configuration that was concentric and involved co-current and counter-current stages. In the co-current stage, water flows over the surface in a downward direction and the steam also passes in a downward direction, coming in direct contact with the cold seawater. Approximately 90% of the steam is condensed at this stage. The remaining steam then passes to a counter-current stage where the cold water continues to flow downward and the steam flows in the opposite direction; approximately 8% of the remaining steam is condensed. The remaining steam and noncondensable gases are exhausted by the vacuum system. Thermal effectiveness in excess of 95% has been routinely achieved (Figure 12).

In 1986, the state-of-the-art for turbines that would be used in OTEC plants was assessed. The closed-cycle plant, Rankine cycle utilizing either ammonia or Freon, is well established and is state-of-the-art technology. Existing steam turbine technology can support the open-cycle OTEC turbine requirements for plants up to about 5 MWe, the limiting factor being the construction of blades longer than 51 in. Beyond that, new blade designs, such as those used in helicopter configurations, are needed. Also evaluated were

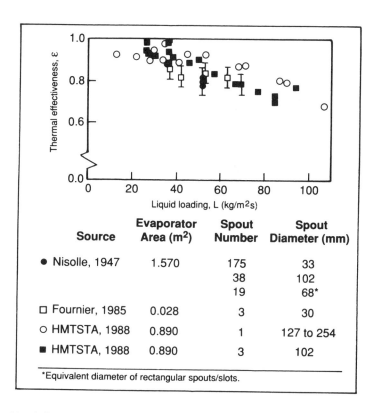

Figure 11 Influence of liquid loading on thermal effectiveness of vertical-spout evaporator. (Figures courtesy of National Renewable Energy Laboratory, *Results of Scoping Tests for Open-Cycle OTEC Components Operating with Seawater*, SERI/TR-253-3561, Golden, CO: Solar Energy Research Institute, Sept. 1990.)

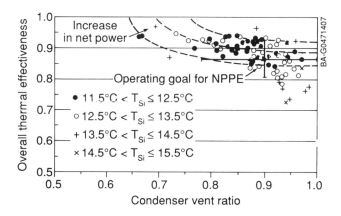

Figure 12 Thermal and venting performance of direct-contact condenser with structured packings. (Figures courtesy of National Renewable Energy Laboratory, *Results of Scoping Tests for Open-Cycle OTEC Components Operating with Seawater*, SERI/TR-253-3561, Golden, CO: Solar Energy Research Institute, Sept. 1990.)

Figure 13 210 kW (gross) OC-OTEC experimental apparatus. (PICHTR photograph.)

axial flow machines, radial inflow machines, radial inflow machines, and mixed flow machines.

Significant advances had been made in the OTEC technology, so that by the late 1980s all necessary validated codes for prediction of the various subsystems performance had been developed. System experiments to incorporate all the data were formulated and culminated in the 210-kW OC-OTEC Experimental Apparatus installed in the NELH in 1992 (Figure 13).

This experimental facility was designed to demonstrate system net power, subsystem performance, prototypical operation, reliability, availability, and production of fresh water. The nominal design conditions for the system are warm water temperatures at 26 °C with a flow rate of approximately 9600 gpm. At the Seacoast Test Facility, the cold water temperature is approximately 6.6 °C at the design point and flows at a rate of 6500 gpm. The overall mixed discharge flow rates will be 13,300 gpm. Operation of the NPPE will be over the full tidal variation of +2 ft to 1.2 ft, with overall power generation projected at 210 kW. Parasitic power objectives are: vacuum system and exhaust compressor, no more than 40 kW; warm water pumping power, no more than 58 kW; and cold water pumping power, no more than 54 kW. The essential auxiliaries are limited to approximately 12 kW, for a total of 148 kW parasitic power, leaving a net of 46 kW. The design is not optimal and pays significant additional penalty required to pump the full 13,000 gal cold water, of which only 6500 gpm is used in the apparatus.

The power block is built around a single, vertical-axis, mixed-flow turbine supported by a concrete vacuum vessel, 7.6 m in diameter and 9.5 m high. The electrical generator is located above the turbine assembly and outside the vacuum enclosure. Steam is produced in an annular flash evaporator at the periphery of the vacuum vessel. The steam flows up from the evaporator and enters the turbine radially inward. The steam exits the turbine axially in the center of the vessel. A conical exhaust diffuser is used for pressure recovery. The condenser is a direct-contact, structured-packing condenser composed of two coaxial stages. The noncondensable gases liberated from the seawater streams, at pressures of 1 to 3% atmospheric, and a small amount of uncondensed steam are compressed and exhausted using a multiple-stage vacuum compression system. All subsystems have been instrumented to measure the parameters required to assess performance.

The design and construction of this facility has been completed.

For OC-OTEC, a summary of the technology and the significant advances are:

1. Theoretical models have been developed and experimentally verified for the evaporator, condenser, and gas desorption systems.
2. Freshwater experimental tests have been conducted.
3. Continued seawater experiments are underway.
4. Good correlation between experiments both in seawater and freshwater and analytical codes have been obtained.
5. Significant technological advances have been made in the identification of inexpensive alloys and materials for utilization in OTEC heat exchangers (both OC- and CC-OTEC applications; see CC-OTEC section).
6. Highly efficient flash evaporators have been developed.
7. Significant advances developing extremely effective direct contact heat exchangers have been completed.
8. A 40-in. polyethylene pipe has been deployed, thereby establishing deployment techniques.

The design of the 210-kW (gross) OC-OTEC Experimental Apparatus currently operating in Hawaii allows for a slipstream of steam (10%) to be diverted to an aluminum surface condenser for the production of fresh water. The freshwater production subsystem is now in the final construction phase and is expected to be *online* by February 1994, producing 7000 gal a day.

Shakedown testing of the OC-OTEC demonstration plant began in December 1992. These tests were programmed to test the functionality of the subsystems in preparation for the acceptance of hardware components. The shakedown testing phase was extended through April 1993 due to problems with the vacuum compression subsystem. In addition, an instability in the turbine generator subsystem was encountered during synchronous operation with the utility power grid. All other subsystems performed according to design conditions.

Performance tests on the OC-OTEC demonstration plant were initiated in April. Within 1 month, a world's record for OTEC had been achieved. The plant, when operating off–grid, had not only met the design conditions but was producing 213 kWe when the surface water was 25.9 °C and the deep water temperature was 5.9 °C. Altogether, the plant consumes approximately 150 kW, yielding a net power of 60 kW. Higher outputs were expected in the summer and fall of 1993; continued subsystem testing of the NPPE through 1993 was planned. Assuming synchronization problems will be solved, long-term grid connected power production tests are planned through 1995. These experiments will provide the basic data on operation of an OC-OTEC plant that produces electrical energy as well as freshwater.

THE BY-PRODUCTS:
FRESHWATER, AIR CONDITIONING, AND NUTRIENTS

FRESHWATER

The condensate of the open cycle and hybrid OTEC systems is desalinated water, suitable for human consumption and agricultural purposes. This water is actually more pure (less saline) than the water provided by the municipal water system in Honolulu. The market value of desalinated water in the Pacific Islands ranges from $1 to $4.60 per kilogallon and may be even higher in locations with no groundwater resources. A 1-MWe plant can readily produce over 1 million gallons of fresh water per day. Calculations indicate that OTEC freshwater production costs are competitive with reverse osmosis.

AIR CONDITIONING

The cold water discharge temperature from an OTEC plant can be between 10 and 16 °C, depending upon electrical energy and freshwater production. Typical temperatures in the chillers of air conditioners are between 7 and 14 °C. Thus, even after electrical energy production and freshwater production, a sufficient temperature differential remains to provide air conditioning potential. If a slipstream (less than 3%) of cold seawater (at 4.5 °C) for a 1-MWe plant was diverted for air conditioning, some 300 hotel rooms could be cooled at less than 25% of the cost of a conventional system.

NUTRIENTS

Early OTEC experiments drew on the deep ocean water as a key element of the energy system. However, enterprising scientists quickly found other applications for this water, which is relatively pathogen-free and has high concentrations of nitrates, phosphates, and silicates (i.e., nutrients that foster the growth of plants and algae). The site for the initial experiments and eventual commercial development has been NELH, located on the western shoreline of the Big Island of Hawaii.

Experiments on the land-based cultivation of marine organisms at NELH have blossomed into major businesses and fledgling new enterprises, many still in the research and development stage. These applications for the cold seawater can play a key role in improving the economic appeal of the total OTEC system. Commercial enterprises are currently growing salmon, trout, opihi, oysters, lobsters, sea urchins, abalone, kelp, nori, and other macro and micro algae. While the ultimate commercial viability of these product lines is yet to be proven, their promise is significant.

OTEC ECONOMICS AND THE MARKET

Significant technological advances have been achieved, yet the question remains — "Is there a market for OTEC?"

Yes, there is a market. This budding market focuses on the Pacific and Caribbean Islands. The estimated market size by the year 2015 is on the order of 5000 MWe, corresponding to an economic industry of some U.S.$18.5 billion.

In 1987, PICHTR investigated the market for OTEC by surveying 26 Southeast-Asia and Pacific Island sites to determine their potential for this renewable energy resource. Total additional demand for power projected for these islands over the base year 1987 for the year 2000 was 2953 MWe and in the year 2015, a cumulative 24,898 MWe. In 2015, this represents a demand for about 250 plants of a 10-MWe size. Projections beyond 2015

Table 1 **OTEC market penetration scenarios**

Nominal Net Power (MWe)	Type	Scenario Requirement	Scenario Availability
1	Land-based OC-OTEC with second stage additional water production	$45/barrel of diesel $1.6/m³ water	South Pacific Island Nations by year 1995
10	Land-based (as above)	$25/barrel of fuel oil $0.85/m³ water –or– $30/barrel with $0.8/m³ water	American Island Territories and other Pacific Islands by year 2000
40	Land-based hybrid (ammonia power cycle with flash evaporator downstream)	$44/barrel of fuel oil $0.4/m³ water –or– $22/barrel $0.8/m³ water	Hawaii, if fuel or water costs double by year 2000
40	Closed-cycle land-based Closed-cycle plantship	$36/barrel $23/barrel	By year 2005

to 2050 assume a tripling of the 2015 demand or at least 75,000 MWe; this would be the equivalent of about 750 plants of 10-MWe size.

The market for OTEC in the Pacific Island countries depends on many factors. One thing was certain: it was the freshwater component and mariculture potential of the OC-OTEC that prompted several Pacific Island governments to requisition PICHTR to conduct studies of the feasibility of this technology in their islands. While the Pacific Islands are presently dependent on imported fuel for generating electricity, fluctuations in the price of oil do not seem a major determinant in their decision to request these studies. Even the availability of an inexhaustible, renewable clean source of energy does not seem to be as attractive as the potential benefits of freshwater and mariculture for their economic development.

While the market for OTEC may be influenced by the availability and cost of its alternatives, it is primarily the costs of switching from diesel-fired power plants to another technology (such as gas turbines with all the accompanying difficulties in the technology) that have supported the Pacific Islands' interest in OTEC as an alternative. The Asian and Pacific Island nations seem to be more influenced by the potentially hazardous impact of current energy resources on the environment. OTEC promises a source of clean energy with little impact on the environment.

In assessing the economic potential of these markets, it is useful to consider certain global indicators when discussing the needs of a given community. Domestic water needs in developed countries are met with about 50 gal per person. In agricultural regions, the use is seven to ten times greater. The electrical power needs for domestic and industrial utilization gives a rule of thumb that for each 1000 to 2000 people, 1 MWe power is needed (approximately 1 kW per person). In less developed countries, five to ten times more people's needs can be met with the same power requirement.

Is OTEC cost effective today? Yes, in limited market areas outlined and under economic conditions noted in Table 1.

At the capital costs ranging from 6000 $/kW to 20,000 $/kW depending on the size, type, and location of the plant, OTEC requires significant, if not prohibitive, up-front investment to realize its long-term potential.

OTEC AND THE ENVIRONMENT

Although OTEC is looked upon as being an environmentally benign renewable energy, there are environmental impacts, believed to be small, that should be considered. Particular attention should be given to construction phases when large displacements of the surrounding environment will occur. Also, the effect of the discharge of large volumes of water, both cold and warm, is not totally known. The behavior of a mixed discharge plume at depth is being studied, but its actual effect has not been measured. There is consideration of visual impacts and the potential impact on shipping and recreational activities. However, OTEC does have significant beneficial aspects.

Operating a 40-MWe OTEC plant instead of a conventional 40-MWe fossil fuel plant can save some 510,000 barrels of oil per year and can eliminate 146,000,000 lb CO_2, 116,000 lb NO_x, and over 3,720,000 lb sulfur per year. These significant savings in emissions, when considered as part of the cost equation, go a long way toward making OTEC a more attractive and cost competitive alternative.

CONCLUSIONS

Great strides have been made, yet technological issues still remain. Proof of commercial scale OTEC performance remains to be accomplished. Long-term data on reliability, operation, and maintenance of OTEC is needed. Scale-up uncertainties exist; improved seawater supply systems require work, development of manufacturing technology for heat exchangers must be undertaken, and further work on power systems, particularly for OC-OTEC, must be done.

There are economic issues: high capital costs continue to plague OTEC; proof of economic viability remains to be shown. As long as fossil fuels are cheap, OTEC or any renewable energy can only marginally compete in the energy market. Unless a thoughtful sustainable energy strategy is developed, no renewable will become a major energy contributor.

Any sustainable energy strategy must be concerned with not only short-term low cost of energy, but also long-term economic, social, energy security, and environmental impacts. It must consider the integration and utilization of renewable technologies into the existing energy mix; the demand as well as supply side economics and planning; and, conservation.

We need dedication, innovation, and aggressiveness. We must be dedicated to developing, improving, and integrating these concepts into the existing energy structure. We need to recognize economic issues, but not let them become the sole determining factor. Quality of life now and for our children must also be considered. National energy security by utilization of indigenous energy resources is worth something. Reductions in CO_2, NO_x, and SO_x result in direct economic benefits. We must quantify them.

We must be innovative; innovative in our application of an environmentally sound and sustained energy policy. Change does not occur overnight.

There is inertia. Inertia of perception, "We have never done it that way before." There is inertia of technology. It takes on the order of 50 years for a technology to fully develop. There is inertia of infrastructure. One cannot utilize a clean technology like hydrogen because the infrastructure for supply, storage, and distribution does not exist. There is inertia of economics; oil, coal, and gas are abundant and cheap today.

We must be aggressive in our pursuit of this environmentally sound, sustainable energy policy. This aggressiveness must transcend national boundaries. CO_2 emission is a worldwide problem, requiring a worldwide solution. Energy companies delivering the resource as well as those using the resource to produce and distribute power must aggressively adopt a long-range perspective rather than a myopic view of bottom-line economics.

Governments must aggressively pursue policies that encourage reduction in fossil fuel consumption and greater utilization of renewable alternatives. They must promote the implementation of supply and demand side management as well as conservation measures. Tax incentives for the use of non-fossil fuels must be implemented. Consideration should be given to a carbon tax, an allowance for utilities to share in savings resulting from supply and demand side management, and more aggressive conservation measures.

The general public must be aggressively taught the need for new approaches and how to achieve them, as well as the consequences for ignoring these needs. We must not wait for the next oil embargo, or Middle East war, or gradual depletion of our fossil reserves, or the hard fact of global warming before we move to solve these problems.

We have solutions to these problems. We need to aggressively pursue them now.

BIBLIOGRAPHY

McHale, F. A. et al., OTEC Cold Water Pipe At-Sea Test Program Phase II, Suspended Pipe Test, Final Report, Hawaiian Dredging & Construction Co. Oceanic and Atmospheric Administration, Department of Commerce, August 1984; and Ocean Thermal Energy Conversion Cold Water Pipe At-Sea Test Program Design Report, July 1982.

Nihous, G. C., Syed, M. A., and Vega, L. A., Conceptual design of an open-cycle OTEC plant for the production of electricity and fresh water in a pacific island, in *Proc. Int. Conf. on Ocean Energy Recovery,* Honolulu, November 1989. American Society of Civil Engineers, New York, 1990.

Penney, T. R. and Bharathan, D., Power from the sea, *Sci. Am.*, 256, 86, 1987.

Rogers, L. and Trenka, A., An update on the U.S. DOE Ocean Energy Program, in *Proc. Int. Conf. on Ocean Energy Recovery*, Honolulu, Hawaii, November 1989, American Society of Civil Engineers, New York, 1990.

Seymour, R. J., Ed., *Ocean Energy Recovery: The State of the Art*, American Society of Civil Engineers, New York, 1992.

Syed, M., Nihous, G., and Vega, L., Use of Cold Seawater for Air Conditioning, *Proc. OCEANS '91 Conf.*, Honolulu, Hawaii, October 1991.

Takahashi, P. K. and Trenka, A. R., "Status of Ocean Thermal Energy Technology in 1989", *Proc. Solar World Congress*, Kobe, Japan, September 1989.

Vega, L. and Nihous, G. C., At-Sea Test of the Structural Response of a Large Diameter Pipe Attached to a Surface Vessel, *Proc. Offshore Technology Conf.*, Houston, Texas, May 1988.

Vega, L., Nihous, G., Lewis, L., Resnick, A., and Van Ryzin, J., OTEC Sea Water Systems Technology Status, in *Proc. Int. Conf. on Ocean Energy Recovery*, Honolulu, Hawaii, November 1989, American Society of Civil Engineers, New York, 1990.

Vega, L. and Trenka, A., "The Near Term Market Potential for OTEC in the Pacific Basin", *Proc. PACON '90 Conf.*, Tokyo, Japan, July 1990.

Vega, L., Economics of OTEC, *Ocean Energy Recovery: The State of the Art*, 7, 152–181, American Society of Civil Engineers, New York, 1992.

Chapter 2

Environmental Compatibility of Geothermal Energy

Marshall J. Reed and Joel L. Renner

CONTENTS

INTRODUCTION

Geothermal energy is one of the cleaner forms of energy now available in commercial quantities. The use of this alternative energy source, with low atmospheric emissions, has a beneficial effect on our environment by displacing more polluting fossil and nuclear fuels. Rapidly growing energy needs around the world will make geothermal energy exceedingly important in several developing countries. In the production of geothermal energy, wells are used to bring hot water or steam to the surface from underground reservoirs. The thermal energy carried in the produced fluid can be used for direct heating in residential, agricultural, and industrial applications; or the thermal energy of higher temperature systems can be used to produce electricity.

Geothermal energy provides an enormous resource for low-temperature applications such as heating and cooling buildings, drying agricultural products, and process heating for industry. For example, geothermal heat pumps can be installed in almost all areas of the U.S. to provide greater efficiency in heating and cooling of buildings and supplying hot water than either all electric systems or systems with air-source heat pumps. Only a modest part of the potential of geothermal energy has been developed because the service industry is small and the price of competing energy sources is low. Electrical power production is the most profitable use of geothermal energy and has grown the most. Our discussion of the environmental aspects of geothermal energy utilization will concentrate on the production of electricity.

The U.S., Japan, and the European Community are continuing experiments in the extraction of thermal energy from high-temperature, subsurface zones with low initial permeability (often called "hot dry rock"). In these investigations, one deep well is used for the injection of water, at high pressure, into artificially fractured rock, the water extracts heat from the fracture surface, and a second deep well produces steam and hot water. This method of energy production is not yet economic, and the presently commercial geothermal operations depend on naturally occurring hydrothermal systems.

The U.S. Department of Energy (DOE) conducted research into the extraction of energy from the geopressured (very high pressured) brines in the Gulf Coast area of Texas and Louisiana, and concluded that even the extraction of methane as a byproduct did not make this energy source economic. Experiments were also conducted by the DOE to investigate the recovery of thermal energy from magma systems. The technique considered for energy extraction involved the injection of water to cool the magma to a fractured glass and then the continued injection and production of water to carry heat to the surface.

Geothermal energy performs a small but important role in the supply of energy for electric power generation in the U.S., and geothermal electricity plays an even greater role in some developing countries (the Philippines, Mexico, Indonesia, El Salvador, Kenya). In 1991, geothermal electrical production in the U.S. was 15,738 GWh (gigawatt hours), and the generation of this electricity provided approximately $1 billion dollars in revenue.[1] This use of geothermal energy displaces the equivalent of over 30 million barrels of imported oil per year.

The U.S. geothermal electric-power industry has grown to be the largest in the world, with over 2100 MW (megawatts electricity) generating capacity operating at over 90% availability. Slightly over half, 1100 MW generating capacity, is from The Geysers geothermal field in California. The magnitude of development in the U.S. is followed by the Philippines with 890 MW, Mexico with 700 MW, Italy with 545 MW, and New Zealand with 460 MW.[1] Iceland has the unique situation of an overabundance of hydroelectric potential, and most geothermal energy is used to provide heating and hot water for commercial and residential customers.

Geothermal energy use avoids the problems of acid rain, and it greatly reduces greenhouse gas emissions and other forms of air pollution. Geothermal reservoirs, either dry steam or hot water, are naturally occurring hydrothermal convection systems. Natural fluids are usually complex chemical mixtures, and geothermal waters exhibit a wide range of compositions and concentrations of solutes. The concentrations of solutes generally increases with the temperature of the geothermal system, and higher concentrations of some elements often require remedial action for protection of the environment. Potentially hazardous elements (Hg, B, As, and Cl) produced in geothermal brines are largely injected back into the producing reservoir. A continuing strong market for geothermal electrical generation is anticipated as a result of the increasing interest in controlling atmospheric pollution and because of the spreading concern about global warming. Geothermal development will serve the growing need for energy sources with low atmospheric emissions and proven environmental safety.

Land use for geothermal wells, pipelines, and power plants is small compared to land use for other extractive energy sources such as oil, gas, coal, and nuclear. Low-temperature geothermal applications are usually no more disturbing of the environment than a normal water well. Geothermal development projects often coexist with agricultural land uses, including crop production or grazing.

GEOTHERMAL ENERGY APPLICATIONS

HIGH-TEMPERATURE ELECTRICAL USE

The production of electricity requires a greater concentration of energy than other applications. Many geothermal systems contain water or steam at temperatures above 175 °C, and temperatures up to 400 °C have been recorded. If hot fluid is available in great enough quantities, a geothermal power plant can be installed that uses the produced steam directly to drive a turbine generator system.

In 1960, The Geysers in northern California became the first U.S. geothermal field to produce electricity, and this remains the only commercial development in the U.S. that is classified as a dry-steam geothermal system. In this low-pressure, single-phase system, dry steam is the pressure-controlling medium filling the fractured rocks. The pressure increases only slightly with depth due to the density of the steam. Initial conditions in The Geysers reservoir at a depth of 1.5 km (kilometers) included temperatures near 250 °C and pressures near 3.3 MPa (megapascals). Over 30 years of production, the pressure has dropped to less than 1 MPa in the areas of production wells, but the temperature has remained constant. Early developers found that this dry-steam type of geothermal system is very rare.

Most geothermal fields are water-dominated, where liquid water at high temperature, but also under high (hydrostatic) pressure, is the pressure-controlling medium filling the fractured and porous rocks. The pressure increases along a hydrostatic gradient in water-dominated reservoirs, and the temperature will often increase along the boiling point (liquid-vapor equilibrium) curve with depth. In water-dominated geothermal systems, water comes into the wells from the reservoir; and, in the flashed-steam power-plant technology, the pressure decreases as the water moves toward the surface, allowing part of the water to boil. Since the wells produce a mixture of flashed steam and water, a separator is installed between the wells and the power plant to separate the two phases. The flashed steam goes into the turbine to drive the generator, and the water is injected back into the reservoir.

The water-dominated geothermal system in Dixie Valley, Nevada has several features that are common to many of the geothermal fields in eastern California, Nevada, and western Utah. The water contains 0.45 weight percent dissolved solutes, and these constituents are in chemical equilibrium with the reservoir rocks. At a depth of 2 km, the temperature is 240 °C and the fluid pressure is 24 MPa; this is the hydrostatic pressure from the overlying column of water.[2] At these conditions, the geothermal fluid is liquid water. The production wells penetrate permeable zones along the active Stillwater fault, which is the physical boundary between Dixie Valley and the Stillwater Range. Steam is allowed to flash in the wells and is separated at the surface to drive the turbines. The separated water is injected to maintain reservoir pressure.

BINARY-PLANT ELECTRICAL GENERATION

Most water-dominated reservoirs below 175 °C are pumped to prevent the water from boiling as it is circulated through heat exchangers to heat a secondary liquid. In these binary power systems, heat is transferred to an organic compound with a low boiling temperature (commonly propane or isobutane), and the resulting organic vapor then drives a turbine to produce electricity. Binary geothermal plants have no emissions because the organic fluid is continuously recirculated in a closed loop, and the entire amount of produced geothermal water is injected back into the underground reservoir. A higher conversion efficiency is required to economically use lower-temperature water for electrical production, and the binary equipment has a higher capital cost to achieve this greater efficiency. The identified reserves of lower-temperature geothermal fluids are many times greater than the reserves of high-temperature fluids, providing an economic incentive to develop more efficient binary power plants.

LOW-TEMPERATURE DIRECT HEAT USE

Warm water, at temperatures above 20 °C, can be used directly for a host of processes requiring thermal energy. Thermal energy for swimming pools, space heating, and domestic hot water are the most widespread uses, but industrial processes and agricultural drying are growing applications of geothermal use. In a 1990 inventory, the U.S. was using over 5×10^{12} kJ (kilojoules) of energy annually from geothermal sources for direct heating of commercial and residential installations.[3] In Iceland, more than 95% of the buildings are supplied with heat and domestic hot water from geothermal systems, and this heat has directly replaced the burning of fossil fuels. The cities of Boise, Idaho; Elko, Nevada; Klamath Falls, Oregon; and San Bernardino and Susanville, California; have geothermal district-heating systems where a number of commercial and residential buildings are connected to distribution pipelines circulating water at 54 to 93 °C from the production wells.[4] There is believed to be a high potential for growth in district heating because numerous geothermal resources are co-located with population centers, especially in the western half of the U.S. The U.S. Department of Energy currently is funding a comprehensive inventory of low-temperature geothermal systems with special emphasis on co-location with population centers. Preliminary results indicate that there may be twice the number of systems that have been identified previously.

Typical direct-use applications are either closed systems with the produced fluids being injected back into the geothermal reservoir, or systems where the produced water is pure enough for beneficial use or disposal to surface waterways. Experience has shown that it is worthwhile to inject as much of the cooled geothermal water back into the reservoir as possible to maintain pressure and production rates. The direct use of geothermal energy for heating offsets the carbon dioxide production from combustion of fossil fuels — usually oil or gas — in a large number of residential or commercial furnaces.

GEOTHERMAL HEAT PUMPS

The use of geothermal energy through ground-coupled heat pump technology has almost no impact on the environment and has a beneficial effect in reducing the demand for electricity. Geothermal heat pumps use the reservoir of constant temperature, shallow groundwater as the heat source during winter heating and as the heat sink during summer cooling. Shallow groundwater is normally about 5 °C above the mean annual air temperature for any locality in the U.S. Because of this constant temperature, the energy efficiency of geothermal heat pumps is about 30% better than that of air-coupled heat pumps and 50% better than electric-resistance heating. Depending on climate, advanced geothermal heat pump use in the U.S. reduces energy consumption and, correspondingly, power plant emissions by 23 to 44% compared to advanced air-coupled heat pumps, and by 63 to 72% compared to electric-resistance heating and standard air conditioners.[5] The need for electrical generation capacity at the central power station is reduced by 2 to 5 kW for each residential installation and by about 20 kW for average commercial installations. Thus, for each 1000 homes with geothermal heat pumps, the utility can avoid the installation of 2 to 5 MW of generating capacity.

Geothermal heat pumps can be used in a variety of installations. A system is comprised of 1) the heat pump mechanical unit, 2) the closed-loop or open-system ground heat exchanger, and 3) the building water loops. In closed-loop systems, water or a mixture of water and an environmentally safe antifreeze solution is circulated through a pipe to remove heat from, or reject heat to the ground. There is thus no contact between the solution in the closed-loop pipe and the groundwater or soil. In a vertical installation, the heat exchanger loop is a U-shaped pipe inserted in a hole 50 to 150 m (meters) deep. In horizontal installations, the heat exchanger loop is either rigid or flexible pipe laid in trenches about 2 m deep. Flexible tubing shaped in a spiral (often called a "slinky") and placed in a trench can be used to increase the effective heat exchange surface area of a horizontal loop and to reduce the length of trenching by 40%. The open vertical system uses a water well to provide groundwater to the heat pump, and, depending on need, the water can be used within the building, can be discharged at the surface, or can be injected in a second well. In single-well, open installations, water can be withdrawn from the bottom of the well, circulated through the heat pump, and returned to the top of the well. This method depends on the open communication with the groundwater system, is often a lower cost option, and is used in large commercial applications where space is limited.

The use of heat pump technology is associated with disturbance of soil during installation; however, since this application is normally associated with the simultaneous construction of homes and industrial buildings, there is only a small and transient surface disturbance. Geothermal heat pumps require less frequent maintenance and repair; refrigerants are installed in sealed systems at the factory (like a refrigerator) and no field connections are required. Equipment has a much longer lifetime since no part of the heat pump is outside the building and exposed to the elements. During operation, there are no emissions from closed-loop systems because the ground-loop heat-exchange fluid (usually water) is contained. If an antifreeze is needed, environmentally compatible antifreeze such as potassium acetate can be used so that there is no risk of accidental release of polluting compounds to ground or surface waters.

ENVIRONMENTAL CONSIDERATIONS

AIR QUALITY

All known geothermal systems contain the equilibrium distribution of carbonate, bicarbonate, and aqueous carbon dioxide species in solution; and, when a steam phase separates from boiling water, carbon dioxide is the dominant (over 90% by weight) noncondensible gas. Most hydrothermal systems have very low oxygen activity, and these systems commonly contain the reduced species H_2S, NH_3, and CH_4 in the steam phase. In most geothermal systems, noncondensible gases make up less than 5% by weight of the steam phase. Thus, for the same output of electricity, carbon dioxide emissions from geothermal flashed-steam power plants are only a small fraction of emissions from power plants that burn hydrocarbons. Binary geothermal power plants do not allow a steam phase to separate, so carbon dioxide and the other gases remain in solution and are reinjected into the reservoir, resulting in no atmospheric emissions. For each megawatt-hour of electricity produced in 1991, the average emission of carbon dioxide by plant type in the U.S. was: 990 kg from coal, 839 kg from petroleum, 540 kg from natural gas, and 0.48 kg from geothermal flashed-steam.[6]

Hydrogen sulfide can reach moderate concentrations in the steam produced from some geothermal fields, and some systems contain up to 2% by weight of H_2S in the separated steam phase. This gas presents a pollution problem because it is easily detected by humans at concentrations of less than 1 ppm in air. Development of technology to remove H_2S was the first major research effort for joint industry-government funding in the National Geothermal Program. H_2S control became a pressing problem at The Geysers because of increasingly more stringent environmental standards promulgated by the California Air Resources Board. Now, either the Stretford process or the incineration and injection process is used in dry-steam and flashed-steam geothermal power plants to keep H_2S emissions below 1 ppb.

The efficiency of these processes in removing over 99.9% H_2S from the air emissions has resulted in Lake County, California (containing part of The Geysers geothermal field) receiving the Outstanding Performance award in 1992 from the California Air Resources Board for compliance with the California Clean Air Standards.[7] Use of the Stretford process in many of the power plants at The Geysers results in the production and disposal of about 13,600 kg sulfur per megawatt of electrical generation per year. Figure 1, based on the diagram of Henderson and Dorighi,[8] shows the typical equipment used in the Stretford process at The Geysers. Some of this sulfur is contaminated with vanadium (the Stretford catalyst) and must be washed before disposal.

The incineration process burns the gas removed from the steam to convert H_2S to SO_2, the gases are absorbed in water to form SO_3^{2-} and SO_4^{2-} in solution, and iron chelate is used to form $S_2O_3^{2-}$.[9] Figure 2, derived from the diagram of Bedell and Hammond,[9] shows the incineration abatement system. The major product from the incineration process is a

24

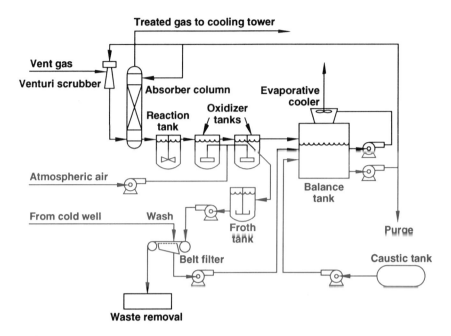

Figure 1 Typical equipment used in the Stretford process for hydrogen sulfide abatement at The Geysers geothermal field. (*From Henderson, J. M. and Dorighi, G. P., Geothermal Resources Council Trans., 13, 593, 1989, with permission.*)

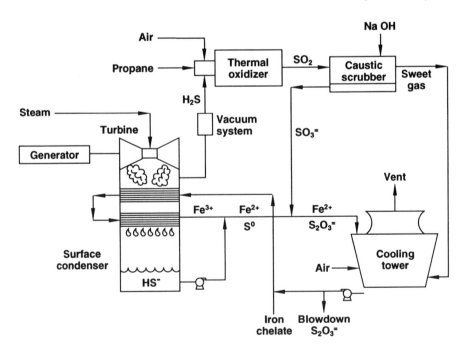

Figure 2 Equipment used in the incineration process for hydrogen sulfide abatement at The Geysers geothermal field. (*From Bedell, S. A. and Hammond, C. A., Geothermal Resources Council Bull., 16, 3, 1987, with permission.*)

soluble thiosulfate that is injected into the reservoir with the condensed water used for the reservoir pressure maintenance program. Recent advances in the use of an oxidizing biocide to remove H_2S from cooling tower circulating water have the potential to decrease cost and increase efficiency of removal.[10]

The environmental effects of H_2S and SO_2 are quite different but, at a distance of 5 km downwind from the source, studies have shown that all of the H_2S has been oxidized by the air to SO_2. For discussions of air emissions, we have converted the H_2S to SO_2. Sulfur emissions from geothermal flashed-steam power plants are only a small fraction of emissions from power plants that burn solid or liquid hydrocarbons. For each megawatt-hour of electricity produced in 1991, the average emission of SO_2 by plant type in the U.S. was: 9.23 kg from coal, 4.95 kg from petroleum, and 0.03 kg from geothermal flashed-steam (from data of Colligan[6]).

Ammonia occurs in small quantities in many geothermal systems; but, in flashed-steam geothermal power plants, the ammonia is oxidized to nitrogen and water as it passes into the atmosphere. Because the high pressures of combustion are avoided, geothermal power plants have none of the nitrogen oxides emissions that are common from fossil fuel plants. For each megawatt-hour of electricity produced in 1991, the average emission of nitrogen oxides by plant type in the U.S. was: 3.66 kg from coal, 1.75 kg from petroleum, 1.93 kg from natural gas, and zero from geothermal (from data of Colligan[6]).

WATER QUALITY

The waters in geothermal reservoirs range in composition from 0.1 to over 25 weight percent dissolved solutes. The compositions and concentrations of geothermal waters depend on the rock type of the reservoir, the temperature, and the pressure. Systems in sedimentary rocks seem to have higher concentrations than those in volcanic or granitic rocks, but there is wide variability within a single reservoir rock type. Temperatures up to 380 °C have been recorded in geothermal reservoirs in the U.S., and many chemical species have a significant solubility at high temperature. For example, all of the geothermal waters are saturated in silica with respect to quartz. As the water is produced, silica becomes supersaturated; and, if steam is flashed, the silica becomes highly supersaturated. Upon cooling, amorphous silica precipitates from the supersaturated solution.

The high flow rates of steam and water from geothermal wells usually prevent silica from precipitating in the wells, but careful control of fluid conditions and residence time is needed to prevent precipitation in surface equipment. Silica precipitation is delayed in the flow stream until the water reaches a crystallizer or settling pond. There the silica is allowed to settle from the water, and the water is then pumped to an injection well. It is necessary to inject the geothermal water back into the reservoir to maintain the pressure and flow rate at the producing wells. Precipitated silica is removed from the water so that the solid material does not clog the injection well or reservoir. The most soluble of the other species in solution remain in solution and are injected. Other species, which have precipitated, are washed from the silica and injected with the wash water. The removed silica requires disposal, but research is underway to find a commercial use for the silica produced. Many of the solids removed from geothermal processes require drying before disposal to reduce both volume and mass.

The Salton Sea geothermal system in the Imperial Valley of southern California has presented some of the most difficult problems in brine handling. Water is produced from the reservoir at temperatures between 300 and 350 °C and concentrations between 20 and 25% solutes by weight. This brine is in equilibrium with the mineral phases in the reservoir, but the concentration that occurs when 20% of the mass is allowed to vaporize leaves the brine supersaturated with respect to several solid phases. To remove solids from the steam, crystallizers are used upstream of the turbines, and to remove solids from the injection water, both clarifier and thickener tanks are needed. Figure 3, modified from

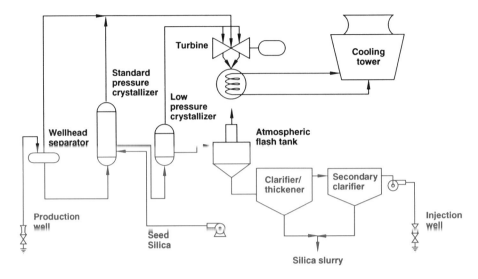

Figure 3 The flow stream for removal of solids from the vapor and brine in typical power plants in the Salton Sea geothermal field. (*From Signorotti, V. and Hunter, C. C., Geothermal Resources Council Bull., 21, 277, 1992, with permission.*)

the diagram of Signorotti and Hunter,[11] shows the flow stream for removal of solids from the vapor and brine. As an alternative, one power plant in the Salton Sea geothermal field uses the addition of acid to lower the pH and keep the solutes in solution.[11] The output from the crystallizers and clarifiers is a slurry of brine and amorphous silica. The methods used to de-water the salt and silica slurry from Magma Energy operations in the Salton Sea geothermal system are described by Benesi.[12]

Some geothermal systems, such as Dixie Valley in Nevada, form a high pH water through the evolution of carbon dioxide from solution.[1] This high pH permits the silica concentration in solution to remain at much higher levels without causing the precipitation of amorphous silica. At high pH, some of the silica in solution forms an ionic complex ($H_3SiO_4^-$) reducing the concentration of the neutral complex ($H_4SiO_4^0$) that controls polymerization and precipitation as amorphous silica.

In the U.S., only the lower-temperature geothermal waters that are of drinking-water quality are allowed to flow into streams or lakes. All other geothermal applications require that the cooled water be injected back into the reservoir. To protect potable ground waters in shallow aquifers, both the production and injection wells are lined with steel casing pipe and cemented to the surrounding rock. This type of well completion prevents the loss of geothermal water to any freshwater aquifers and confines the injection to the geothermal reservoir. Repeated examination of casing and cement, using sonic logging instruments, assures that no leakage occurs.

The production and injection system for geothermal water also prevents any contamination of surface waters. Water injection in the hotter geothermal systems does not require any pump pressure at the surface, since the cold injection water drops under the influence of gravity into the less dense, hot water of the reservoir. Cooler geothermal systems or those with rocks of lower permeability will require some pump pressure to inject the water into the reservoir. Geothermal power plants in the U.S. use cooling towers to condense the turbine exhaust fluid (either steam or organic fluid), and no waste heat is dumped into rivers or the sea. Waste heat disposal from fossil and nuclear power plants can cause disruption of the biota in local water bodies.

LAND USE

The actual land used in geothermal operations is fairly small, and other applications such as crop growing or grazing can exist in proximity to the roads, wells, pipelines, and power plants of a geothermal field. The average geothermal plant occupies only 400 m^2 (square meters) for the production of a gigawatt hour over 30 years.[13] If the entire life cycle of each energy source is examined, the energy sources based on mining such as coal and nuclear require enormous areas for the extraction and processing in addition to the area of the power plant. The disturbed surface from open pit mining is an area with no plant life to participate in the carbon cycle or in evapotranspiration to replenish the water in the atmosphere.

ACKNOWLEDGMENTS

The authors express their appreciation for useful suggestions and critical comments in reviews from Phillip M. Wright and John E. Mock. This study was supported by the Geothermal Division of the U.S. Department of Energy, partially under DOE Idaho Operations Office contract DE-AC07-76ID01570 with EG&G Idaho, Inc.

REFERENCES

1. **McLarty, L. and Reed, M. J.,** The U.S. geothermal industry: three decades of growth, *Energy Sources*, 14, 443, 1992.
2. **Reed, M. J.,** Thermodynamic calculations of calcium carbonate scaling in geothermal wells, Dixie Valley geothermal field, U.S.A., *Geothermics*, 18, 269, 1989.
3. **Lund, J. W., Lienau, P. J., and Culver, G. G.,** The current status of geothermal direct use developments in the United States, update: 1985 - 1990, *Geothermal Resources Council Trans.*, 14, 277, 1990.
4. **Rafferty, K.,** A century of service: the Boise Warm Springs water district system, *Geothermal Resources Council Bull.*, 21, 339, 1992.
5. **L'Ecuyer, M., Zoi, C., and Hoffman, J. S.,** *Space Conditioning: The Next Frontier*, U.S. Environmental Protection Agency, EPA430-R-93-004, Washington, D.C., 1993.
6. **Colligan, J. G.,** U.S. electric utility environmental statistics, in *Electric Power Annual 1991*, U.S. Department of Energy, Energy Information Administration, DOE/EIA-0348(91), Washington, D.C., 1993.
7. **Anderson, D. E., Ed.,** Lake County lauded for cleanest air, *Geothermal Resources Council Bull.*, 22, 50, 1993.
8. **Henderson, J. M. and Dorighi, G. P.,** Operating experience of converting a Stretford to a Lo-Cat(R) H$_2$S abatement system at Pacific Gas and Electric Company's Geysers unit 15, *Geothermal Resources Council Trans.*, 13, 593, 1989.
9. **Bedell, S. A. and Hammond, C. A.,** Chelation chemistry in geothermal H$_2$S abatement, *Geothermal Resources Council Bull.*, 16, 3, 1987.
10. **Gallup, D. L.,** "BIOX" A new hydrogen sulfide abatement technology for the geothermal industry, *Geothermal Resources Council Trans.*, 16, 591, 1992.
11. **Signorotti, V. and Hunter, C. C.,** Imperial Valley's geothermal resource comes of age, *Geothermal Resources Council Bull.*, 21, 277, 1992.
12. **Benesi, S. C.,** Dewatering of slurry from geothermal process streams, *Geothermal Resources Council Trans.*, 16, 577, 1992.
13. **Flavin, C. and Lenssen, N.,** Designing a sustainable energy system, in *State of the World, 1991, A Worldwatch Institute Report on Progress Toward a Sustainable Society*, W. W. Norton, New York, 1991, 21.

Chapter 3

Total Fuel Cycle Emissions Analysis of Biomass-Ethanol Transportation Fuel

Cynthia J. Riley and K. Shaine Tyson

CONTENTS

ABSTRACT

In 1991, the U.S. Department of Energy (DOE) unveiled its National Energy Strategy (NES), a framework of policy initiatives to increase energy efficiency and reduce U.S.

dependence on imports and fossil fuels. The strategy endorsed a particular methodology, the total fuel cycle analysis (TFCA), as a tool to describe and quantify the environmental, social, and economic costs and benefits associated with energy alternatives. A TFCA should quantify inputs and outputs, their impacts on society, and the value of those impacts that occur from each activity involved in producing and using fuels. New fuels and energy technologies can be consistently evaluated and compared using TFCA, providing a sound basis for ranking policy options that expand the fuel choices available to consumers.

DOE has chosen ethanol produced from lignocellulosic biomass as a high priority option for research and development. At the request of DOE, a fuel cycle analysis was completed to quantify the inputs and outputs of a hypothetical biomass-ethanol industry in the year 2010 that produces a 95% ethanol, 5% gasoline fuel product (E95). A comparison of the results to a similar study of reformulated gasoline (RFG) was made.

Five regional biomass-ethanol fuel cycles were examined to evaluate the impact of different energy crop mixes on the levels of inputs and outputs. The technology of producing ethanol from biomass was based on engineering designs, research trends, past industrial experience, and expert opinion. Projections of future crude oil mixes, refining product outputs, and organizational structure were used to characterize the future RFG industry. Each fuel cycle is represented by a flow chart of activities based on a model industry. From this, an inventory of inputs (electricity, chemicals, materials, etc.) and outputs (fuel, emissions, wastes, etc.) was created for each fuel cycle. Only the operational phase of the fuel cycles was examined. The industrial activities for each fuel cycle are divided into five stages: feedstock production, feedstock transportation, fuel production, fuel distribution, and end use. This convention is used to describe the fuel cycles and the results. The discussion of results focuses on the gaseous, solid, and liquid fuel cycle emissions because the major issues impacting fuel use today are the environmental implications.[1,2]

This chapter in an excerpt summary of selected results from the final study *Fuel Cycle Evaluations of Biomass-Ethanol and Reformulated Gasoline Fuels, Volume I*.[3]

The conclusions drawn from this study are:

- R&D in vehicle technology can produce substantial benefits in terms of reduced emissions because the majority of emissions are produced in the end-use stage.
- E95 fuel cycles can produce 90% less carbon dioxide (CO_2) emissions compared to the RFG fuel cycle.
- E95 fuel cycles produce less nitrogen oxides (NO_x), sulfur dioxide (SO_2), CO_2, and particulate matter (PM) than RFG, when emissions associated with electricity production are included in the fuel cycle analyses.
- Ethanol fuels can extend our fossil fuel resources in the transportation sector because they require much fewer fossil fuel resources per Btu of fuel to produce.
- This study can be used to rank fuels based on selected criteria, such as CO_2 emissions, but impact and valuation analyses are required to conclude that one fuel is preferred to another.

INTRODUCTION

The National Energy Strategy (NES) presents a road map of policies that could lead to reduced dependence on imported fuels, more efficient use of domestic resources, economic growth, and a cleaner environment. To help reach these goals, the NES recommended the total fuel cycle analysis (TFCA) as the methodology for the U.S. Department of Energy (DOE) and its agencies to use for evaluating fuels and energy technologies.

One of the specific options identified by DOE is "Enhanced Transportation Biofuels Production R&D," which proposes to accelerate the research and development of biofuel technologies in the hope that they may become commercial sooner, and thus provide more benefit to the American public. DOE's Office of Energy and Efficiency and Renewable Energy (DOE/EERE), which funds biofuels technology development, wanted to enhance its capability to conduct credible evaluations of alternative fuel options by applying TFCA to biomass-ethanol and RFG fuels. This chapter summarizes the findings of the TFCA for these fuels. The information presented here is an excerpt summary of selected results from the final study *Fuel Cycle Evaluations of Biomass-Ethanol and Reformulated Gasoline Fuels, Volume I and II*.[4]

These fuel cycle analyses focused on measuring the amounts of inputs and outputs produced by two transportation fuels: E95, a blend of ethanol and 5% gasoline, and RFG. The ethanol is made from lignocellulosic feedstocks, trees, and grasses, using an experimental technology. Ethanol made from grain is not discussed. The time frame for the analyses is 2010.

The fuel cycles examined are snapshots in time. Technology and industry are constantly changing. The technologies used to model the biomass-ethanol industry represent researchers' best assumptions about how this industry might function. These fuel cycle analyses focused on measuring the inputs and outputs of two fuel cycles, similar to a mass and energy balance. This report provides the information necessary to rank fuels by specific criteria, such as CO_2 emissions. This report also provides the information required to conduct impact studies, but does not include impact studies or estimates of the costs associated with impacts.

These fuel cycle analyses provided a number of benefits:

1. Helped formulate future research agendas to answer questions that arose during this study and to provide data that did not exist for this study
2. Organized existing information
3. Improved the existing engineering design for biomass-ethanol production
4. Created a better understanding of how the biomass-ethanol industry may operate
5. Created a database of emissions for site-specific impact studies
6. Established a basis for future cost-benefit studies

The remainder of the report consists of several sections. The following subjects are discussed: 1) the TFCA methodology and its implementation (including the rationale behind the choices of fuels evaluated); 2) the industrial systems and technologies used to produce, deliver, and utilize the fuels; 3) the findings of the analysis; and 4) the conclusions drawn from the analysis and their implications.

TOTAL FUEL CYCLE ANALYSIS METHODOLOGY

TFCA provides a systematic approach for evaluating fuel resources and technologies. The fuel cycle analysis is determined by the following tasks:

1. Define the fuels or fuel cycles to be analyzed.
2. Define the fuel cycle boundaries that will limit the analysis.
3. Define the types of fuel cycle impacts to be analyzed (social, economic, technological, and environmental).

The following discussion of boundary conditions and assumptions is critical to understanding how the results provided should be used and for understanding the lessons learned from applying TFCA.

FUEL CYCLES

The two transportation fuels for the fuel cycle study are:

- E95, 95% ethanol manufactured from energy crops, trees, and grasses, with 5% gasoline added as denaturant in 2010.
- Reformulated gasoline (RFG) with methyl tertiary butyl ether (MTBE) in 2010.

These fuels were chosen because of their prominence in policies proposed by DOE and the Environmental Protection Agency (EPA).

Producing ethanol from lignocellulosic biomass is not a commercial technology today. However, by 2000 a number of facilities could be operating using low-cost feedstocks such as municipal solid waste (MSW) and by 2010, cellulosic crop technologies (often referred to as energy crops) should be commercially available. In addition, the biomass-ethanol industry will rely on energy crops as its primary source of feedstock because the unused supply of cellulosic waste materials may dwindle as demand for these materials increase (recycled paper, electric power, ethanol, etc.). The ethanol referred to in this study is produced from lignocellulosic biomass (such as trees and grasses), using an experimental technology. Ethanol from grain is not discussed.

The NES projected that nearly all gasolines will be reformulated by 2000 (U.S. DOE 1991 b). RFG using MTBE was selected because it is the most common RFG produced today.

The CAAA of 1990 requires the use of RFG containing oxygenates (Title II), and clean fuels in fleets in serious, severe, and extreme ozone nonattainment areas and in serious carbon monoxide (CO) nonattainment areas. Deadlines for adopting and using these fuels depend on the specific area and fuel considered. Specific clean fuels are not mandated but several alternative fuels are listed, including natural gas, methanol, ethanol (if the methanol and ethanol content of the fuel equals or exceeds 85% by volume), electricity, liquefied petroleum gas, RFG or reformulated diesel, and hydrogen.

The CAAA requires all fuels in the year 2000 to meet CAAA Tier I standards in motor vehicles (Title II, Section 203). By 2010, Tier II standards will be promulgated with stricter limitations on air emissions from vehicles. Cleaner burning fuels will be required and ethanol is listed in the CAAA as a clean fuel alternative. Thus, the fuel cycle for 2010 assumes that ethanol is produced from energy crops and is consumed as a denatured fuel in dedicated ethanol vehicles.

E95 is ethanol denatured with 5% gasoline; neat ethanol has to be denatured according to existing regulations of the Bureau of Alcohol, Tobacco and Firearms, to control the collection of taxes on alcohol purchased for consumption and to discourage human consumption of fuel ethanol. Gasoline is a common denaturant today, although other denaturants are available.

Both fuels are consumed by light-duty passenger vehicles. E95 is consumed in dedicated ethanol vehicles with optimized technology; dedicated ethanol vehicles are assumed to be available by 2010, according to the NES and other industry sources. In 2010, ethanol vehicles are projected to get 28.25 miles per gallon (mpg) and RFG vehicles are projected to get 35.6 mpg. The results of this study—the fuel cycle inventories—are presented in grams or gallons of outputs for every mile traveled by a light-duty passenger vehicle.

The data inventory was managed by the Total Emission Model for Integrated Systems (TEMIS). TEMIS is an accounting tool and does not optimize or project variables. It does allow for a wide array of sensitivity analyses by altering major parameters such as engine efficiencies or crop yields to determine the effects on the total inventories. The input and output characteristics of each activity in the fuel cycle and the magnitude of the activity are part of the basic database. TEMIS is used to link the various activities and adjust the relative magnitudes of the activities to reflect a consistent basis for the evaluation.

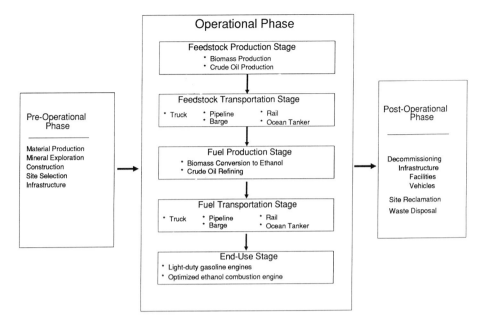

Figure 1 Fuel cycle boundaries.

No attempt was made to optimize technologies or markets represented by the fuel cycles based on economic or social criteria. Future economic parameters, such as costs and profits, will be affected by environmental issues, costs of environmental controls, and regulations. The industry structure examined is reasonable given what we know today about existing or similar industrial structures.

FUEL CYCLE BOUNDARIES
Only the operational phase of a fuel cycle (e.g., activities directly associated with producing and consuming the fuels) is documented in this study (Figure 1). Emissions associated with construction and decommissioning of the infrastructure required to produce, deliver, or consume the fuels are not included in the inventories. Drilling and other activities associated with exploration for crude oil were not included in the fuel cycle analysis because these activities are generally one-time occurrences that resemble construction and development more than daily operational activities.

A number of previous studies were examined to determine the effect of excluding pre- and post-operational phases. Deluchi (1992) constructed ethanol and RFG fuel cycles to estimate energy consumption and greenhouse gas emissions. His analysis showed that 10 to 15% of the total energy inputs of the fuel cycles are used in the production of the materials used in construction of the infrastructure and the vehicles.

Deluchi assumed that 2 to 3% of the energy content of the end-use fuel is used in exploration, production, and drilling for onshore and offshore oil. The DOE Handbook (1983) estimates that the energy used to produce on-shore oil in the lower 48 states is 1.5% of the energy in the crude produced, with about half of that used in development drilling and half used for oil production.

The exclusion of construction activities may be a significant issue but would require more information on future biomass-ethanol industrial development than is currently available. The future size and location of the biomass-ethanol industry has yet to be established and is controversial. This study was limited to the operational phase because it can be defined based on engineering principals and published information.

Biomass-Ethanol as E95

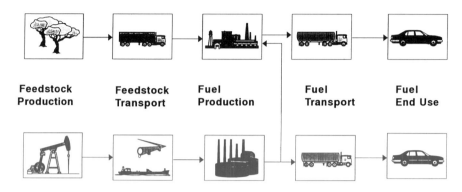

| Feedstock
Production | Feedstock
Transport | Fuel
Production | Fuel
Transport | Fuel
End Use |

Benchmark - Reformulated Gasoline of 1990 CAAA

Figure 2 Fuel cycle stages

The operational phase of the fuel cycle is divided into five stages: feedstock production, feedstock transportation, fuel production, fuel distribution, and end-use, which is primarily the combustion of fuels in light-duty passenger vehicles (Figure 2). Table 1 summarizes the major activities included in each stage of the fuel cycles examined in this report. Figure 3 provides a flow diagram for each of the two fuel cycles, showing how the outputs from one activity become the inputs to the next. The results reported later in this document have been allocated between co-products. The descriptions of the fuel cycles go into detail on allocation assumptions.

This study uses a three-part approach to evaluate a fuel cycle: (1) present detailed descriptions of the engineering systems that produce, transport, convert, and consume feedstocks and fuels; (2) construct a model industry that incorporates the activities defined in 1; and (3) build inventories of inputs and outputs for the fuel cycles. Estimates of fuel cycle inputs and outputs are based on theoretical engineering designs of the fuel cycles studied. The future petroleum industry is assumed to be nearly identical to the existing petroleum industry. The biomass-ethanol industry is created from a hypothetical set of assumptions based on existing agricultural practices, transportation infrastructure, and engineering designs. Outputs include estimates of air pollutants, solid wastes, water effluent, and energy products such as fuel, electricity, and heat. Inputs include labor, electricity, feedstocks (crude oil and biomass), chemicals, water, fuels, and equipment.

Six individual fuel cycles were created. These cases consist of five energy crop-E95 fuel cycles and the RFG fuel cycle (Table 2). The fuel cycle scenarios are limited to characterizing the domestic industry, although the RFG fuel cycle includes imported crude oil.

Five sites for biomass-ethanol production were chosen to reflect characteristics found in the surrounding regions (Figure 4). Regional variation in energy crop production inputs and outputs is very likely. Climate, soil characteristics, and other natural parameters affect which crops are produced, their yields, and agronomic practices, and thus affect the level of inputs and outputs of biomass production. Different mixes of energy crops affect the yield of ethanol, and thus affect the inputs and outputs of the fuel production stage. The five sites selected are: Peoria, IL; Lincoln, NE; Tifton, GA; Rochester, NY; and Portland, OR. Biomass production and conversion (fuel production) are located in the vicinity of these cities. Fuel was assumed to be consumed in the local area surrounding these cities.

Table 1 **Fuel cycle stages and activities**

Fuel Cycle Stage	E95	RFG
Feedstock production	Prepare land for planting; plant, tend, and harvest biomass crops and store on farm. Biomass crops: perennial grasses annual grasses short rotation trees	Crude oil production from domestic sites, on-site processing and storage; imported crude oil production same as domestic. By-products: natural gas
Feedstock transportation	Load biomass into trucks, rail, or barge for transportation to ethanol conversion facility; unload.	Transport crude oil via truck, pipeline, barge, and tanker in U.S. boundary waters to storage facilities; store; deliver crude to refineries via pipeline, barge, and tanker; unload and store at refinery.
Fuel production	Lignocellulosic crops converted to E95 using 2010 technology. Gasoline fuel cycle inventory included (5% denaturant) in this stage. By-product: electricity	Crude oil converted to reformulated gasoline and other products. MTBE production is excluded; MTBE is treated as input. By-products: non-gasoline products
Fuel distribution	E95 stored at conversion plant, loaded into railcars, transported to dedicated bulk tanks in bulk terminals at major metro areas in region and unloaded, loaded into tank trucks and delivered to retailers, unloaded and stored at retail facilities, pumped into dedicated vehicles.	Reformulated gasoline is transported in pipelines, barges, tank trucks, and tankers to bulk terminals, stored, loaded into tank trucks for retail delivery, unloaded into retail storage, and pumped into passenger vehicles.
End use	Combustion in a light-duty passenger car, dedicated ethanol engine.	Combustion in a light-duty passenger car, conventional gasoline engine.

In the RFG fuel cycle, we have assumed that imported crude has the same production emission characteristics as domestic crude oil production. This assumption can overestimate or underestimate actual inputs and outputs associated with international oil production, but the scope of estimating actual values was beyond this study. The emissions from transporting imported crude from the 200-mile economic trade boundary to U.S. ports are included but not the emissions that occur before the oil reaches the 200-mile boundary. The lack of readily available data and the modeling requirements involved to simulate crude oil transportation limited the treatment of this activity. The location and volumes of domestic crude oil production were taken from NES projections; refining and fuel

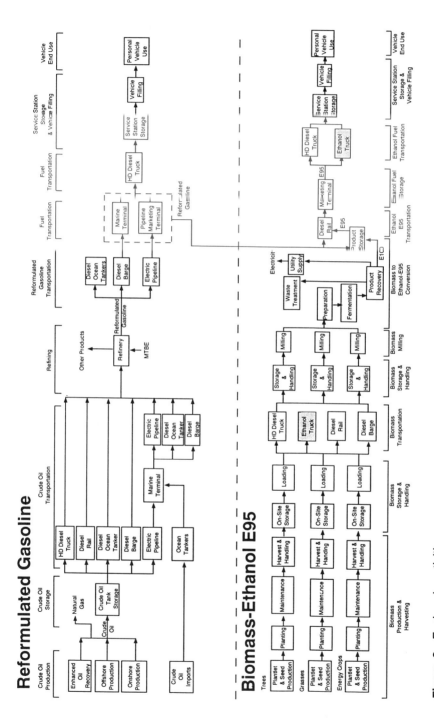

Figure 3 Fuel cycle activities.

Table 2 **Descriptions of fuel cycle cases**

Year	Descriptions
2010	RFG with inputs/outputs of crude oil production, transportation, and refining allocated between RFG and other products. Imported crude oil is assigned the same inputs/outputs as domestic crude oil production.
2010	E95 from Tifton biomass, includes inputs/outputs of the RFG case for the 5% gasoline content. Feedstock conversion, transportation, and production characteristics allocated between ethanol and electricity products.
2010	E95 from Peoria biomass, includes inputs/outputs of the RFG case for the 5% gasoline content. Feedstock conversion, transportation, and production characteristics allocated between ethanol and electricity products.
2010	E95 from Lincoln biomass, includes inputs/outputs of the RFG case for the 5% gasoline content. Feedstock conversion, transportation, and production characteristics allocated between ethanol and electricity products.
2010	E95 from Portland biomass, includes inputs/outputs of the RFG case for the 5% gasoline content. Feedstock conversion, transportation, and production characteristics allocated between ethanol and electricity products.
2010	E95 from Rochester biomass, includes inputs/outputs of the RFG case for the 5% gasoline content. Feedstock conversion, transportation, and production characteristics allocated between ethanol and electricity products.

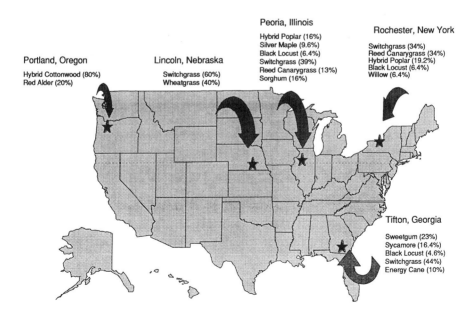

Figure 4 Ethanol fuel cycle locations.

consumption are assumed to be similar to patterns that exist today. All biomass and ethanol production is assumed to occur in the U.S.

FUEL CYCLE SCENARIOS

This section summarizes the major activities of the fuel cycles.

BIOMASS-ETHANOL FUEL CYCLES

The five biomass-ethanol scenarios are differentiated by feedstock and location. The feedstocks and plant locations are designed to bracket a range of potential scenarios that could lead to variations in environmental outputs from feedstock production, transportation, and conversion. The only difference between the scenarios is the choice of feedstock, such as blends of dedicated energy crops.

Each bioethanol production facility was assumed to require 1814 dry metric tons (2000 short tons) of feedstock per day to provide consistent capacities for comparative purposes. The ethanol plants produce between 295 and 320 million liters (78 and 85 million gallons) per year of E95. The ethanol yields varied according to feedstock composition. Fuel distribution varies between cases; ethanol fuels are distributed among regional cities based on a weighted average of population distribution in the region.

The feedstock production and transportation stages of the fuel cycle are described first, followed by a summary of biomass-ethanol conversion and fuel distribution.

Dedicated Energy Crop Supply

Biomass production, transportation, conversion, fuel distribution, and end-use were assumed to occur in the vicinity of the five locations selected. Biomass crops produced at each location were selected based on soil characteristics, climate, harvesting time schedules, storage characteristics, and available data from field trials. Data from field trials were projected to year 2010 based on recent trends. These projections involved yield estimates, input requirements, and cultural practices possible by year 2010. Researchers assumed that farmers will be employing more low-impact, environmental practices by 2010.

Crop establishment, cultural management, harvesting, and storage operations vary among the three broad classes of cellulosic energy crops: woody crops, perennial herbaceous crops, and annual herbaceous crops. Farmers in different regions were assumed to use similar practices for each type of crop.

The land available for energy crop production includes the counties within a 100-mile radius of each of the five ethanol manufacturing facilities, with the conversion facilities located in the approximate center of the areas. The total acreage used for energy crops is limited to a maximum of 7% of the suitable land across all land quality designations. This assumption would make energy crop production the fifth most important crop in each area and minimizes land competition.

Energy crop yields were expected to grow over time as scientists select and breed energy crops for desirable traits, and hybridize and propagate exceptional plant material (genetic research). Moreover, breeding superior crops is also expected to reduce management requirements; faster growth will reduce the frequency of weed control, and greater tolerance to stresses will reduce the need for pest control. Estimates of future yields were solicited from energy crop researchers in several regions. These estimates are based on expert opinion and are believed to be conservative. Soil conservation practices (such as reduced tillage methods) are assumed to be sufficiently advanced so that biomass crops maintain high survival rates and yields. Reduced tillage will minimize soil erosion in the early years of tree crop establishment and will reduce soil losses associated with annual crops.

A unique characteristic of energy crop production systems is that through photosynthesis they capture carbon dioxide from the atmosphere, release oxygen, and convert much of the carbon to useable energy feedstocks. Some of the carbon sequestered is returned to the atmosphere through the decomposition of a portion of the biomass harvesting residues, storage losses, leaf litter, and small roots that die each year. Some of the carbon initially captured by the growing biomass accumulates as organic matter in the soil until an equilibrium condition is reached, which may take 30 to 50 years. The net change of carbon in the soil and in above-ground tree stems and branches represents pools of carbon that are "sequestered" or removed from the atmosphere for relatively longer periods of time; thus, they represent a benefit of the biofuels system.

Harvested energy crops are stored on the farm until they are transported to an ethanol facility. Trees and thin-stemmed grasses are baled and can be stored covered or uncovered. Thick-stemmed grasses are harvested as forage and stored in silage facilities. Varying harvest schedules allow energy crops to be delivered to the ethanol facility year-round, minimizing conflicts with local demands for harvesting equipment and labor. Storage losses are accounted for in the transportation stage of the fuel cycle. Transportation distances depend on the distribution of cropland, geography, and available routes. In some cases, bulk commodity transportation modes (such as rail and barge) are available, while other sites rely exclusively on truck transportation.

Ethanol Production

The conceptual design for the lignocellulosic biomass-to-ethanol production process is based on research and process development work sponsored by the DOE Biofuels Program. The major drawback in this design is the lack of actual experimental data that would support the estimates of processing inputs, system efficiency, and system outputs. The inventory characteristics used in this study are the result of a mass and energy balance. Experimental data are used for specific assumptions or to model specific processes; however, the effects of running the process on a totally integrated basis (i.e., running all the process steps in series using effluent from one step as the feed to the next step) is uncertain.

Feedstock compositions and the material and energy balance consequences cause the major differences among the five cases. The compositions of the various energy crops were estimated based on data from the literature.

Energy crops enter the plant and are stored and processed in the feedstock handling area. After size reduction, the biomass is treated with a dilute sulfuric acid solution. This step increases the digestibility of the cellulose fraction and hydrolyzes the hemicellulosic fraction into sugars. This solution is neutralized and prepared for fermentation. Enzymes are used to hydrolyze the cellulose into glucose; then, microorganisms ferment the sugars to ethanol and carbon dioxide. The hydrolization and fermentation is combined into one system, called the "simultaneous saccharification and fermentation process," which is the foundation of this engineering design. Other designs are possible, and each different design would produce different fuel cycle inventories.

Ethanol produced from the fermentation steps is recovered, dehydrated, denatured with 5% (by volume) gasoline, and sold as fuel grade ethanol (E95). The fuel cycle inventory associated with gasoline production is added to the ethanol inventories in this stage. Thus, inventory characteristics for the ethanol production stage include full fuel cycle inventory characteristics for gasoline.

Solid wastes from fermentation and ethanol recovery are dewatered and sent to a fluidized bed boiler where high-pressure steam is generated. The recovered solids are mostly lignin and insoluble protein that entered the plant as part of the feedstock. These components have substantial heating value and are a major source of fuel for the boiler. Other liquid and gaseous waste streams are also sent to the boiler for energy recovery. The

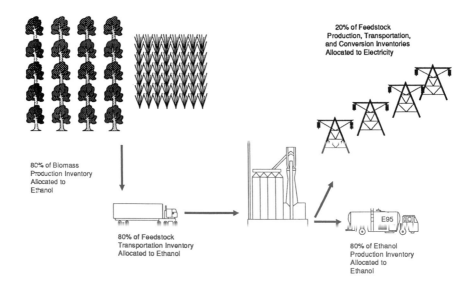

Figure 5 E95 fuel cycle allocation diagram.

high-pressure steam is let down through a steam turbine, which generates electricity for the plant and provides lower pressure steam for internal process users. Excess electricity is produced and sold to the local utility grid. The capacity of the cogeneration facilities ranges from 13 to 21 MW.

Liquid separated from the solids after ethanol recovery is processed in a wastewater treatment system. The wastewater is assumed to be treated to the standards required for industrial wastewater pretreatment; effluent is assumed to be sent to a publicly owned treatment works (POTW) facility. The exact nature of the effluent is unknown, although it is believed to be substantially similar to effluent from corn-ethanol plants.

Ash and uncombusted material from the boiler is recovered as a solid waste that requires disposal. It is assumed to be nonhazardous and therefore suitable for disposal in a licensed landfill. The ash should be similar to ash from power plants fired by wood and agricultural residues.

For each of the cases evaluated, a detailed material and energy balance was estimated, complete with utility summaries and chemical summaries. In all cases, the biomass production, transportation, and conversion inventory inputs and outputs were divided between two products: ethanol and electricity. Although this apportionment varied in each case, on the average 80% of all the inputs and outputs of the conversion stage and the previous stages was allocated to the ethanol product, and 20% was allocated to the electricity product (Figure 5). Similar methodology is used for the refinery allocation in the reformulated gasoline scenarios to account for the fact that multiple products are derived from crude oil.

Ethanol Fuel Distribution

We have assumed that the gasoline transportation and storage facilities could be used for ethanol with minor modifications. To simplify the types of transportation available and types of fuels used in them, all locomotives and trucks are assumed to be identical and use #2 diesel fuel. Fuel pumps at bulk facilities and retail terminals are assumed to be all electric. While we recognize that the industry is more complex and uses a variety of equipment and fuels, these simplifications were necessary for this analysis. Transportation mode efficiencies are based on published statistics. Vapor losses from storage tanks

an assumption of uniform tank design and size. All storage tanks and tank
pped with vapor recovery systems to reduce volatile organic compounds
;ions.
ution stage begins at the ethanol plant when the E95 is loaded onto rail cars.
plants are located in regions that support a railroad infrastructure, allowing
to be transported to the surrounding major cities by rail. The rail cars unload
ilk storage plants located at or near the rail line in major cities located in a
lius around the ethanol plant. E95 from the bulk plant is loaded into tank
elivered directly to retail stations. The average truck travels 50 miles round-
the bulk plant and the retail stations. Finally, E95 is unloaded into retail or
storage tanks, where it is pumped on demand into customers' cars.

ILATED GASOLINE FUEL CYCLE

sumes that RFG will be the primary fuel used by the year 2000. The RFG fuel
ucted for this study assumes that the future gasoline industry is substantially
similar to the gasoline industry today. The RFG in these fuel cycles has a composition that
is consistent with CAAA standards for an RFG containing 2% oxygen by weight. MTBE
is the oxygenate used in the RFG fuel cycle. This fuel cycle study assumes that RFG based
on MTBE is the only gasoline produced by the petroleum industry despite contrary
projections. We did not attempt to model the future petroleum industry with all of its
infinite variations.

The NES provides a recent forecast of the petroleum oil industry for the year 2010.
The strategy scenario, used for this fuel cycle study, includes advances in petroleum
production and utilization technologies and enough information to construct hypothetical
slates of crude oil qualities and refinery characteristics.

Crude Oil Production

Crude oil production begins with the wellhead. Exploration and drilling are assumed to
be pre-operation activities and are not included in this analysis. Conventional crude oil
production technology will remain essentially similar to current technologies through the
year 2010. Speculative resources, such as oil shale or gas hydrates, are not included
because their economic exploitation is considered unlikely by 2010, given the expected
economic conditions and anticipated technological development. The NES assumption
that controversial resources, such as the Arctic National Wildlife Refuge and Outer
Continental Shelf areas, will be developed and producing by 2010 is incorporated into this
analysis.

The techniques that produce crude oil vary according to the properties of the crude,
the geology of the underground reservoir, the age of the field, and its location (onshore,
offshore). Most of the current domestic production of crude oil is from onshore oil fields
using primary recovery technologies. However, these methods are expected to shift
toward secondary and tertiary techniques as fields age. Secondary and tertiary techniques
are more energy intensive than primary methods and employ gases, steam, and mechanical means of enhancing the flow of crude oil from the reservoir as the field becomes
depleted. By 2010, heavier crudes will be produced, and secondary and tertiary production methods will account for a larger portion of the total production. Thus, the characteristics of the hypothetical blend of crude oils available to refineries and the inputs and
outputs associated with crude oil production are projected to change over time.

The inputs and outputs associated with crude oil production are allocated between the
two co-products produced from a wellhead—natural gas and crude oil—on a contained-
Btu basis (Figure 6). Thus, only 58% of the emissions created during crude oil production
are assigned to the crude oil that is transported to the refinery. See the refining description
for a discussion of other allocation assumptions.

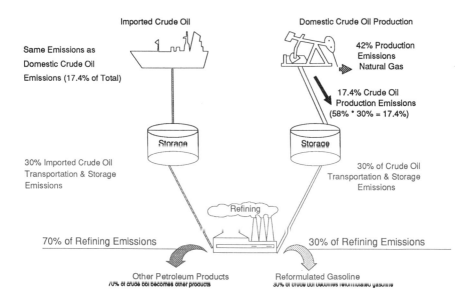

Figure 6 RFG fuel cycle allocation diagram.

Imported crude oil characteristics are added to the fuel production stage. Even with the domestic oil production incentives present in the NES, more than 37% of the oil demanded by refineries will be imported by the year 2010. Estimating foreign oil production characteristics is the best approach to the RFG fuel cycle inventory; however, collecting this information was beyond the scope of this study. The case constructed for the RFG fuel cycle assumes that imported oil is assigned the same production characteristics as domestically produced oil.

Crude Oil Transportation
Domestic crude oil is stored in tankage near the wellhead; then it is transported to crude storage tanks at the refineries. Offshore and Alaskan crude is assumed to be transported by pipeline to a marine tank storage facility; from there it is transported by ocean tanker to coastal refineries or to refinery storage facilities. Current transportation patterns are assumed to be relatively stable throughout the next two decades. National average statistics of the portion of crude oil transported in each mode are used to derive weighted average transportation estimates.

Only the characteristics associated with transporting imported crude oil from the 200-mile economic boundary to the port are included in the fuel cycle study. Transportation characteristics for the beginning of the journey are not included. Imported crude oil is unloaded into storage tanks at existing port facilities. The majority of the imported oil is transported by pipeline to refineries. Because most refineries that depend on imported crude oil are located at ports, imported crude oil is not transported the same distances as domestic crude oil.

The inventory characteristics for crude oil transportation are subject to an allocation assumption described in detail in the following section on refining.

Refining
The petroleum refining industry is the link between crude oil and finished products. The major variables that affect refinery operations with respect to the production of RFG are: (1) crude oil characteristics, (2) crude oil refining technology, and (3) RFG specifications.

The characteristics of the crude oil slate available to refineries will influence U.S. refinery operations. Similarly, the specifications for the major refinery outputs will also affect refinery operations.

For the purposes of this study, a simplifying assumption was made that the U.S. crude refining system can be characterized by two geographical components: one east of the Rocky Mountains that encompasses crude oils processed in the Petroleum Administration for Defense Districts (PADDs) I through IV, and the other west of the Rockies encompassing refining in PADD V. Average crude slate API (American Petroleum Institute) gravities and sulfur contents were forecast for both geographical regions. Two refinery scenarios for the year 2010 were investigated:

- West Coast (PADD V)
- United States less West Coast (PADDs I through IV)

The second step was to define the RFG product specifications. The following list describes the average RFG composition and property characteristics expected.

- Aromatic content: 25% by volume
- Benzene content: 1.0% by volume
- Olefin content: 15% by volume
- Oxygen content: 2.0% by weight
- Summer RVP (Reid vapor pressure): 8.5 psi
- Sulfur content: 100 ppm

The study's approach formulates the gasoline pool to meet these specifications on a nationwide average basis using a plausible scenario based mainly on changes to catalytic reforming operations. MTBE is assumed to be the oxygenate in the U.S. gasoline pool; 11% MTBE corresponds to 2% oxygen. MTBE may be manufactured in a refinery, but for purposes of this study, MTBE is considered a separate input to the gasoline refining process, and no environmental releases associated with its production were calculated. As a result, the fuel cycle inventory provided in this report underestimates total fuel cycle inputs and outputs.

National average refining and blending scenarios are developed based on the two individual refinery scenarios listed previously, along with projected crude production rates, API gravities, sulfur content, and reformulated gasoline product specifications. The scenarios developed assumed that more than 98% of the fuel is produced by complex/integrated refineries. The scenarios proposed are not an attempt to achieve the optimum, but are intended to be plausible on an average nationwide basis. In reality, each refinery will try to achieve an optimum strategy for its individual situation. The refining scenarios evaluated in this study include:

- Reducing reformate severity and therefore reformate volume
- Reducing alkylate and butane volumes in the pool
- Diverting butanes to maximize production of isobutylene, used to make MTBE
- Increasing FCC light olefins production in 2010 (up to that date, the U.S. may be able to import worldwide supplies of isobutylene or MTBE)
- Extracting benzene from reformate
- Eliminating deliberate blending of other aromatics
- Increasing the manufacture of hydrogen to make up for reduced production of catalytic reforming hydrogen

At the same time, the scenarios include increased vacuum distillation and coking volumes to contend with the trend toward heavier crude oils. They also include increased hydrotreating and caustic washing to contend with higher sulfur contents of crude oils.

Environmental releases (air emissions, water releases, and solid wastes) are based on published factors (release/barrel throughput). Environmental releases are calculated by multiplying the annual throughput volumes for each refining step by the emission factors. Major inputs to the refinery include the crude oil, natural gas, electricity, and MTBE for blending with the final gasoline product. Although there are many other chemical inputs to a refinery, they were not included in this study because characterization was difficult and it was expected that the impact on conclusions would be small. Major outputs include the reformulated gasoline stream blended with MTBE and other refinery products that are grouped in this study as other products. These include LPG (liquid petroleum gas), aviation gasoline, benzene, kerosene, jet fuel, heating oil, diesel fuel, fuel oil, coke, and miscellaneous specialty oils and waxes.

All the fuel cycle characteristics for the crude oil production, transportation, and refining stages reported are weighted by the ratio of the gasoline base (gasoline without MTBE) to total refinery product based on the energy content of the product streams. Only 33% of the fuel cycle characteristics associated with crude oil production, transportation, and refining are assigned to RFG. As the characteristics of the crude oil slate and the product slate change, the ratio of gasoline to total refinery output changes. U.S. production of gasoline is projected to fall from 7 million bpd in the year 2000 to 6.3 million bpd in 2010; whereas, crude oil demand is projected to increase from 12.3 million bpd to 13.7 million bpd between 2000 and 2010.

Air emissions are estimated using factors for criteria pollutants, aldehydes, and ammonia obtained from AP-42 (EPA 1985) and modified when appropriate to include control technologies expected to be in place by 2010. The emission factors for greenhouse gases such as carbon dioxide and methane are derived from energy consumption and combustion data.

RFG Distribution

The RFG transportation infrastructure in 2010 is expected to resemble the existing infrastructure because major changes are not considered in the NES. RFG can be transported via pipeline, barge, rail, and truck from the refinery to bulk terminals or marine terminals. From bulk terminals the fuel is usually transported to bulk plants in local metropolitan areas using tanker trucks. Trucks are used to transport the fuel from bulk plants to retail outlets. Fuel consumption for transporting gasoline is reported for the nation as a whole. Thus, it is not necessary to develop detailed estimates of how much gasoline is transported by each mode for any given distance. The lack of distances could be confusing, but keep in mind that if national estimates of fuel use in gasoline transportation are available, they are preferred to detailed modeling of a complex system. The assumption is made that the percentage of fuel that travels through the various transportation modes remains constant.

Number 2 diesel is assumed to be the only fuel used in trucks, rail, and inland barges. Number 6 diesel is assumed to be the only fuel used in ocean tankers and barges. Pipeline pumps and pumps at storage facilities are assumed to be all electrically driven.

The primary sources of emissions are vehicle emissions, primarily from rail and trucks because pipeline pumps are assumed to be electric. Vapor recovery controls are assumed to be universally employed with an recovery efficiency of 95%. Vapor recovery systems are assumed to be used at the pumps in all retail stations.

FUEL END USE CHARACTERISTICS

E95 and RFG are consumed in light-duty, spark-ignition passenger vehicles that represent technology available in 2010. Fuel composition and vehicle performance are estimated using an engineering analysis based on the technical literature. The emission values are generated from published EPA data. Changes in emission levels expected from vehicles using ethanol fuels are projected from identified changes in emissions from vehicles using reformulated gasoline. Ethanol vehicle performance is based on a theoretical analysis of the physical and chemical property differences between RFG and ethanol fuels. The theoretical analysis is then supported through a comparison with empirical data on actual engine performance measurements presented in the literature.

Vehicle emissions from RFG are based on a scenario of proposed CAAA Tier II standards being met by 2010. Evaporative emission standards have not been proposed by the EPA, and therefore, they are projected to equal the exhaust VOC levels as currently observed. Carbon dioxide and sulfur dioxide emissions are based on fuel carbon and sulfur content, respectively, and on projected fuel economy for each fuel. The fuel economy projections are based on NES estimates for a compact vehicle. Fuel economy projections for RFG are based on changes in fuel energy content resulting from the hydrocarbon distribution in an RFG.

E95
By the year 2010, fully optimized engines for ethanol fuels should be available. They could take the form of dedicated-fuel, high-compression engines designed to run specifically on E85 or E95, or they could be variable-fuel, variable-compression engines with highly sophisticated engine control systems able to optimize engine performance for a variety of fuels. The theoretical analysis suggests a 15% efficiency advantage for ethanol over gasoline, including the effect of greater tank and fuel weight. On a proportional basis this would translate to a 14% advantage for E95. Insufficient data are available to confirm these percentages experimentally. On a constant compression ratio basis the theoretical advantage for ethanol would be 7%. The available data indicates an assumption that a 15% advantage for an optimized engine is a reasonable estimate of future potential. This theoretical value is assumed as the correct measure of potential by 2010. Because of its lower energy density, light-duty passenger vehicles are assumed to get 28.25 miles per gallon on E95 and 35.6 miles per gallon on reformulated gasoline.

RFG
The CAAA requires that RFG be sold in the nine worst ozone nonattainment areas starting in 1995. States or cities can also elect to use RFG to satisfy local environmental goals. The NES projects that RFG will replace conventional gasoline by the year 2000. Future vehicle efficiency projections are based on the NES projections of new-car efficiency ratings for the year 2010 of 37.1 miles per gallon based on 1990 gasoline. The estimated energy density of RFG containing 15% MTBE, plus enough added alkylate to replace aromatics and olefins, is approximately 4% less than the energy density of 1990 gasoline. Converting the NES data to miles per million Btu yields a fleet average mileage projection of 35.6 miles per gallon in the year 2010 using RFG. This corresponds to 244 miles per million Btu.

ALLOCATION METHODOLOGY
Fuel cycle characteristics for a stage or activity were divided among the co-products of that stage or activity in three areas: crude oil production, crude oil refining, and ethanol production. In addition, prior activities were also subjected to the allocation. Analysts assigned inventory characteristics on the basis of the ratio of energy in the final product compared to the energy of the total outputs.

Co-production of Crude Oil and Natural Gas
Natural gas is often produced with crude oil. It is referred to as associated gas. If all the inventory characteristics of producing crude oil are assigned to crude oil, the natural gas produced is "free" to society; it has no costs associated with its production. Indeed, this

is how it is viewed by some analysts. The RFG case assumes that the inventory associated with crude oil production is divided, between the natural gas and crude oil produced. Crude oil is assigned 58% of the production characteristics and natural gas is assigned the remaining 42%.

Co-production of Multiple Refinery Products

Crude oil is transformed into RFG and numerous other products including jet fuel, fuel oil, fuel gas, diesel, propane, petrochemicals, coke, and asphalt. In the RFG case, the refinery characteristics are divided between RFG and "all other products" based on a Btu equivalent value of total output. In the year 2010, 30% of the refinery emissions are assigned to RFG.

The characteristics of the crude oil production and transportation stages are similarly allocated. The remaining crude oil production, transportation, and refining inventory characteristics are assigned to "other petroleum products." Therefore, only 30% of the crude oil transportation emissions are assigned to RFG, only 17.4% of the crude oil production inventory is reflected in the RFG fuel cycle (0.58 × 0.30).

Biomass-Ethanol Conversion Process

The biomass conversion facility produces two products: E95 and electricity. The characterization of the activities that produce, transport, and convert biomass need to reflect only that portion that actually contributes to ethanol production, rather than electricity. Therefore, the results reflect an allocation of the characteristics of feedstock production, transportation, and conversion based on the ratio of energy content in the ethanol to that of the total products. Each regional case is slightly different, because different feedstocks yield different proportions of ethanol and electricity. The average of the allocation characteristics of the five 2010 cases is 80% to ethanol, 20% to electricity.

FINDINGS

The discussion of results focuses on the gaseous, solid, and liquid emissions because the major issues revolving around fuel use today are their environmental implications.[2a] The data inventories of all of the activities involved with producing fuel to power a car for a common distance are aggregated into totals for each stage of the fuel cycle and for the fuel cycle as a whole. Any common basis may be used. In this report, emissions are reported on a grams or milliliters per light-duty passenger vehicle mile traveled (VMT) basis.

E95 AND RFG FUEL CYCLES

There is little difference in emission characteristics from each stage of the five E95 fuel cycles (Table 3). The differences that do occur among the E95 cases are caused by different types of feedstocks and different feedstock transportation characteristics.

CO emissions are 6 to 8% higher for E95 compared with RFG. NO_x emissions for E95 range from 97 to 104% of NO_x emissions from RFG, and SO_2 emissions are 60 to 80% lower for E95 fuels. Particulate emissions are 100 to 150% higher for E95, and VOC emissions (excluding biogenic emissions) are 13 to 15% less than RFG. E95 produces less than 10% of the CO_2 emissions that RFG produces. All of the emissions associated with producing and transporting feedstocks and producing fuel have been allocated between the various products as described earlier.

Carbon Monoxide Emissions (CO)

An average of 92% of the CO emissions from the E95 fuel cycles and 98% of the CO emissions from the RFG fuel cycle come from the passenger vehicle in the end-use stage (Figure 7). Vehicle emissions are 1.7 g CO per mile for both fuels, based on the

Table 3   E95 and RFG fuel cycle emissions (milligrams per VMT)

Emission	Fuel	End Use	Fuel Distrib.	Fuel Prod.	Feedstock Transport.	Feedstock Prod.	Total
CO	E95	1695.9	2.2	99.4[a]	7.2	43.7	1848.4
	RFG	1700.0	2.7	7.3	9.1	6.4	1725.5
NO_x	E95	199.4	6.5	68.3[a]	11.1	43.7	329.0
	RFG	199.6	4.5	65.3	20.9	37.2	327.5
PM	E95	0.0[b]	0.1	4.5[a]	0.1	4.5	9.2
	RFG	0.0[b]	0.2	2.1	1.0	0.7	4.0
SO_2	E95	3.7	0.2	21.1[a]	0.8	2.0	27.8
	RFG	40.0	0.3	4.0	0.9	4.5	85.6
CO_2[c]	E95	15.1	0.9	3.6[a]	2.5	5.8	27.8
(grams)	RFG	243.0	1.0	26.9	4.1	14.7	289.7
VOC[d]	E95	159.7	17.2	18.8[a]	2.0	10.1	207.8
	RFG	179.6	35.4	3.6	11.8	12.7	243.1
Wastewater	E95	n/a	n/a	490.0[a]	n/a	0.0	490.0
(ml)	RFG	n/a	n/a	57.0	n/a	91.0	148.0
Solid	E95	n/a	n/a	16,010.0[a]	n/a	0.0	16,010.0
Wastes	RFG	n/a	n/a	544.0	n/a	91.0	635.0

[a] Includes gasoline fuel cycle emissions for gasoline added to ethanol in this stage.

[b] Particulate emissions from passenger vehicles not available for E95 or reformulated gasoline.

[c] Fossil CO_2, does not include CO_2 sequestered in biomass or released from fermentation or ethanol combustion.

[d] VOC totals, excluding biogenic emissions.

assumption that vehicles and fuels will be designed for cars to ensure that the proposed Tier II standards of the CAAA are met. Technologies, such as improved catalytic converters and other pollutant traps, could benefit both fuels.

E95 fuel cycles produce 6 to 8% more CO than the RFG fuel cycle because of the combustion of solid wastes in the boiler of the ethanol production facility. Refineries were assumed to purchase excess power needs, and the emissions associated with that electricity are not included in the base cases; however, they are included in the electricity sensitivity cases. Although biomass combustion is perceived as a mature technology, many technological advances in boiler efficiency are under examination by NREL and others. More efficient biomass boilers and/or improved emission controls could be developed by the year 2010, which could diminish boiler emissions.

Nitrogen Oxides Emissions (NO_x)

There is no significant difference in the amount of NO_x produced by either fuel cycle; the emissions from the average E95 fuel cycle and the RFG fuel cycle are roughly the same for each stage (Figure 8). NO_x emissions for crude oil transportation are higher than those of biomass transportation because of the longer distances involved.

The passenger vehicles, in the end-use stage, produce about 61% of the NO_x emissions in both fuel cycles. Vehicle emissions were 0.2 g NO_x per mile for both fuels. Analysts assumed that both fuels and vehicles are designed to meet the proposed Tier II standards of the CAAA.

Fuel production is the second largest NO_x source for both fuel cycles, producing 20% of the total emissions. NO_x is produced during the combustion of the waste biomass in the ethanol plant's boilers and the combustion of petroleum by-products in the refinery.

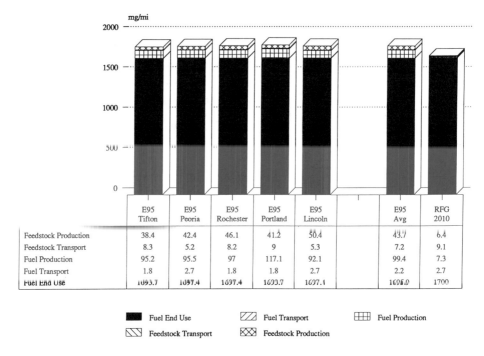

Figure 7 Fuel cycle emissions of carbon monoxide.

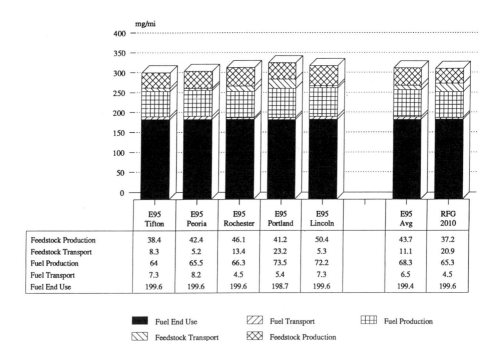

Figure 8 Fuel cycle emissions of nitrogen oxides.

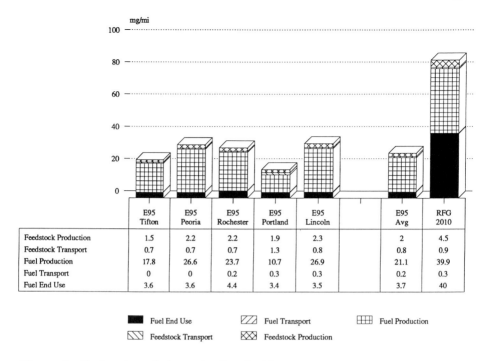

	E95 Tifton	E95 Peoria	E95 Rochester	E95 Portland	E95 Lincoln		E95 Avg	RFG 2010
Feedstock Production	1.5	2.2	2.2	1.9	2.3		2	4.5
Feedstock Transport	0.7	0.7	0.7	1.3	0.8		0.8	0.9
Fuel Production	17.8	26.6	23.7	10.7	26.9		21.1	39.9
Fuel Transport	0	0	0.2	0.3	0.3		0.2	0.3
Fuel End Use	3.6	3.6	4.4	3.4	3.5		3.7	40

Fuel End Use ▨ Fuel Transport ⊞ Fuel Production
◨ Feedstock Transport ⊠ Feedstock Production

Figure 9 Fuel cycle emissions of sulfur dioxide.

Analysts assumed that ammonia injection is used to control NO_x emissions from the ethanol plant's boiler. The NO_x emissions from the boilers are a combination of thermal NO_x, and the combustion of the nitrogen in the protein portion of the solid waste stream.

The other major NO_x source is feedstock production. NO_x emissions are produced by farm vehicles using diesel fuel. Farm vehicle use is correlated with biomass yields (lower yields require more land under cultivation and more diesel fuel, and the types of biomass grown) and some management and harvesting activities are more energy intensive than others. Because land quality affects biomass yields and the management practices required, it is difficult to draw any conclusions about specific crops having a major influence on the level of NO_x emissions. The variability in NO_x emissions for the feedstock transportation stage is due to different modes of transportation: truck, rail, and barge. NO_x emissions are higher when rail and barge are used to move feedstocks (Portland, Oregon and Rochester, New York, respectively). The other cases relied exclusively on truck transportation.

Sulfur Dioxide Emissions (SO₂)

SO_2 is produced from two sources: transportation vehicle emissions (diesel-fueled and passenger) and stationary sources, such as the conversion facility and the refinery (Figure 9). Even if the level of sulfur in RFG is reduced from 350 to 50 ppm—reducing emissions in the end-use stage by 86%—total fuel cycle SO_2 emissions from the RFG fuel cycle will still exceed those from E95 fuel cycles.

Pure ethanol does not contain sulfur; however, the denaturant gasoline contains sulfur. Since the denaturant represents only 5% by volume, E95 provides a significant reduction in SO_2 emissions from passenger vehicle exhaust over RFG.

More than 75% of the SO_2 produced in the E95 fuel cycles results from combusting wastes at the conversion facility. The proteins in biomass contain sulfur, which is the

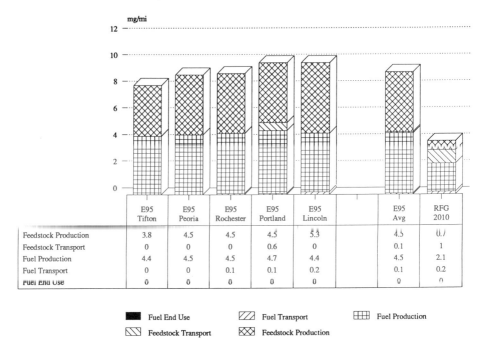

mg/mi	E95 Tifton	E95 Peoria	E95 Rochester	E95 Portland	E95 Lincoln		E95 Avg	RFG 2010
Feedstock Production	3.8	4.5	4.5	4.5	5.3		4.5	0.7
Feedstock Transport	0	0	0	0.6	0		0.1	1
Fuel Production	4.4	4.5	4.5	4.7	4.4		4.5	2.1
Fuel Transport	0	0	0.1	0.1	0.2		0.1	0.2
Fuel End Use	0	0	0	0	0		0	0

■ Fuel End Use   ▨ Fuel Transport   ⊞ Fuel Production
◫ Feedstock Transport   ⊠ Feedstock Production

Figure 10   Fuel cycle emissions of particulate matter.

source of SO_2 emissions from the boiler. Most of the regional variation in SO_2 production in the E95 fuel cycles is the result of differences in the protein content of the feedstocks used. The Portland, Oregon, conversion facility produces the least SO_2 because it uses exclusively wood feedstocks that do not contain high levels of protein; the Lincoln, Nebraska, plant produces the most SO_2 because it uses exclusively grass feedstocks that contain relatively high levels of protein. SO_2 emissions from the conversion facility boilers at other facilities fall between these extremes because feedstocks are composed of mixtures of wood and grass biomass.

Sulfur contained in the crude oil is the source of SO_2 emissions from the refinery. Refineries may be required to reduce their SO_2 emissions in the future, resulting in lower SO_2 emissions than presented in this study.

Feedstock production and transportation activities create SO_2 from diesel fuel used in tractors, trucks, and other equipment. Reducing the sulfur content of diesel will affect the total SO_2 emissions from both fuel cycles in direct proportion to the amount of diesel fuel consumed in both fuel cycles.

Particulate Matter Emissions (PM)

Approximately half of the particulates produced in the E95 fuel cycles are tailpipe emissions from diesel-fueled farm and feedstock transportation vehicles; the other half are emissions from the ethanol conversion facility (Figure 10). In the RFG fuel cycle, over 50% of the particulates are produced from the refinery, followed by another 25% from crude oil transportation (diesel use in tankers, railroads, etc.); the remainder is produced from production and processing equipment at the wellhead and RFG distribution. Data on the quantity and composition of particulates from passenger vehicles fueled by E95 or RFG were not available and therefore are not included in the total emission levels presented.

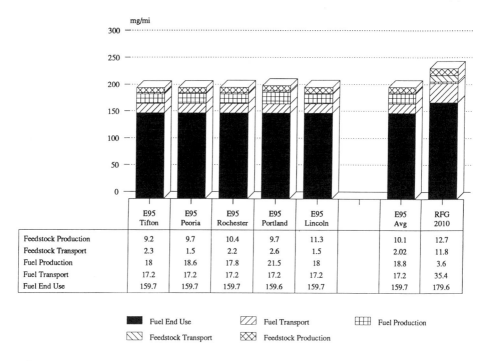

The following data accompany the figure:

	E95 Tifton	E95 Peoria	E95 Rochester	E95 Portland	E95 Lincoln		E95 Avg	RFG 2010
Feedstock Production	9.2	9.7	10.4	9.7	11.3		10.1	12.7
Feedstock Transport	2.3	1.5	2.2	2.6	1.5		2.02	11.8
Fuel Production	18	18.6	17.8	21.5	18		18.8	3.6
Fuel Transport	17.2	17.2	17.2	17.2	17.2		17.2	35.4
Fuel End Use	159.7	159.7	159.7	159.6	159.7		159.7	179.6

Legend: ■ Fuel End Use　▨ Fuel Transport　⊞ Fuel Production　◹ Feedstock Transport　⧅ Feedstock Production

Figure 11　Fuel cycle emissions of volatile organic compounds.

The particulate emissions from the conversion facility are divided equally between boiler fly ash emissions and dust from the feedstock handling and preparation activities. The fly ash emissions are a function of the quantity and composition of the material fed to the boiler.

Particulate emissions from feedstock and fuel transportation are positive but very low. In most cases, these estimates are shown as zero. The exception of the E95 Portland fuel cycle is caused by transporting biomass feedstock by rail, which is responsible for the relatively high levels of particulate emissions in the feedstock transportation stage.

If airborne soil erosion, fertilizers, and pesticides are included in the accounting of particulates, total particulate emissions in the E95 fuel cycles increase dramatically. An impact analysis is required to determine if some or all of these airborne farm emissions would have occurred in the absence of a biomass-ethanol industry, and if so, how much of these emissions are the direct result of the biomass-ethanol industry.

Volatile Organic Compound Emissions (VOC)

VOC emissions were divided into two source categories: (1) biogenic VOC emissions produced by growing plants, and (2) nonbiogenic VOC emissions produced during the use or combustion of fossil fuels and volatile chemicals. This allows us to compare the quantities of nonbiogenic VOC emissions of the two types of fuel cycles—E95 and RFG. RFG fuel cycles do not produce any biogenic VOC emissions.

Approximately 75% of the nonbiogenic VOC emissions produced from the E95 and RFG fuel cycles are evaporative and exhaust emissions from the passenger vehicles used in the end-use stage (Figure 11). Exhaust emissions were assumed to be identical for both fuels—0.09 g per mile. Evaporative engine losses were less for E95 (0.07 g per mile) compared to RFG (0.09 g per mile). This difference caused end-use emissions from

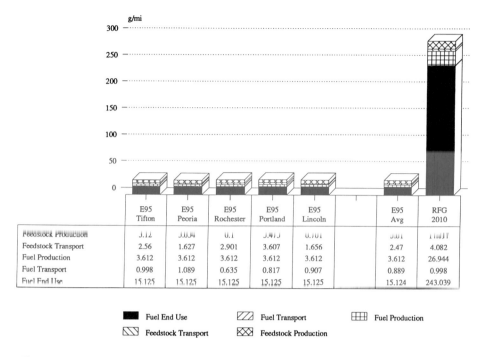

	E95 Tifton	E95 Peoria	E95 Rochester	E95 Portland	E95 Lincoln		E95 Avg	RFG 2010
Feedstock Production	3.12	3.034	0.1	3.413	0.101		3.01	1.037
Feedstock Transport	2.56	1.627	2.901	3.607	1.656		2.47	4.082
Fuel Production	3.612	3.612	3.612	3.612	3.612		3.612	26.944
Fuel Transport	0.998	1.089	0.635	0.817	0.907		0.889	0.998
Fuel End Use	15.125	15.125	15.125	15.125	15.125		15.124	243.039

- ■ Fuel End Use
- ▨ Fuel Transport
- ▤ Fuel Production
- ◺ Feedstock Transport
- ▩ Feedstock Production

Figure 12 Fuel cycle emissions of carbon dioxide.

dedicated passenger vehicles using E95 to be 11% less than emissions from vehicles using RFG.

The remaining VOC emissions are produced from the combustion of diesel fuel in equipment used to produce and transport feedstocks and fuels. The biomass conversion processes also produce significant amounts of VOCs from the boilers.

If biogenic VOC emissions are included in the VOC accounting framework, total VOC emissions in the E95 fuel cycles increase 600 to 1600% depending on the proportion of trees produced in the biomass feedstock mix. Deciduous trees produce nearly 10 times more biogenic VOCs than any other agricultural crop except corn.[5] Analysts assumed that herbaceous biomass crops did not produce biogenic VOC emissions, although it is likely that these emissions will be produced in small quantities. The extent that tree crops displace corn and other crops will determine the net changes in localized biogenic VOC emissions. This net analysis should be undertaken in the future.

Not enough data exist to completely define the components of the biogenic and nonbiogenic VOC emissions in sufficient detail to perform ozone impact studies. Each specific VOC compound has a different reactivity and chemical signature in the atmosphere. Some decompose rapidly and others have complex reaction chains. The differences in the composition of VOC emissions will influence the timing, persistence, and impacts of ozone creation in a locality.

Carbon Dioxide Emissions (CO₂)

E95 fuel cycles produce an average of 27.9 g net CO_2 per VMT. The RFG fuel cycle produces 290 g CO_2 (Figure 12). CO_2 emissions from the E95 fuel cycles are positive because diesel vehicles that burn fossil fuel are used in transportation, farming, and other minor activities, and because 5% of E95 consists of gasoline. Thus, a portion of the RFG fuel cycle is added to the E95 fuel cycle, reflecting the fuel cycle emissions associated with the denaturant. Displacing gasoline with ethanol fuels is a policy option

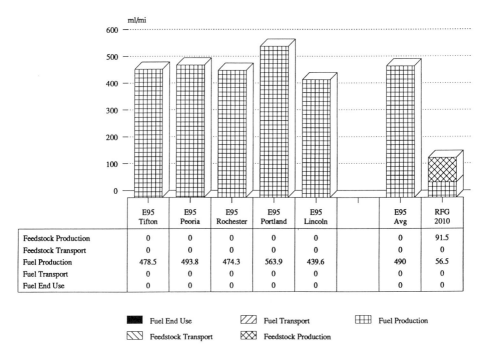

	E95 Tifton	E95 Peoria	E95 Rochester	E95 Portland	E95 Lincoln		E95 Avg	RFG 2010
Feedstock Production	0	0	0	0	0		0	91.5
Feedstock Transport	0	0	0	0	0		0	0
Fuel Production	478.5	493.8	474.3	563.9	439.6		490	56.5
Fuel Transport	0	0	0	0	0		0	0
Fuel End Use	0	0	0	0	0		0	0

Fuel End Use   Fuel Transport   Fuel Production
Feedstock Transport   Feedstock Production

Figure 13 Fuel cycle emissions of wastewater.

that appears to have a substantial impact on transportation-related CO_2 emissions. More than 90% of the CO_2 emissions associated with RFG can be avoided by replacing gasoline with E95.

Wastewater Emissions
The E95 fuel cycles produce 490 milliliters (ml) of wastewater per VMT, on average, compared with only 148 ml per VMT in the RFG fuel cycle (Figure 13). The wastewater produced in the E95 fuel cycle comes from the conversion facility, except for the water that is reflected in the 5% gasoline contained in E95. The wastewater in ethanol plants could be reduced by as much as 60% with more sophisticated water recycling designs.

The process water from ethanol production can be treated by city sanitation plants to produce potable water. The wastewater stream is an optimal environment for growing organisms and as such is suited to other agricultural uses.

Most of the wastewater produced in the RFG fuel cycle is formation water that is produced during oil production. It commonly contains salts, metal, oil, radionuclides, and other hazardous materials. Most of the formation water is reinjected into the oil reservoir or other geological zones. The formation water reinjected and the process water used for enhanced oil recovery (EOR) is not considered wastewater. If they were, estimated wastewater produced during crude oil production would be approximately 20 times higher than the total reported here.

Solid Waste Emissions
The E95 fuel cycles produce 11.7 to 22.2 g solid waste per VMT. Of this waste, about half is gypsum produced from neutralizing sulfuric acid used in the pretreatment process, and half is the ash remaining after the nonfermentable residues are combusted (Figure 14). If another method of biomass pretreatment could be used that did not require acid prehydrolysis, solid waste production could be cut by half. The solid waste produced by

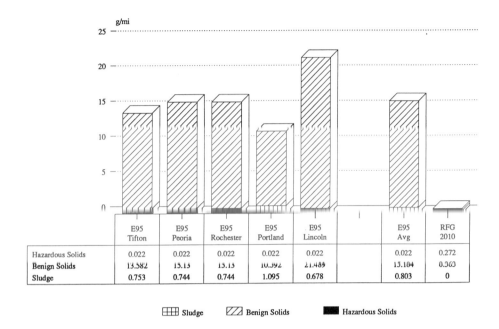

	E95 Tifton	E95 Peoria	E95 Rochester	E95 Portland	E95 Lincoln		E95 Avg	RFG 2010
Hazardous Solids	0.022	0.022	0.022	0.022	0.022		0.022	0.272
Benign Solids	13.582	15.13	15.13	10.392	21.489		13.104	0.363
Sludge	0.753	0.744	0.744	1.095	0.678		0.803	0

⊞ Sludge   ▨ Benign Solids   ▓ Hazardous Solids

Figure 14   Fuel cycle emissions of solid wastes.

an ethanol plant is not considered hazardous. Currently, biomass ash from combustion boilers is in demand as a landfill amendment to control acidity.

Approximately half of the 0.64 g solid waste per VMT produced in the RFG fuel cycle is considered dangerous: hazardous, toxic, carcinogenic, etc. Future waste reduction technologies, high-temperature combustion, and other alternatives are being explored that could reduce petroleum industry wastes.

Impact of Adding the Secondary Emissions Associated with Electricity

If one assumes that the by-product electricity sold by the ethanol plant offsets or partially eliminates the need for a utility company to produce electricity, then the avoided emissions can be viewed as emission "credits" for the electricity produced from the ethanol plant. Similarly, when electricity is consumed in a fuel cycle, the emissions associated with producing that electricity should be included in the fuel cycle. For the six cases discussed previously, NO_x, SO_2, CO_2, particulates, and solid waste emissions per kilowatt-hour (kwh) were subtracted as credits when electricity was produced and added to the fuel cycle inventories (debits) when it was consumed. Table 4 shows the incremental changes in the base cases caused by the sensitivity analysis.

The ethanol fuel cycles are regional. Some stages of the RFG fuel cycle have activities in them that are regionally concentrated (like refining and oil production), whereas other stages are national in character (fuel distribution). Utilities also have regional characteristics, depending on local resource endowments and environmental air quality regulations. Therefore, analysts estimated regional electricity generation emissions characteristics for each region.

Characteristic electricity generation emissions are added to the fuel cycle where electricity is consumed and credited against emissions where electricity is produced. For crude oil production, transportation, and refining, the activities are apportioned to various regions, depending on where they occur today. Thus, the emission debits and credits for

Table 4 **Changes in emissions from base cases[a] from electricity emission debits and credits (mg per VMT)**

Emission	Fuel Cycle	Fuel Distribution	Fuel Production	Feedstock Transport	Feedstock Production	Total Change
NO_x	E95	24	−71	0	0	−47
	RFG	28	5	10	5	48
SO_2	E95	23	−83	0	0	−60
	RFG	31	5	6	3	45
CO_2	E95	7.6	−28.9	0	0	−21.3
(grams)	RFG	10.2	2.5	3.9	2.0	18.6
PM	E95	1.8	−5.3	0	0	−3.5
	RFG	2.3	0.5	0.7	0.4	3.9
Solid	E95	1430	−3170	0	0	−1740
Waste	RFG	1900	360	640	270	3170

[a] The numbers shown should be added to the values presented in Table 3 to calculated TFC emissions for these sensitivity analyses.

these stages of the fuel cycle were weighted by the proportion of the activity that occurred in each region.

National average emissions are applied to electricity consumption in the RFG fuel cycles for fuel distribution. This may not accurately portray actual emissions if specific electric usage for distribution fuel is examined regionally. However, since national average statistics were used to estimate electricity consumption in fuel distribution, using the national average of electric generation emissions was appropriate for this study.

All the emissions considered in the electricity sensitivity analysis of the E95 fuel cycles are reduced because more electricity is produced by the ethanol facility than is consumed in the entire fuel cycle. In some cases, the electricity production credit offsets more than the total amount of SO_2 and CO_2 produced throughout the entire fuel cycle, including emissions associated with the electricity consumed.

All the emissions examined in the electricity sensitivity analysis of RFG increase because a large amount of electricity is consumed in refining and fuel distribution, and no electricity is produced. Total NO_x emissions increase 15%, SO_2 emissions increase more than 50%, particulate emissions nearly double, and solid waste emissions increase 500%. CO_2 emissions increase 6%.

Each mile traveled on E95 produces only 6.5 g CO_2 when the electricity credits are included. This is a reduction of more than 75%. Total NO_x emissions are reduced by about 14%. Solid waste is reduced by more than 10%. Particulate emissions are cut by more than one third and SO_2 emissions are offset by a factor of 2.1.

When the results of the electricity sensitivity cases are compared for E95 and RFG, E95 provides a net benefit to society by significantly reducing the amount of air pollutants produced by its fuel cycle compared with RFG. This type of analysis is the primary reason that total fuel cycle analysis is important to policy makers because it provides a mechanism in which the many costs and benefits associated with a fuel can be compared equally. This comparison is limited to an inventory of selected physical inputs and outputs. Economic and social impacts should be included in the future for a complete analysis.

ENERGY RATIOS

This study uses three methods to address the issue of fuel cycle energy efficiency: 1) process energy requirements, 2) fossil fuel use (depletable resources), and 3) total energy

Table 5 **Total energy requirements**

	E95	RFG 2010
Process Energy Inputs (Btu VMT)		
Feedstock production	167.8	34.8
Feedstock transportation	31.3	121.5
Fuel production	81.0	190.7
Fuel distribution	150.7	194.9
Subtotal process energy inputs	430.8	541.9
Feedstock Energy Inputs and MTBE (Btu VMT)		
Biomass feedstock	4659.6	n/a
Crude oil feedstock	245.4	3105.8
MTBE	0	293.5
Subtotal feedstock energy and MTBE	4905.0	3399.3
Fuel Including Fuel Additives, Ethanol or MTBE (Btu VMT)		
End-use fuel energy value	2752.0	3108.0
Energy Ratios		
Nonfeedstock inputs/fuel output	0.157	0.174
Total fossil inputs/fuel output	0.246	1.27
Total inputs/fuel output	1.94	1.27

(Table 5). Throughout the energy analysis, lower heating values are assumed. Biomass heating values are estimated on a dry weight basis. The heat rate of 10,400 Btu per kWh for electricity captured the inefficiencies of electricity production. Energy embodied in fertilizer, chemicals, and electricity is included.

Process Energy Requirements

Process energy is energy required to operate equipment in each of the four stages of the fuel cycle: feedstock production, feedstock transportation, fuel production, and fuel distribution. The end-use stage is not included in this category since the only operation that occurs in that stage is the combustion of the fuel to provide mobility; it is shown below under *Fuel including fuel additives*. Process energy does not include feedstocks (not even the feedstocks consumed to provide process energy in refineries and ethanol production facilities, e.g., shrinkage) or fuel additives such as MTBE.

E95 fuel cycles are slightly more efficient than RFG, consuming fewer Btus of process energy inputs per Btu of output. On the whole, the differences in process energy consumed per Btu of energy output is relatively similar; however, some interesting differences among the stages are noteworthy.

Feedstock production is almost three times more energy intensive (Btu energy consumed per Btu energy feedstock produced) for E95 than for RFG. This is the result of producing a relatively diffuse, low-Btu fuel. Half of the energy required in feedstock production for E95 is used to fuel farm equipment (diesel) and half is embodied in the production of nitrogen fertilizer.

The energy consumed in feedstock transportation is four to five times higher for RFG than for ethanol fuels on a basis of Btu of energy consumed per Btu of feedstock moved. Nearly 60% of the energy requirements in crude transportation are electricity inputs for pipeline transportation. The remainder is diesel for tanker, barge, rail, and truck transportation. Crude oil is transported longer distances (average 615 miles) compared with biomass (26 to 48 miles), which offsets the benefits of moving a more condensed energy product.

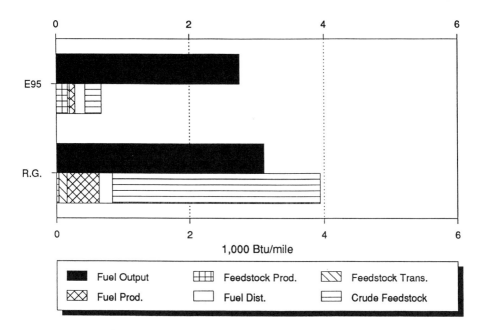

Figure 15 Total energy requirements without biomass feedstocks.

Crude oil refining is more energy intensive per Btu of final product (RFG without MTBE) than biomass conversion to E100 (pure ethanol without denaturants), when only the process energy inputs are considered. Part of this conclusion results from not accounting for internally produced and consumed process energy in either fuel cycle (by-products of refining and ethanol conversion that are combusted for process heat and power).

The E95 fuel cycles consume less energy in the distribution stage compared with the RFG fuel cycles because RFG distribution is based on national average transportation distances and E95 distribution is based on regional distribution infrastructure patterns.

Fossil Fuel Energy
Focusing on fossil energy inputs provides an insight into the effects of an ethanol fuel industry on our depletable resources. The total impact of consuming fossil fuels is examined by adding the crude feedstocks to the first energy balances; this includes the crude feedstocks that are transformed into gasoline and added to the ethanol fuel cycles in the conversion stage. Figure 15 provides a breakdown of process energy inputs and outputs by stage with crude oil feedstocks as a separate input.

Only 0.24 Btu fossil energy is required to produce 1 Btu E95, whereas 1.27 Btu fossil energy is required to produce 1 Btu RFG. Clearly, a biomass-ethanol industry could extend our fossil fuel supply over a longer period of time if the ethanol is used as a dedicated fuel to augment or displace future gasoline demand.

All Energy Sources
The third method of calculating energy ratios reflects the sum of all of the inputs, fossil and renewable, associated with fuel production. Ethanol fuel cycles appear to be less efficient than RFG fuel cycles. Nearly twice as much total energy is used to produce E95 than is contained in the fuel itself; however, 87% of that energy is renewable. The lower energy efficiency of the E95 fuel cycle is primarily the result of converting only a fraction

of the feedstock into a condensed liquid fuel and using a low-Btu boiler fuel in the ethanol plant.

In Table 5, only a fraction of total energy inputs are shown in each of the fuel cycles—the portion required to produce, transport, and convert feedstocks into liquid fuel. The allocations discussed earlier have been applied.

CONCLUSIONS AND DISCUSSION

This study presents data on environmental emissions produced by two fuel cycles: E95 and RFG, which can be used to support impact studies, cost/benefit studies, and economic analyses. Providing the best possible estimates of the quantities of emissions created by an industry is necessary to conduct credible and useful studies of environmental impacts and their benefits or costs. This study has focused on providing quality information for further analysis.

The results of this work can be used to evaluate limited policy objectives. If decision makers need to reduce a particular emission, such as carbon dioxide, then this report provides information that can be used to evaluate the benefits of substituting gasoline with E95 and RFG. For example, this report indicates that E95 reduces CO_2 emissions, which could reduce or forestall global warming, if substituted for RFG. However, we have only quantified CO_2, and not necessarily included other greenhouse gases such as nitrous oxide and methane. The information summarized here and described in more detail by Tyson[4] is a powerful tool, but not the only tool needed to evaluate policy options for transportation.

Each fuel examined in this report has some advantages that the other fuels do not; e.g., reduces CO, VOC, or other emissions. No one fuel examined can be characterized as better or worse than its alternatives based on the results of this study alone because benefits of reducing some emissions are offset by increases in other emissions. Future analysis of economic, environmental, and health impacts of the emissions volumes reported are required to support this type of conclusion.

The analysis revealed a number of interesting results:

- Vehicle emissions create the bulk of most of the gaseous emissions.
- Increasing use of E95 is a promising option for reducing CO_2 emissions from the transportation sector because E95 fuel cycles produce less than 10% of the CO_2 emissions produced by the RFG fuel cycle.
- When emissions from electricity generation are added to the fuel cycle analysis, E95 fuels produce significantly less NO_x, SO_2, particulates, and CO_2 emissions than RFG.
- Ethanol fuels can extend our fossil fuel resources in the transportation sector until a permanent solution is found for our dependency on petroleum, since ethanol fuels require fewer fossil fuel resources than RFG.
- Assumptions concerning technology performance, particularly emission control equipment, environmental regulation, and allocation assumptions, heavily influence the results of this study.

Vehicle emissions dominated total fuel cycle gaseous emissions in all the fuel cycles. Improvements in engine performance, catalytic converters, and other vehicle emission controls will benefit both fuels. CAAA standards for vehicle emissions will play a central role in determining the emission characteristics of the fuel cycles. Owing to the lack of data on ethanol fuel emissions, many emission estimates are based on the assumption that fuel and auto manufacturers will design systems to meet regulations. Thus, these regulations are critical focal points of the analysis.

E95 is a promising option for reducing CO_2 emissions from the transportation sector. Used in sufficient quantities, fuel substitution policies can be effective policy tools for mitigating global warming, because most of the CO_2 produced from the ethanol fuel cycle is recycled each year in new growth of trees and grasses.

There are further benefits of E95 substitution when the electricity from the ethanol facility is considered and soil carbon accumulation is included. Soil carbon accumulation was not accounted for in the base cases because it was treated as a long-term investment rather than a short-term operational characteristic. If electricity offsets are included in the analysis, E95 fuel cycles produce 5 g CO_2 per VMT. If E95 vehicles captured 10% of the passenger vehicles market by the year 2010, U.S. CO_2 emissions could be reduced by 35 million tons per year.

In addition, E95 can reduce the production of SO_2 emissions in the utility sector. If E95 is substituted for RFG, the U.S. production of SO_2 will fall. Similarly, when E95 is substituted for RFG, NO_x and particulates are also reduced. It is clear that E95 fuels provide substantial environmental benefits in emission reductions once the electricity produced from the ethanol facility is factored into the analysis.

This last result further emphasizes the impact of assumptions on fuel cycle estimates. Excluding secondary emissions, such as the emissions from producing electricity, can underestimate the total impact a fuel creates on society. The readers should keep this in mind and recall that the emissions of other inputs, such as fertilizer and MTBE, are not included in this study except in the energy balance discussion. Previous work by Deluchi[1] indicates that electricity credits and fertilizer emissions offset each other in biomass-ethanol fuel cycles (Deluchi only examined greenhouse gases). If this remains true, then the cases shown in this report are accurate estimates of emission inventories.

Approximately 87% of the energy inputs used to make E95 are renewable. The fossil fuel energy consumed in the E95 fuel cycle is less than 20% of the total energy consumed in the RFG fuel cycle. Ethanol fuels can prolong our limited petroleum resources and reduce our dependence on imported oil. Each year the world's consumption of petroleum increases and the exhaustible reserves shrink. Therefore, the fact that ethanol made from crops and trees does not require large amounts of fossil fuels to produce and is made from a renewable resource, will become a large benefit in the future. A renewable resource is not limitless; our future production of ethanol will be limited by land use policies. However, this country will be able to produce a constant amount of fuel, whether it is 10% or 50% of the gasoline demand, year after year after year. The availability of a fuel that can be substituted for gasoline could be very important for our future generations.

The issues at stake often become obscured by more immediate concerns, such as clean air, more fuel-efficient vehicles, testing and demonstration of new technologies, and the current economic condition. Information is needed to address all these concerns and provide a basis for trade-offs. The TFCA methodology has proven to be a useful analytical tool for DOE, to the extent that it can be used to develop detailed estimates of emission inventories. This information can be used to rank future technologies or fuels in a consistent manner based on specific emission criteria.

In addition, the TFCA demonstrated that useful information can be collected and organized in a manner that provides insights concerning both the development of new technologies and their environmental implications. The process in which scientists and engineers were asked to develop their best estimate of one specific combination of technologies needed to produce ethanol from biomass, and estimate the required inputs and wastes produced, led to many questions concerning the technologies selected and improvements in the overall design. Several new lines of research were developed as a result of this work.

This study describes only one unique combination of technologies used to produce ethanol from biomass. Many others are possible. Inasmuch as the biomass-ethanol technology has been examined in detail, changes to the process can be integrated into the database developed for this project, and the impacts of new technologies or engineering designs can be examined in the future.

We believe that the TFCA methodology has been a useful tool for DOE and will provide the department with the type of environmental information needed to assess future technologies and alternative fuels.

ACKNOWLEDGMENTS

This project was conducted for the U.S. Department of Energy at the request of the Assistant Secretary of Energy Efficiency and Renewable Energy, with funding from the Office of Transportation Technologies and the Office of Planning and Assessment. The work was performed by:

- National Renewable Energy Laboratory (formerly the Solar Energy Research Institute)
- Oak Ridge National Laboratory
- Pacific Northwest Laboratory
- Meridian Corporation
- Radian Corporation
- E. A. Mueller
- CH_2M Hill
- J. E. Sinor Consultants, Inc.

REFERENCES

1a. Clean Air Act Amendments of 1990, 42 United States Code, Title II, Sections 201 – 235. DeLuchi, M. A., Emissions of Greenhouse Gases from the Use of Transportation Fuels and Electricity. Volume I. Main Text; and Volume II. Appendices. Center for Transportation Research, Argonne National Laboratory. ANL/ESD/TM-22, 1991.

1b. Environmental Protection Agency, Compilation of Air Pollutant Emission Factors, Volume I. Stationary Point and the Areas Sources. (AP-42). 4th ed., Research Triangle Park, NC, 1985.

2a. U.S. Department of Energy, *National Energy Strategy*, U.S. Government Printing Office, DOE/S-0082P, February 1991a.

2b. U.S. Department of Energy, *National Energy Strategy: Technical Annex 2. Integrated Analysis Supporting the National Energy Strategy: Methodology, Assumptions, and Results*, (NES2), U.S. Government Printing Office, DOE/S-0082P, February 1991b.

3. **Tyson, K. S., Riley, C. J., and Humphreys, K. K.,** Fuel Cycle Evaluations of Biomass-Ethanol and Reformulated Gasoline Fuels, Volume I, Summary report (draft), National Renewable Energy Laboratory, Golden, CO, 1992.

4. **Tyson, K. S., Ed.,** Fuel Cycle Evaluations of Biomass-Ethanol and Reformulated Gasoline Fuels, Volume II, Appendices (draft), National Renewable Energy Laboratory, Golden, CO, 1992.

5. **Pierce, T. E., Lamb, B. K., and Van Meter, A. R.,** Development of a Biogenic Emissions Inventory System for Regional Scale Air Pollution Models, presented at the 83rd Annu. Air and Waste Management Assoc. Meet. Exhibition, June 24-29, 1990, 90-94.3.

Chapter 4

Organic Carbonyl Compounds in Albuquerque, New Mexico, Air: A Preliminary Study of the Effects of Oxygenated Fuel Use

Carl J. Popp, Lin Zhang, and Jeffrey S. Gaffney

CONTENTS

ABSTRACT: A suite of inorganic and organic species was analyzed for four 2-4 day time periods over a year in Albuquerque, New Mexico, to determine baseline conditions for organic pollutants under the current air pollution control parameters. Concentrations of low molecular weight carbonyl compounds were relatively high compared with areas such as Los Angeles. Formic acid concentrations in air samples were significant even in winter. In addition, ratios of peroxypropionyl nitrate to peroxyacetyl nitrate are higher than expected and may be related to the use of oxygenated fuels which are used to mitigate CO concentrations. The number of CO violations in Albuquerque has decreased steadily since 1982 and the downward trend has continued since 1989 when oxygenated fuel use was mandated. It is, therefore, difficult to correlate directly the drop in CO violations to the use of oxygenated fuels when such factors as fleet turnover, wood burning controls, emissions testing, and meteorological conditions also may be playing significant roles. More detailed studies are needed to determine the specific relationship between the use of oxygenated fuels and the air quality in Albuquerque, New Mexico, and similar urban areas in the western U.S.

INTRODUCTION

Urban air pollution has developed into a serious problem in many areas of the U.S. with mobile sources of pollutants playing a major role. Pollutants presenting potential health hazards include carbon monoxide (CO) and ozone (O_3) as well as more exotic species such as peroxyacetylnitrate (PAN) and peroxypropionylnitrate (PPN). In addition, O_3,

PAN, and PPN have phytotoxic properties[1] and organic species can contribute to increased levels of O_3. Oxides of nitrogen ($NO_x = NO + NO_2$) can contribute to acidic rains and fogs, and particulates and NO_2 can cause visibility problems often resulting in the "brown cloud" phenomenon common in many cities during wintertime temperature inversions, especially prevalent in the western U.S. The inversions trap not only components responsible for visibility degradation, but all pollutants resulting in a series of complex chemical reactions perpetuating the formation of higher concentrations of pollutants and subsequent transport from the pollution source.

In order to mitigate the problems associated with increased pollutant emission rates in urban centers, many municipalities have adopted various measures such as required emission testing of motor vehicles, bans and limitations on activities such as wood burning, and the mandated use of oxygenated fuels in the wintertime. As of 1991, six large urban areas in the mountain western and southwestern U.S. (Albuquerque, NM; Denver, CO; Phoenix, AZ; Las Vegas, NV; Reno, NV; and Tucson, AZ) have adopted the use of oxygenated fuels in the winter as one step toward reducing CO pollution[2] and many more municipalities are expected to follow suit. As outlined by Gaffney and Marley,[3] a number of issues and unanswered questions have been raised regarding emissions of precursors of organic carbonyl compounds including organic acids, aldehydes, and PAN, and the ultimate net effect the use of alternative fuels has on the chemistry of urban atmospheres. Typical strategies for using oxygenated fuels containing such additives as ethanol, methanol, and ethers [methyltertiarybutyl ether (MTBE) as an example] are directed at reducing both CO and O_3 emissions. Using such blended fuels (or pure alcohols) has the potential to increase atmospheric concentrations of low molecular weight, photochemically active organic compounds[4,5] that may result in increased concentrations of ozone, hydrogen peroxide, and/or organic oxidants. In this study, the concentrations of air pollutants including CO, O_3, NO_x, PAN, PPN, and formic acid were monitored for several periods during one calendar year in Albuquerque, NM, where use of oxygenated fuel is mandated during the winter. The species CO and NO are direct emission products, while the species NO_2, O_3, PAN, PPN, and formic acid are more likely to arise as secondary pollutants. Greater than 95% of the oxygenated fuel currently in use in Albuquerque is ethanol-enhanced,[6] which is apparently an unusually high percentage. For comparison, data are reported for Socorro, NM, and a remote mountain site which are located away from the urban influence of Albuquerque and where the use of oxygenated fuel is not required.

METHODS, PROCEDURES, AND SAMPLE SITES

STUDY AREA, SAMPLE SITES, AND GENERAL SAMPLING PROTOCOL

The primary study area is the city of Albuquerque, which has a metropolitan population of about 500,000 and is located along the Rio Grande Valley at an altitude of 5000 ft (1524 m) in central New Mexico, while Socorro (altitude = 1400 m) has a population of ~9000 and is located 75 miles (120 km) south of Albuquerque (Figure 1A). The region is high desert (Upper Chihuahuan) averaging less than 9 in (<23 cm) of precipitation yearly. The remote, mountain site is located at Langmuir Laboratory which is operated by New Mexico Tech and is at an altitude of 10,500 ft (3300 m) in the Magdalena Mountains west of Socorro (Figure 1A). Two sampling sites in Albuquerque were selected to coincide with air quality sampling sites operated by the Albuquerque Air Quality Control Division (AAQCD) and are shown as 1 (Site 2R) and 2 (Site 2ZM) on Figure 1B. Site 2ZM is located near the center of the highest traffic density in Albuquerque (35,000 to 55,000 vehicles per day)[7] and should be representative of direct emissions, while site 2R is located in the more rural south valley with much lower traffic densities and considerably more vegetation due to agricultural use and the presence of a cotton-

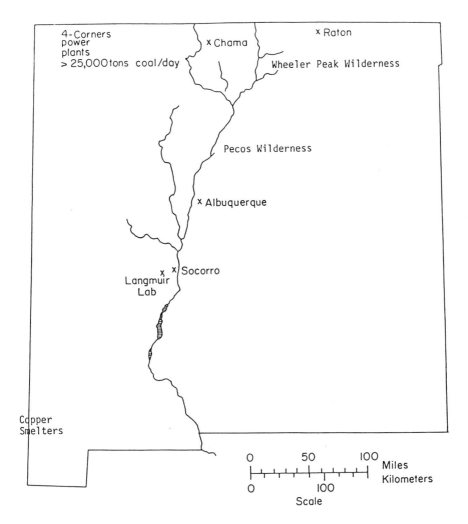

Figure 1A Map of New Mexico showing locations of Albuquerque, Socorro, and Langmuir Laboratory.

wood forest along the Rio Grande flood plain. The distance from 2ZM to 2R is 9 miles (14.4 km).

In order to determine any differences in the atmospheric composition of trace pollutants, samples were taken during wintertime when oxy-fuels use is mandated in Albuquerque and during summer and spring when the use of such fuels is not required. All sampling times in Albuquerque involved simultaneous measurement of NO_x, O_3, CO, formic acid, PPN, and PAN. Samples from Albuquerque were collected in August and December 1991, in February and May 1992, and at similar times in Socorro. Samples at Langmuir Laboratory were collected in the summer (June/July) of 1991 and 1992.

NO_x, CO, AND OZONE ANALYSIS

NO_x and O_3 were analyzed using standard chemiluminescent oxides of nitrogen analyzers and ultraviolet ozone analyzers. Instruments include Dasibi 1003 O_3 analyzers and Monitor Labs 8440 and Thermoelectron Corp. NO_x Analyzers. New Mexico Tech's NO_x analyzer was calibrated with certified cylinders of NO in nitrogen that are checked yearly

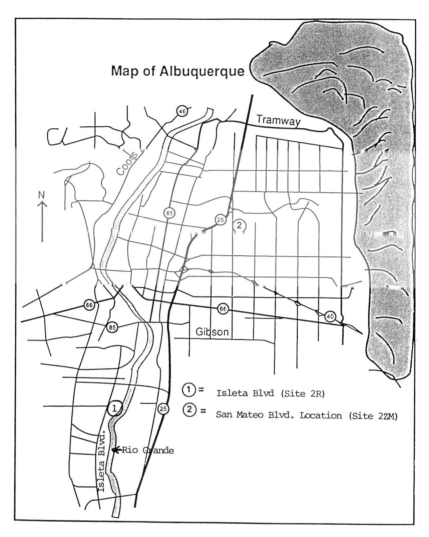

Figure 1B   Map of Albuquerque showing location of sampling sites.

at the National Oceanic and Atmospheric Administration Aeronomy Lab in Boulder, CO. To check conversion efficiencies of the NO_x analyzer, NO_2 permeation tubes were used. Instruments operated by the City of Albuquerque regularly are subject to U.S. EPA compliance checks. Ozone analyzers were calibrated using a Thermo Environmental Instruments, Inc. Model 49PS ozone calibrator or similar equipment. Analyses were averages for 1-h periods.

PAN AND PPN ANALYSIS

PAN and PPN were analyzed by gas chromatography using an electron capture detector and gas chromatograph (Shimadzu Mini II) equipped with a chromosorb W-HP/10% carbowax column and automatic sampling valve and timer to collect and analyze samples every half hour. The analysis procedure and calibration follows that described by Tanner[4] and Gaffney.[1] Calibration of PAN was attained by analyzing acetate from the base hydrolysis of PAN using ion chromatography (Dionex 2000 Si/AS 10 column) and also by splitting the PAN and PPN flow from a diffusion tube through NaOH with subsequent

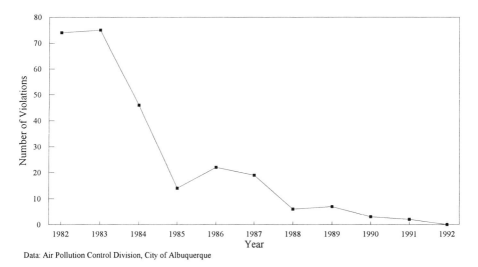

Data: Air Pollution Control Division, City of Albuquerque

Figure 2 Number of carbon monoxide violations by year for all sites in Albuquerque, NM.

spectrophotometric analysis of nitrite.[8] PAN and PPN were also calibrated using a chemiluminescent NO_x analyzer with flow split from a PAN or PPN loaded diffusion tube. PAN and PPN were prepared by the acid nitration of the appropriate peracids as described by Gaffney.[1] The detection limit for both PAN and PPN was 30 pptv.

FORMIC ACID

Organic acids were scrubbed from the air using a fine water mist following the procedure described by Cofer et al.[9] A 1-μ pore size Teflon®* prefilter was placed upstream from the nebulizer (DeVilbiss Model 40) to trap particulates and air was pulled through the nebulizer at about 1 l/min for about 2 h. Flow rates for air sampling were monitored by rotameters calibrated using a 2-l bubble meter. The distilled water in the nebulizer (~7 ml) was diluted to 25 ml in a volumetric flask and stored under refrigeration after several drops of chloroform were added to inhibit bacterial degradation of the organic acids. Analysis was accomplished using suppressed ion chromatography (Dionex 2000 Si) and an AS10 column with NaOH (40 mM) eluent. Calibration standards were prepared from the sodium salt (99+%) of formic acid. The detection limit for formic acid was 0.2 ppbv for the 2-h sample.

RESULTS AND DISCUSSION

AMBIENT AIR QUALITY IN ALBUQUERQUE, NM, AND CO VIOLATIONS

In recent years, the City of Albuquerque has implemented several policies to address air quality issues. Woodburning advisories were instituted in 1988 and were issued in winter when meteorological conditions (temperature inversions) were imminent. A motor vehicle emissions testing program was also begun in 1988. In 1989, the use of oxygenated fuels was mandated for Albuquerque and for Bernalillo County in which the city is located. The oxygenate could be either ethanol or MTBE. Initially, both additives were used extensively, but as of 1991–1992, the fuel mix was dominated by ethanol (>95%.)[6] The number of CO violations of the 8-h U.S. EPA standard since 1982 are plotted in Figure 2.[10] The most dramatic drop in violations occurred between 1983 and 1985 (from

---

* Registered Trademark of E. I. du Pont de Nemours and Company, Inc., Wilmington, DE.

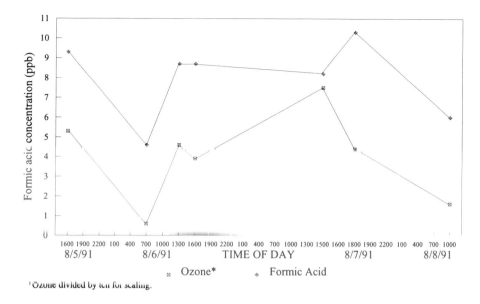

Figure 3A Concentrations of gas-phase formic acid and O_3 during August 5, 6, 7, and 8, 1991 at site 2ZM at Albuquerque.

72 to 12 violations), reaching a low of zero in the past winter (1992–1993). Inasmuch as most of the drop in the number of violations happened well before air pollution control measures were introduced, it is difficult to assign credit to the measures taken by the City of Albuquerque. Some of the decrease is undoubtedly due to fleet turnover and improved catalysts resulting in cleaner-burning engines. However, given that the number of CO violations have decreased considerably, this study seeks to address issues related to the potential for changes in organic species.

FORMIC ACID

Gas-phase formic acid was analyzed beginning in the summer of 1991 and plots of concentration vs time from Albuquerque (August 1991, March and February 1992) and the mountain laboratory are shown in Figures 3 to 6. Average values for Albuquerque, Socorro, and the mountain site are summarized in Table 1. Formic acid concentrations peaked from midday to late afternoon during all sampling periods at both the Albuquerque and remote sites, corresponding to times of increased photochemical activity and increases in oxidants such as O_3, and suggest direct emissions and/or rapid formation from precursors (Figures 3A to 6). This contrasts with results in Los Angeles, which often showed high concentrations of formic acid at night.[11] Potential precursors to the formation of formic acid are formaldehyde and plant emissions such as isoprene and terpenes[12] (Jacob and Wofsy 1988) that undergo photolytic and/or chemical decomposition to form formaldehyde, which then may be oxidized to formic acid. The potential also exists for the direct emission of formic acid by plants. These sources and cycling effects are summarized in a review by Keene and Galloway.[13] The time necessary for these processes is on the order of hours. The highest concentrations of formic acid (10 to 17 ppbv) in Albuquerque occurred in August at the South Valley site (2R, Figure 3B), where vegetation probably contributes to relatively high biogenic emissions. The two Albuquerque sites exhibited parallel behavior in August; but at the high traffic density site (2ZM, Figure 3A), formic acid concentrations were lower, again possibly related to lower biogenic emissions, either of direct emissions and/or precursors such as isoprene. In

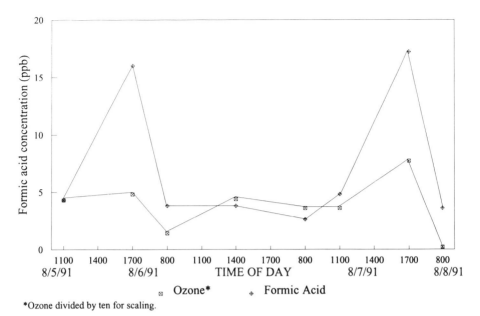

Figure 3B Concentrations of gas-phase formic acid and O₃ during August 5, 6, 7, and 8, 1991 at site 2R at Albuquerque.

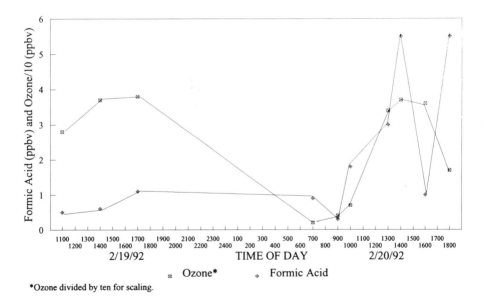

Figure 4 The concentrations of formic acid during February 19–20, 1992 at 2ZM, Albuquerque, NM.

contrast to the summertime samples, the highest formic acid concentrations in December (Table 1) and February (Figure 4) were 11.1 ppbv and 5.5 ppbv, respectively. The December samples averaged 8.9 ppbv for both sites in Albuquerque, which was surprisingly high because for several days the city was subject to an unusual period of continual cloudiness, fog, and drizzle with no sunshine during the sample period. As a consequence,

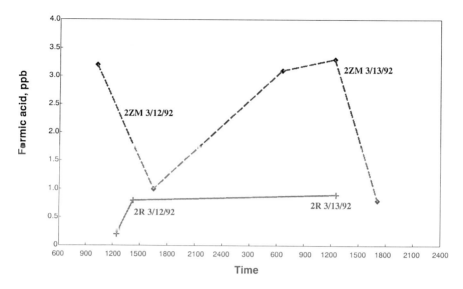

Figure 5 Concentrations of gas-phase formic acid during March 12 and 13, 1992 at 2ZM and 2R at Albuquerque.

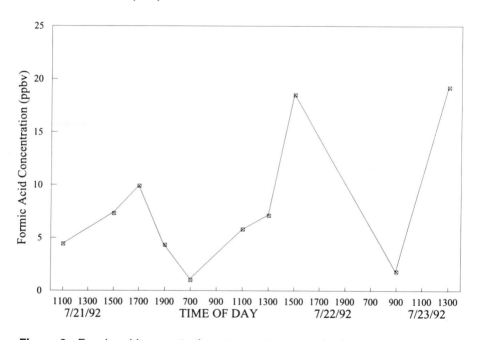

Figure 6 Formic acid concentration at a remote mountain site, July 21–23, 1992.

O_3 was extremely low and air quality was generally very good. This together with almost no photochemical activity apparently led to a buildup of formic acid in February (Figure 4) and March (Figure 5). The formic acid concentrations peaked at 5.5 and 3.5 ppbv in February and March, respectively, at site 2ZM and the averages were 2.1 ppbv (February) and 2.0 ppbv (March), which were much less than found in August or December. In comparison with Albuquerque, formic acid concentrations in Socorro were also high in

Table 1 **Average values for the concentrations of formic acid (FA) in Albuquerque, Socorro, and Magdalena mountain air from this study compared with FA concentrations from other studies**

Site	Date	FA Concentration[a]	Ref.
Albuquerque, NM	Summer 1991-92	6.9 (14)	
	Winter 1991-92	4.1 (16)	
	February	2.1 (11)	
	December	8.9 (5)	
Socorro, NM	Summer 1991-92	1.6 (6)	
	Winter 1991-92	6.8 (11)	
	February	NA[b]	
	December	6.9 (6)	
Magdalena mountains	Summer 1991-92	6.0 (19)	
	Winter 1991-92	NA[b]	
	February	NA[b]	
	December	NA[b]	
Tucson, AZ	Winter 1979-1980	2.0	19
Amazon (Brazil)	July 1985	1.6	20
Southern California	September 1988-1989	2.8–6.1	11
Southern California	1989	1–13	21
Central Pennsylvania	Summer 1991	2.5	22

[a] Concentrations are in the gaseous phase, expressed in ppbv and the number of samples averaged is shown in parentheses.

[b] Not analyzed.

winter but were low in summer, with average concentrations of 6.8 ppbv in the winter and 1.6 ppbv in the summer (Table 1). The biogenic emissions in summer might be expected to be similar in Socorro and Albuquerque but much greater anthropogenic organic emissions in Albuquerque might account for higher concentrations of formic acid precursors in the city. It is not clear why wintertime formic acid concentrations are relatively high in Socorro, as they were in Albuquerque, except that it is possible that wood burning may contribute which might explain high wintertime concentrations in Socorro when temperature inversions trap smoke from stoves and fireplaces.

The summertime average of formic acid at the remote mountain site was 6.0 ppbv, with peak values of 20 ppbv for samples collected in both the summer of 1991 and 1992 (Table 1 and Figure 6). The source of the formic acid at this site was probably plant emissions (direct emissions and/or oxidation of hydrocarbon emissions by plants). As with daily variations shown in Albuquerque, the formic acid concentrations reached their peak values in the afternoon when both plant emissions and photochemical activity would be highest (Figure 6).

PEROXYACETYL AND PEROXYPROPIONYL NITRATES

PAN and PPN concentrations are shown on a diurnal basis for two of the sampling periods (August 1991 and March 1992) in Figures 7 and 8, respectively. PPN concentrations are lower than PAN concentrations but concentrations of the two species correlate well, indicating similar sources of organic precursors and similar photochemical behavior. The lowest PAN concentrations (0.05 ppbv) occurred during the December sampling; this correlates with the cloudy and rainy weather conditions resulting in relatively low NO_x and O_3 concentrations. When sunlight and high O_3 are present, PAN concentrations reached maxima near 6 ppbv in August 1991 (Figure 6). These high values in summer

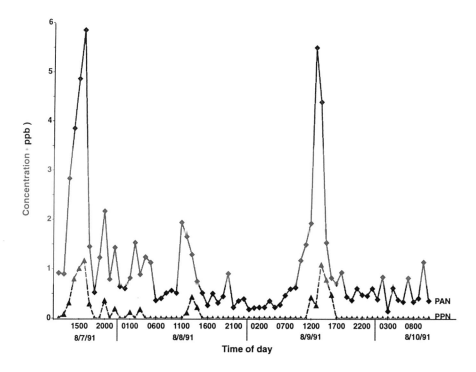

Figure 7 PAN and PPN concentrations during August 7–10, 1991, Site 2ZM, Albuquerque.

were surprising considering the instability of PAN at high temperatures. The only historical PAN data for Albuquerque was obtained in February 1988.[14] The average PAN concentration for the February 1988 sampling was 0.74 ± 0.52 ppbv, and in February 1992 an average value of 1.1 ± 0.5 ppbv was obtained in this study. This slight increase in PAN concentrations for data obtained before and after the implementation of the oxygenated fuels program is insufficient evidence to suggest that levels of PAN precursors may be increasing. However, it has been suggested[4,11,15] that the PPN/PAN ratio may change when the mix of organic precursors changes. In Table 2, PPN/PAN ratio data are summarized. All data are normalized to take into account the fact that some authors assumed that the electron capture detector (ECD) responses to PAN and PPN were similar and often only the PAN response was calibrated. Grosjean[16] has shown that PPN is about 82% as sensitive as PAN. Additionally, our calibrations showed a similar factor of 78% when both PAN and PPN were subject to individual calibration. For comparison purposes, all data from other authors in Table 2 (except those of Grosjean[11,15] and Shepson[17]) are corrected to take the ECD response into account.

A large increase in the percentage of PPN/PAN appears to have occurred between February 1988 and February 1992 ($9.3 \rightarrow 28.7$), suggesting an increase in primary emissions of propionyl precursors relative to acetyl precursors. The effect is more evident in February (PPN/PAN $\times$ 100 = 28.5) than in March (17.0) or August (19.3) (see Table 2). Because the percentage of PPN/PAN averages for August and March are also lower than the values found in February, the data may corroborate the theory that oxygenated fuels (mandated in February but not in August or March) may be playing a role in causing high ratios such as those found by Tanner et al.[4] in Brazil where ethanol-enhanced fuel was also used and in Southern California where ratios of 14 to 28 have been found by Grosjean

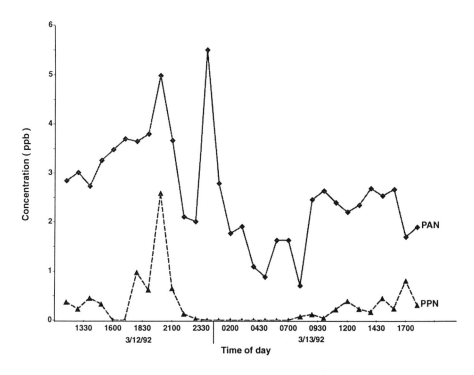

Figure 8 PAN and PPN concentrations during March 12–13, 1992, Site 2ZM, Albuquerque.

Table 2 **PPN/PAN × 100 at various sites (ratios are in ppbv units)**

Site (date)	PPN/PAN × 100	Ref.
Albuquerque (February 1988)	9.3	14
Albuquerque (February 1992)	28.7	This study
Albuquerque (August 1991, March 1992)	18.0	This study
Albuquerque (overall average for 4 dates, 172 samples)	19.3	This study
Rio de Janiero	25.6	4
Urban Eastern U.S.	16.5	23
Urban Western U.S.	6.4	23
Remote Western U.S.	7.7	23
Southern California Mountains	28.0	11
Southern California Mountains	14–19	15
Ontario, Canada	8.9	17

et al.[11,15] However, because the PPN/PAN ratios seem to be generally higher on a year-round basis in this study, other anthropogenic PPN source(s) may also be important. Although PPN has been thought to originate only from the oxidation of anthropogenic hydrocarbons, Grosjean et al.[18] have recently shown that a biogenic precursor, *cis*-3-hexen-1-ol, is capable of producing PPN in laboratory experiments designed to simulate atmospheric conditions. These phenomena warrant further investigation especially when the use of ethanol-enhanced fuel would be expected to increase PAN (2-C chain) rather than PPN (3-C chain). Concentrations of PAN in Socorro were below the detection limit

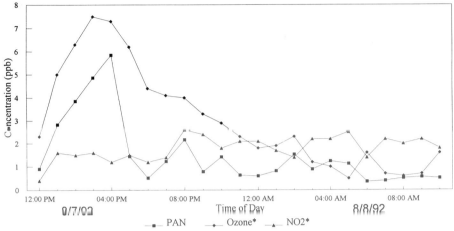

*Ozone and NO2 divided by ten for scaling

Figure 9 PAN, ozone, and NO₂ concentrations during August 7–8, 1991, Site 27M, Albuquerque.

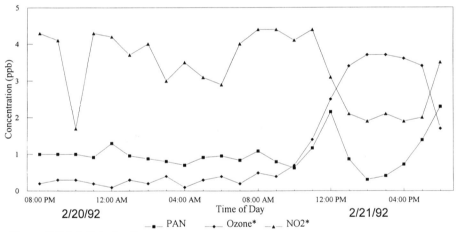

*Ozone and NO2 divided by ten for scaling

Figure 10 PAN, ozone, and NO₂ concentrations during February 20–21, 1992, Site 2ZM, Albuquerque.

of 30 pptv in the winter. Gaffney et al.[14] report PAN values averaging less than 0.4 ppbv for forested areas near Los Alamos in north-central New Mexico, with maximum values no greater than 2 ppbv.

PAN, O₃, AND NO₂ CORRELATION

PAN concentrations correlate reasonably well with O_3 concentrations, and $PAN/O_3/NO_2$ plots for two of the sampling periods are shown [August, 1991 (Figure 9) and March, 1992 (Figure 10)] and are typical for the Albuquerque samples. In general, when temperatures cool and the evening rush hour begins, PAN again increases in the presence of high O_3 and NO_2 concentrations.

CONCLUSIONS

The concentrations of PAN, PPN, and formic acid have been determined in Albuquerque, NM, and two comparison sites as a function of season. These compounds exhibit higher concentrations in Albuquerque than found at a control site in Socorro, NM, and peak values similar to such areas as Los Angeles, suggesting that anthropogenic sources in the urban environment are contributing to levels of organic compounds. These organic compounds may eventually lead to increases in urban ozone concentrations. PAN concentrations have increased slightly since 1988 but, of possibly greater significance, has been a large increase in the PPN/PAN ratio that suggests an increasing primary source of organic pollutants. Further, detailed studies are necessary to ascertain specific cause-and-effect relationships. More long-term, diurnal measurements need to be performed during summer when the amount of oxygenated fuels used is minimal and in winter when the use of such fuel is mandated. In addition, analysis of non-methane hydrocarbons and aldehydes needs to be performed in conjunction with data reported in this study to help establish the role of plant emissions and other reactive carbonyls in the processes.

ACKNOWLEDGMENTS

The authors gratefully acknowledge the Albuquerque Air Quality Control Division for logistic support and access to their sampling sites and data. Specific thanks go to Steve Walker, Mike Wehrle, Ed Peyton, and Barbara Ortega. A number of New Mexico Tech students worked on this project (Casey Caddell, Michelle Cash, Kirk Ferdig, Matt Hind, and Melissa York), and without their assistance the project would not have been completed. Matching funds from Dr. Allan Gutjahr through the Air Quality section of the Geophysical Research Center, New Mexico Tech, Research and Economic Development Office are also acknowledged. This work was performed in collaboration with the U.S. DOE's Office of Health and Environmental Research, Atmospheric Chemistry Program. Financial support of the project came primarily from the New Mexico Water Resources Research Institute Chino Mines Fund, Project No. 01423946.

REFERENCES

1. **Gaffney, J. S., Marley, N. A., and Prestbo, E. W.,** Peroxyacyl nitrates (PANs): Their physical and chemical properties, *The Handbook of Environmental Chemistry*, 4B, 1, 1989.
2. **Miller, S. S.,** Winter's unhealthy air, *Environ. Sci. Technol.*, 26, 45, 1992.
3. **Gaffney, J. S. and Marley, N. A.,** The search for clean alternative fuels: There's no such thing as a free lunch, *Atmos. Environ.*, 24A, 3105–3107, 1990.
4. **Tanner, R. L., Miguel, A. H., deAndrade, J. B., Gaffney, J. S., and Streit, G. E.,** Atmospheric chemistry of aldehydes: Enhanced peroxyacetyl nitrate formation from ethanol-fueled vehicular emissions, *Environ. Sci. Technol.*, 22, 1026, 1988.
5. **Grosjean, D., Miguel, A. H., and Tavares, T. M.,** Urban air pollution in Brazil: Acetaldehyde and other carbonyls, *Atmos. Environ.*, 24B, 102, 1990.
6. **Walker, S.,** City of Albuquerque, Air Pollution Control Division, personal communication, April 1993.
7. **M.R.G.C.G. (Middle Rio Grande Council of Governments),** 1988 Traffic Flows to the Greater Albuquerque Area, 629 Lomas NE, Albuquerque, N.M., 87102, 1988.
8. **Saltzman, B. E.,** Colorimetric microdetermination of nitrogen dioxide in the atmosphere, *Anal. Chem.*, 26, 1949, 1954.
9. **Cofer, W. R., Collins, V. G., and Talbot, R. W.,** Improved aqueous scrubber for collection of soluble atmospheric trace gases, *Environ. Sci. Technol.*, 19, 557, 1985.

10. **Wehrle, M.,** City of Albuquerque, Air Pollution Control Division, personal communication, April 1992.

11. **Grosjean, D.,** Ambient levels of formaldehyde, acetaldehyde, and formic acid in Southern California: Results of a one-year base-line study, *Environ. Sci. Technol.*, 25, 710, 1991.

12. **Jacob, D. J. and Woffsy, S. C.,** Photochemistry of biogenic emissions over the Amazon Forest, *J. Geophys. Res.*, 93, 1477, 1986.

13. **Keene, W. C. and Galloway, J. A.,** The biogeochemical cycling of formic and acetic acids through the troposphere: An overview of current understanding, *Tellus*, 40B, 322, 1988.

14. **Gaffney, J. S., Marley, N. A., and Prestbo, E. W.,** Measurements of peroxyacetyl nitrate (PAN) at a remote site in the southwestern United States: Tropospheric implications, accepted for publication, *Environ. Sci. Technol.*, 27, 1905, 1993.

15. **Grosjean, D., Williams, II, E. L., and Grosjean, E.,** Peroxyacetyl nitrates at Southern California mountain forest locations, *Environ. Sci. Technol.*, 27, 110, 1993a.

16. **Grosjean, D.,** Peroxypropionyl nitrate at a Southern California mountain forest site, *Environ. Sci. Technol.*, 25, 653, 1991.

17. **Shepson, P. B., Hastieg, D. R., So, K. W., and Schiff, H. I.,** Relationships between PAN, PPN and O_3 at urban and rural sites in Ontario, *Atmos. Environ.*, 26A, 1259, 1992.

18. **Grosjean, D., Williams, II, E. L., and Grosjean, E.,** A biogenic precursor of peroxypropionyl nitrate: Atmospheric oxidation of *cis*-3-hexen-1-ol, *Environ. Sci. Technol.*, 27, 979, 1993b.

19. **Dawson, G. A., Farmer, J. C., and Moyers, J. C.,** Formic and acetic acids in the atmosphere of the Southwest U.S.A., *Geophys. Res. Lett.*, 7, 725, 1980.

20. **Andreae, M. O., Talbot, R. W., Andreae, T. W., and Hariss, R. C.,** Formic and acetic acid over the Central Amazon Region, Brazil dry season, *J. Geophys. Res.*, 93, 1616, 1988.

21. **Grosjean, D.,** Organic acids in Southern California air: Ambient concentrations, mobile source emissions, *in situ* formation and removal processes, *Environ. Sci. Technol.*, 24, 77, 1989.

22. **Martin, R. S. H. W., Allwine, E., Ashman, L., Farmer, J. C., and Lamb, B.,** Measurement of isoprene and its atmospheric oxidation products in a central Pennsylvania deciduous forest, *J. Atmos. Chem.*, 13, 1, 1991.

23. **Singh, H. B. and Salas, L. J.,** Measurements of peroxyacetyl nitrate (PAN) and peroxypropionyl nitrate (PPN) at selected urban, rural and remote sites, *Atmos. Environ.*, 23, 231, 1989.

Chapter 5

The Effects of Oxygenated Fuels on the Atmospheric Concentrations of Carbon Monoxide and Aldehydes in Colorado

Larry G. Anderson, Pamela Wolfe, Regina A. Barrell, and John A. Lanning

CONTENTS

INTRODUCTION

The goal of air pollution control activities is to reduce the concentrations of pollutants to which people are exposed. A common method of assessing the effectiveness of air quality control programs is to determine the change in emissions that result from implementing some control procedure. The effects are determined for test systems, which may be a small fleet of vehicles or a small set of test processes. The effectiveness of reducing emissions for the test system is assumed to be appropriate for all systems that are affected by the control program. This leads to information on the expected emissions reductions.

0-87371-978-6/94/$0.00+.50
© 1995 by CRC Press, Inc.

75

Often this emissions information is used to calculate expected changes in ambient concentrations. This is a valuable approach when trying to justify the implementation of a new air quality control program. Prior to implementation, all one can do is to calculate the expected effects. However, once ambient data become available after the implementation of the program, it is crucial that the predicted and actual outcomes be compared.

The effectiveness of ongoing air quality control programs should be assessed by analyzing atmospheric concentrations of the appropriate pollutants. In addition, the assessment must include a search for other effects, whether expected or unexpected, whether positive or negative. In this chapter, we discuss the effects of oxygenated fuels use in Colorado on the atmospheric concentrations of carbon monoxide and aldehydes, and present a statistical time series analysis that addresses the extent to which the use of oxygenated fuels has reduced ambient carbon monoxide (CO) along Colorado's Front Range.

As a portion of our attempts to assess the effects and the effectiveness of Colorado's Oxygenated Fuels Program, we have been monitoring the atmospheric concentrations of formaldehyde (HCHO) and acetaldehyde (CH$_3$CHO) in downtown Denver since December 1987. We have collected 4-h average concentration data, 24 h a day for every winter since oxygenated fuels use began, as well as portions of several summers. This is probably the most extensive data set for formaldehyde and acetaldehyde concentrations at a single site, and is certainly the largest data set for aldehyde concentrations in an area that uses oxygenated fuels.

COLORADO'S OXYGENATED FUELS PROGRAM

PROGRAM DESCRIPTION

The first mandated use of oxygenated fuels to reduce the atmospheric concentration of CO occurred in a portion of Colorado. In 1987, the Colorado Air Quality Control Commission adopted a regulation requiring that oxygenated fuels be sold along much of the Colorado Front Range beginning on January 1, 1988. All automotive fuel sold in the region was required to contain 1.5% oxygen by weight from January 1, 1988 to March 1, 1988. During the first year of the program, the oxygen requirements were met with about 95% of the fuel sold being an 8% by volume mixture of methyltertiarybutylether (MTBE) in gasoline. The remainder of the fuel sold was a 10% by volume blend of ethanol and gasoline.[1] The program required 2.0% oxygen content by weight during the following two winter seasons, from November 1, 1988 to March 1, 1989 and from November 1, 1989 to March 1, 1990. These oxygenate requirements were met in 1988–1989 by 94% of the fuel sold being an 11% by volume blend of MTBE[2] and in 1989–1990 by 92% of the fuel being an 11% MTBE blend.[3] In each case, the remaining fuel sold was a 10% ethanol blend.

The program was expanded for subsequent years. Beginning November 1, 1990, 2.0% oxygen content by weight was required for all automotive fuels; the oxygen content requirement was increased to 2.6% for all grades except the premium unleaded grades beginning December 1, 1990. These levels of oxygen content in the fuels extended through March 1, 1991. These oxygenate requirements were met in 1990–1991 by 87% of the fuel sold being a 14% MTBE blend[4] and in 1991–1992 by 80% of the fuel sold being a 14% MTBE blend.[5] The remaining fuel sold was a 10% ethanol blend. The program has been further revised to comply with the 1990 Clean Air Act Amendments. From November 1, 1992 to March 1, 1993, all automotive fuels contained at least 2.7% oxygen content by weight. Oxygenated gasoline containing 15% by volume MTBE is considered equivalent.[6]

MODELED PROGRAM EFFECTIVENESS

The Colorado Department of Health (CDH) evaluates the effectiveness of the oxygenated fuels program in Colorado by model calculations. Data from a variety of emissions tests with oxygenated fuels are used to assess the reduction in carbon monoxide emissions from motor vehicles using these fuels. These data are used in MOBILE 3 or 4, a fleet emissions model, to calculate the reduction in carbon monoxide emissions from the entire motor vehicle fleet as a result of using oxygenated fuels. Emissions results are then used in the AIRSHED model to calculate the effects of emissions reductions on ambient concentrations of CO. During the first year of the program, 1988, it was reported to have "resulted in an 8 to 11% reduction, as modeled, in ambient air carbon monoxide levels. ... At downtown Denver's CAMP monitoring station, the Oxygenated Fuels Program reduced ambient carbon monoxide levels by 9.36%."[1] "The 1988–1989 program was modeled to have reduced carbon monoxide concentrations by an average of 12% (with an uncertainty of $\pm 3.5\%$) along the Front Range (including Metro Denver and Colorado Springs)."[2] Due to the 1989–1990 program, "motor vehicle tailpipe emissions of carbon monoxide were reduced this year by 15.7% $\pm$ 3.3% (derived from EPA MOBILE 4 Emissions Factor Model). This is a similar tailpipe exhaust reduction to last year's 15.3% $\pm$ 3.2% (derived from the EPA MOBILE 3 Emissions Factor Model). Ambient air concentrations were reduced 12% $\pm$ 3.5% (derived from AIRSHED Model) as a result."[3] "Because of the higher minimum oxygen requirement in most grades of gasoline this year [1990–1991] (2.6% vs 2.0%), and high sales of ethanol blended gasoline this year (13% vs 8%), carbon monoxide exhaust emissions were calculated to have been reduced 19%, compared to 15% the year before."[4] "As a result of this winter's [1991–1992] Oxygenated Gasoline Program, carbon monoxide exhaust emissions were calculated to have been reduced by 23%. This reduction is based on the MOBILE 4.1 emissions model."[5]

AREAS OF CONCERN
Emissions Modeling Uncertainties

Anderson and co-workers[7,8] have shown that the ambient concentrations of CO at CAMP in 1987 were about 70% of the concentrations observed during 1982. On the other hand, emission data and MOBILE modeling suggested that the CO emissions were reduced to about 83% of their 1982 values by 1987. AIRSHED modeling suggested that the ambient concentration of CO at CAMP should be 80% of its 1982 values by 1987.[9] The ambient concentration data suggest that the CO emissions have decreased considerably more than the MOBILE modeling suggests.

During 1991–1992, PRC Environmental Management[10] conducted on-road emissions test studies on an exit ramp from I-25 near downtown Denver. This study suggested that there was a 25% reduction in CO emissions from motor vehicles during the winter oxygenated fuels period, compared to the nonoxygenated fuels periods before and after. PRC calculated the mass reduction in CO emitted from the %CO emitted. They calculated the average emission rate for nonoxygenated fuels of 256 g CO/gallon of fuel and 194 g CO/gallon of fuel using oxygenated fuels. Using the winter fuel consumption data from PRC, the daily CO emissions from motor vehicles not using oxygenated fuels are expected to be 742 metric tons/day (816 tons/day), while it should be 562 metric tons/day (620 tons/day) using oxygenated fuels. This assumes that there is no change in fuel economy as a result of using oxygenated fuels. The PRC study concluded that there was a 1 to 3% fuel economy loss due to the use of oxygenated fuels, based upon a scientific literature survey. The CO emissions calculated by this technique should be the average of all of the CO emissions from motor vehicles that would use oxygenated fuels, throughout the entire Colorado Front Range.

Table 1 **Emissions results for the Colorado Department of Health,
1987 Air Toxics Study**

| | Percent Change in Emissions from Base Gasoline | | | | | |
Pollutant	Gasohol ncat	MTBE ncat	Gasohol cat	MTBE cat	Gasohol cl	MTBE cl
CO	–20.6%	–11.6%	–32.5%	–11.9%	–11.5%	–3.8%
NO_x	–1.0%	–4.5%	5.5%	4.0%	–2.4%	0.6%
Total Aldehydes	–0.8%	–18.3%	37.2%	22.4%	50.4%	38.3%
Formaldehyde	–8.1%	–17.3%	27.8%	22.7%	73.4%	76.4%
Acetaldehyde	100.1%	–3.6%	154.4%	23.3%	122.7%	7.2%

Note: This study measured emissions from 20 vehicles using oxygenated fuels and various emissions control technologies, including four non-catalyst equipped vehicles, nine catalyst equipped vehicles, and seven vehicles using closed-loop emissions control technology.

Gasohol = 10% ethanol blend (3.7% oxygen by weight), MTBE = 11% methyl tertiary butyl ether blend (2.0% oxygen by weight); and ncat = non-catalyst, cat = catalyst, cl = closed-loop

The Colorado Department of Health has constructed an emission inventory for 1990 on-road mobile sources in the Denver modeling area.[11] This inventory suggests that 1063 metric tons/day (1172 tons/day) CO should be emitted for vehicles using oxygenated fuels. The emission inventory area excludes more than 20% of the population that is subject to the use of oxygenated fuels. This mobile source of CO is calculated to be about 80% of the total CO source for the Denver modeling area.[11]

The results of the on-road emissions study and the emission inventory are considerably different, suggesting that one or both of these emissions rates is incorrect. Among the possibilities are: 1) the emission inventory could be incorrect; 2) the conversion from %CO to grams CO/gallon of fuel could be incorrect; or 3) the average %CO in emissions measured under the I-25 and Speer Boulevard driving conditions is lower than the true CO emissions averaged over all driving conditions. Undoubtedly, the emissions measured under these particular driving conditions are not the same as the average emissions under all driving conditions. Hence, the %CO reduction under this set of driving conditions should not be expected to represent the %CO reduction that would be found by averaging over all driving conditions. This discussion is intended to illustrate the fact that there are substantial uncertainties in this type of evaluation.

The uncertainty in the MOBILE model used to estimate motor vehicle contributions to CO emissions has been extensively reviewed.[12,13] Studies have shown that CO emissions may be underestimated by a factor of two to three when compared to atmospheric measurements. Among the major areas of uncertainty in vehicle emissions modeling are: 1) differences in driving patterns from those represented by the Federal Test Procedure (combinations of accelerations, decelerations, cruises, etc.); 2) contributions of high emitters (caused by emissions control tampering, malfunctioning of emission control systems, etc.); and 3) effects of ambient temperature on cold-start emissions.[13]

Potential Adverse Effects
Table 1 shows the results of one set of emissions measurements using oxygenated fuels. This study was conducted by the Colorado Department of Health, and measured CO, nitrogen oxides (NO_x), and aldehyde emissions from 20 vehicles when using gasoline, 10% by volume ethanol blended fuels, and 11% by volume MTBE blended fuels.[7] These emissions data show a decrease in NO_x emissions from 4.5% to an increase of 5.5%, depending on the emissions control technology and the fuel being used. There may be a

small increase in NO_x emissions when oxygenated fuels are used instead of normal gasoline.[7] These same data suggest that there is a substantial increase in formaldehyde emissions from catalyst and closed-loop vehicles when either MTBE or ethanol blended fuels are used. There is also a substantial increase in acetaldehyde emissions from all vehicles using ethanol blended fuels. Hoekman[14] has presented emissions data from a set of vehicles using 11% by volume MTBE as compared to gasoline. These data show that formaldehyde emissions increase by about 18% and acetaldehyde emissions increase by about 5.5% as an average for the vehicles tested in that study.[14] These data sets are relatively small, but are intended to indicate the nature of the changes in emissions that might be expected as a result of using oxygenated fuels.

Some of the early results related to studies of the effects on NO_x and aldehyde concentrations have been presented.[7] The trend analysis in NO_x concentration data has shown no significant effect of using oxygenated fuels. The lack of an effect on ambient NO_x was expected, since the effect of oxygenated fuels on NO_x emissions is smaller than the effect on CO emissions, and no effect on CO concentrations was detected.

Winter formaldehyde concentrations measured in Denver were found to be reasonably high, the maximum 4-h average concentration observed in the earlier study was 32 ppbv,[7] which is as high as almost any measurements reported for Southern California.[15,16] The winter formaldehyde concentration data were highly correlated with the CO concentration, suggesting that motor vehicles are a major source of formaldehyde during the winter.[7] The slope of the formaldehyde vs CO linear regression for the winter "high pollution months" matched quite well the formaldehyde-to-CO ratio measured in the emissions from motor vehicles.

Concerns about the effects of oxygenated fuels on carbon monoxide,[7] nitrogen oxides,[7] and aldehydes,[7,17,18] as well as the direct health effects of MTBE,[19] have been expressed widely across the country. Regardless of the time, effort, and expertise incorporated in the development of new air quality control programs, modeling uncertainties remain. Clearly, such programs must be carefully monitored and evaluated after they are implemented. We present several statistical methods in this chapter for analyzing ambient concentration data to determine trends and to assess the effect of a particular intervention, in this case, the addition of oxygenates to gasoline.

Toxicity of Vehicle Emissions

The U.S. EPA recently released a report dealing with the toxicity of motor vehicle emissions.[20] The focus of the study was on the carcinogenic risk of vehicle emissions. For purposes of comparison, we will use the results for the 1995 base emissions control with a baseline gasoline (1.53% benzene, 32% aromatics, 0% oxygen, and Reid Vapor Pressure - RVP 8.7 psi) and a reformulated fuel (1.0% benzene, 25% aromatics, 2.0% oxygen, and RVP 8.1 psi). In this cancer risk assessment, four specific compounds were considered that are affected by fuel reformulation; these are benzene, 1,3-butadiene, formaldehyde, and acetaldehyde. On switching from the base fuel to reformulated gasoline, the individual cancer risk factor goes from 1.7×10^{-7} to 1.4×10^{-7} for benzene (an 18% decrease), from 8.1×10^{-7} to 8.0×10^{-7} for 1,3-butadiene (a 1.2% decrease), from 1.1×10^{-7} to 1.2×10^{-7} for formaldehyde (a 9.1% increase), and from 1.4×10^{-8} to 1.4×10^{-8} for acetaldehyde (no change). The EPA report[20] makes it clear that the reduction in risk due to benzene exposure is caused by the reduction in benzene and aromatic content of the reformulated fuel, and not directly to the increase in oxygen content in the fuel. There is no significant change in the cancer risk factor for 1,3-butadiene or acetaldehyde related to the use of reformulated fuels (with MTBE as the oxygenate). The cancer risk related to formaldehyde does increase with the use of reformulated fuels.

The emission factors for both formaldehyde and acetaldehyde depend on the vehicle's emission control technology and the quantity and chemical form of the oxygenate added to the fuels.[20] With 15% MTBE blended fuels, formaldehyde emissions increase by factors of 1.67 when three-way catalysts are used, by 1.27 with three-way plus oxidation catalysts, by 2.02 with oxidation catalysts, and by 1.53 for non-catalyst vehicles. The acetaldehyde emission factors for 15% MTBE blended fuels increase by factors of 1.08 with three-way catalysts, by 1.01 with three-way plus oxidation catalysts, by 1.21 with oxidation catalysts, and by 1.44 with non-catalyst vehicles. When 10% ethanol blended fuels are used, formaldehyde emissions increase by factors of 1.48 with three-way catalysts, by 1.23 with three-way plus oxidation catalysts, by 1.24 with oxidation catalysts, and by 1.10 for non-catalyst vehicles. The acetaldehyde emission factors for 10% ethanol blended fuels increase by factors of 2.14 with three-way catalysts, by 2.25 with three-way plus oxidation catalysts, by 2.96 with oxidation catalysts, and by 2.14 with non-catalyst vehicles. The emissions of both of these carbonyls increase when either oxygenate is used, but the formaldehyde emissions increase is larger for MTBE blended fuels, and the acetaldehyde emissions increase is much larger for ethanol blended fuels.

These discussions of toxicity have only considered the carcinogenic risk for these exhaust-related compounds. There are a variety of other health effects of these pollutants. Inhalation reference concentrations (RfCs) are being developed for each of these compounds.[20] These are intended to estimate the continuous exposure to the human population that is likely to have no deleterious effects during a lifetime. Currently, a final RfC exists only for acetaldehyde at 9×10^{-3} mg/m^3 over a lifetime.[20] This corresponds to a continuous exposure to only 5 ppbv acetaldehyde. As we will see later in this chapter, the average outdoor concentration for acetaldehyde throughout our entire measurement program was 2.2 ppbv, about one half the inhalation reference concentration.

Importance of Formaldehyde

The current EPA carcinogenicity assessment for lifetime exposure on the Integrated Risk Information System[21] classifies formaldehyde as B1, a probable human carcinogen, based on limited evidence in humans and sufficient evidence in laboratory animals. Formaldehyde is also a potent irritant of the eyes and mucous membranes in humans exposed to low concentrations.[22] Upper respiratory (nose and throat) irritation has been documented in numerous human exposure studies to occur in the range of 0.1 to 3.0 ppmv.[22] The National Institute for Occupational Safety and Health has a recommended maximum formaldehyde exposure level for an 8-h period averaging 16 ppbv.[23] In addition, formaldehyde is photochemically active and plays an important role in photochemical air pollution problems.

CO DATA ANALYSIS AND INTERPRETATION

We apply statistical techniques to answer the most basic question about the oxygenated fuels program: Was there a reduction in ambient concentrations of CO, and if so, can it be attributed to the oxygenated fuels program? This requires a relatively long series of observed CO concentrations, measured consistently over the entire series. The data we use were recorded as hourly averages of values measured at 1-min intervals at several sampling stations maintained by CDH in compliance with EPA requirements. As expected, these time series data exhibit serial correlation. We use statistical techniques that take into account this violation of the usual assumption of independence.

CHARACTERISTICS OF THE DATA

Quantile-to-quantile plots of hourly averaged ambient CO suggest the data are lognormally distributed. In addition, they are highly serially correlated, exhibit a nonlinear trend, and seasonal fluctuations. Exponential smoothing of monthly, weekly, and daily averaged CO data, and monthly, weekly, and daily maximum 8-h averaged CO data for the downtown Denver concentration data identified a strong downward trend in CO that existed prior to the use of oxygenated fuels and suggested that the effect of using oxygenated fuels was much less than predicted by emissions and air quality modeling.[8] We have used alternative approaches to determine the statistical significance of the trend and the downward shift due to oxygenated fuels.

TIME SERIES ANALYSES
ARIMA Analysis

Where the underlying structure in a univariate time series is determinable, the techniques set forth by Box and Jenkins[24] in 1976 are broadly applicable, and we were able to fit an ARIMA(p,d,q)(P,D,Q) to the data, where p is the number of autoregressive parameters, d is the number of differences required for stationarity, and q is the number of moving average parameters; lower case is for nonseasonal parameters, upper case for seasonal. The results are presented in Table 2. We were able to detrend the series by taking one first difference and one seasonal difference to arrive at a stationary series. Using the ARIMA to assess the effectiveness of oxygenated fuels for CAMP for the time period from January 1981 through December 1992 suggests that there was a 3.2% decrease in CO that could be attributed to the use of oxygenated fuels, but this change was not significantly different from zero. We have not used this analysis technique extensively because the long-term trend that was apparent in the smoothing approach is lost in this analysis. However, it does provide a useful test of alternative analysis techniques.

Table 2 ARIMA (2,1,0)(2,1,0) analysis using log transformed monthly averaged CO data for CAMP with one binary explanatory variable, OXY = 1 when oxygenated fuels are in use, 0 otherwise

Parameter	Estimate	t-ratio	Pr > |t|
AR1	−0.594	−6.393	0.000
AR2	−0.240	−2.590	0.011
SAR1	−0.542	−5.390	0.000
SAR2	−0.342	−3.405	0.001
OXY	−0.032	−0.709	0.480
Constant	0.000	0.118	0.907

N = 132

Loglikelihood = 100.347

Structural Time Series Analysis

Our aim in this approach is to present the structure of the data, "the 'stylized facts' … in terms of a decomposition into components such as trend, season, and cycle."[25] In its most general form, a structural time series model is a regression model whose independent variables are functions of time and the parameters are time varying.[25] The model we present below is the simplest form; the parameters are linear and time invariant.

A structural time series equation that will allow us to explicitly evaluate the effectiveness of the oxygenated fuels program, also referred to as the intervention, while allowing us to estimate the long-term trend in the log-normally distributed ambient CO concentrations, is

$$Y = e^{\alpha} e^{\Sigma \beta_i t_i} e^{\Sigma \beta_j m_j} e^{\beta_k oxy} e^{\varepsilon} \tag{1}$$

$$\ln Y = \alpha + \sum_{i=1}^{3} \beta_i t_i + \sum_{j=4}^{14} \beta_j m_j + \beta_{15} oxy + \varepsilon \tag{2}$$

where Y is monthly averaged CO or monthly maximum 8-h averaged CO, α is the intercept term, the t_i values are centered and orthogonalized trend components (see Appendix A), the m_j values are binary variables for the months (they effectively deseasonalize the data), oxy is a binary variable set to 1 when oxygenated fuels are in use and 0 otherwise, ε is a vector of random errors, and the β values are parameters to be estimated. We chose the two variants of the raw data, monthly maximum 8-h averaged CO and monthly average CO, because the first is closest to the measure used by EPA to determine attainment of air quality standards and the second is a transformation that reduces the effects of day-to-day variability to allow assessment of long-term trends. There is less loss of information due to the averaging in this case than there would be if the observations were independent (see Appendix B).

For the monthly averaged CO, the log transformation gives normally distributed, serially correlated error terms. We use the AREG procedure in SPSS/PC[26] to correct for first-order serial correlation. For the monthly maximum 8-h average CO concentration, the log transformation results in error terms that show no serial correlation. The error terms appear to be distributed closer to normal than we expected. To assess the impact on the distribution of the β values we use the bootstrap,[27] a resampling technique to estimate the distribution of the β values. The skewness and kurtosis were not statistically different from a normal distribution, so we present our results based on the assumption that the β values are normally distributed. (See Appendix C).

Table 3 shows a summary of the oxygenated fuels effects that result from the time series analysis using Equation 2. The results presented in the table are for the monthly average and monthly maximum 8-h averaged concentrations of CO measured at several urban sites along the Colorado Front Range. The sites for which data are presented include three sites in Denver, Continuous Air Monitoring Program (CAMP) at 21st and Broadway, Carriage at 23rd and Julian, and National Jewish Hospital (NJH) at 14th and Albion, as well as a site in Boulder at 2320 Marine and one in Colorado Springs at I-25 and Uintah.

For the monthly averaged CAMP CO data, all of the trend parameters are statistically significant at the 5% level. For the monthly maximum 8-h averaged CO at CAMP, only the cubic portion of the trend is not significant at the 5% level. For both the monthly averaged CO and the monthly maximum 8-h averaged CO, the oxy parameter, $\hat{\beta}_{15}$, is not statistically different from zero. Furthermore, it should be noted that for both CAMP data sets, more than 88% of the variance in the CO data is explained by the cubic trend and seasonal components of the analysis. Also shown in the table is the percent reduction in CO due to oxygenated fuels that would make the oxyfuels parameter significant at the 5% level. For the monthly averaged CAMP data, the oxyfuels parameter would have had to correspond to about a 7.6% decrease in CO to be significant at the 5% level. Similarly, oxyfuels would have had to reduce the monthly maximum 8-h averaged CO by about 13% to be significant at the 5% level. As was suggested earlier, based upon emissions modeling results, the oxygenated fuels program should have reduced CO emissions between 11 and 23% throughout these years. In short, had there been even a 10% reduction in ambient concentrations attributable to the oxygenated fuels program, we would have a better than 80% chance of detecting it.

Table 3 also shows the power of the test, that is, the probability of rejecting the null hypothesis, that there is no change in CO concentrations, ($\hat{\beta}_{15} = 0$) when in fact the alternative hypothesis is true. We chose the alternative hypothesis H_a: $\hat{\beta}_{15} = -0.105$, is based on the expected reduction in ambient CO of at least 10%.[1-5] The power of the test is 0.83 for the monthly averaged CAMP data and 0.43 for the monthly maximum 8-h averaged CO at CAMP (See Appendix D).

Figure 1 shows a plot of monthly averaged CO and monthly maximum 8-h averaged CO concentrations at the CAMP station, along with the fitted values and projections through 1993, and upper and lower 95% confidence limits on the fits and projections. From these plots it is apparent that there has been a strong downward trend in averaged CO concentrations that began in about 1983. This downward trend appears to have

Table 3 **Time series analysis results for monthly averaged and monthly maximum 8-h averaged CO concentrations at Denver's CAMP, Carriage, National Jewish Hospital, and Boulder and Colorado Springs (I-25 and Uintah)**

	Denver CAMP	Denver Carriage	Denver NJH	Boulder	Colorado Springs
Monthly Averaged Data					
$\hat{\beta}_{15}$ (oxyfuels parameter)	0.0048	−0.0629	−0.0755	−0.0400	0.1137
SE $\hat{\beta}_{15}$	0.0401	0.0594	0.0567	0.0703	0.0748
Student's t	0.1187	−1.0579	−1.334	−0.5680	1.5195
p-value for Student's t	0.9057	0.2975	0.1853	0.5710	0.1311
% Reduction in CO corresponding to $t_{.975,130}$[a]	7.6	11.1	10.6	13.0	13.8
Power of the Test[b] (Ha: $\hat{\beta}$ = −0.105)	0.83	0.55	0.58	0.44	0.39
R^2	0.90	0.89	0.81	0.76	0.53
N	145	124	116	145	145
Monthly Maxima Data					
$\hat{\beta}_{15}$ (oxyfuels parameter)	−0.0217	−0.1011	−0.1042	−0.1164	0.0153
SE $\hat{\beta}_{15}$	0.0711	0.0867	0.0861	0.1237	0.0917
Student's t	−0.3056	−1.1667	−1.2100	−0.9410	0.1664
p-value for Student's t	0.7604	0.2459	0.2292	0.3484	0.8681
% Reduction in CO corresponding to $t_{.975,130}$[a]	13.1	15.8	15.7	21.7	16.6
Power of the Test[b] (Ha: $\hat{\beta}$ = −0.105)	0.43	0.33	0.33	0.21	0.31
R^2	0.88	0.85	0.78	0.71	0.74
N	145	124	116	145	145

[a] Minimum change due to oxygenated fuels that would be significant at the 5% level in a two-tailed test.

[b] Probability of rejecting the null hypothesis when the alternative hypothesis (a 10% reduction due to oxygenated fuels) is true.

flattened during the last few years. The monthly maximum 8-h averaged CO concentration exceeds the 9-ppm air quality standard in the winter of 1992–1993 and is predicted to exceed the standard during the winter of 1993–1994.

The results of the analysis of the monthly averaged CO data for CAMP suggest that the CO concentration at CAMP increased by 0.5% as a result of using oxygenated fuels; however, this parameter is not statistically significant. The reason an increase is suggested is understandable. Figure 1 shows that the monthly averaged CO for December 1992 was unusually high, outside the 95% confidence bands on the fitted values. This high value during the winter, coincident with the oxygenated fuels program, forced the estimated oxygenated fuels parameter to become positive. This high CO in December 1992 is most likely due to unusual meteorological conditions occurring during the month. A long-term change in vehicle miles traveled will be embedded in the trend terms that are in effect year-round. Increased vehicle miles traveled would be confounded with the oxygenated fuels parameter only if there was an increase in the winter miles traveled, and not those traveled year-round.

Table 3 shows a comparison between the time series analysis results for monthly average CO concentrations and monthly maximum 8-h averaged CO concentrations

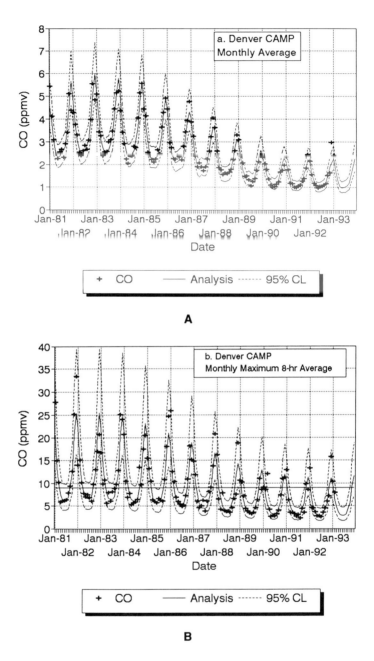

Figure 1 Monthly averaged (a) and monthly maximum 8-h averaged (b) concentration data and trend analysis results for carbon monoxide at Denver's CAMP air monitoring station.

measured at both the NJH and Carriage sites. The results were quite consistent with each other, and suggestive of a 6 to 8% decrease in the monthly average CO concentration and about a 10% decrease in the monthly maximum 8-h averaged CO concentration due to the use of oxygenated fuels, but neither of these was statistically significant. The adjusted R^2 ($\overline{R^2}$) for these two analyses were 0.81 and 0.89, respectively. For the monthly averaged CO data, about an 11% decrease in CO would have been required for statistical significance, and the power of the test was about 0.55 at both sites. For the monthly maximum 8-h averaged CO, an even larger decrease of about 16% would have been required for statistical significance, and the power of the test was about 0.33.

Figure 2 shows the results of the analysis for the Carriage site. The downward trend in CO is apparent in both the monthly average and monthly maximum 8-h averaged CO concentration data. The winter concentrations for CO do not show the continued decrease apparent in the summer data. The significance of the trend suggests that the CO concentrations should continue to decrease. Figure 3 shows the results of the analysis for the NJH site. At NJH, the trend appears to have dropped more in the most recent years, however both of these sites have had 8-h averaged CO concentrations that exceeded 9 ppm for a least one winter month during the last two winters. Although the fitted values for both sites show that the predicted monthly maximum 8-h averaged CO is below this level, the 95% confidence limits do include 9 ppm for both of these sites.

Table 3 shows that the analysis of the Boulder data is also reasonably consistent with that for the NJH and Carriage data. The $\overline{R^2}$ for these data is slightly lower, about 0.75. This analysis suggests that monthly averaged CO measured in Boulder decreased by about 4%, while the monthly maximum 8-h averaged CO decreased by about 11% due to the use of oxygenated fuels. These results are not statistically different from zero at the 5% significance level. For the monthly averaged CO data, a 13% decrease in CO would be required for statistical significance and the probability of rejecting the null hypothesis when the alternative hypothesis, that CO decreased by 10%, is true is only about 0.44. The lower $\overline{R^2}$ and power of the test for the Carriage, NJH, and Boulder data is largely because CO concentrations at these sites are lower than those at CAMP. Figure 4 shows the results of the analysis for the Boulder monitoring site. Again, the downward trend in CO is apparent for both the monthly averaged and the monthly maximum 8-h averaged CO data. The trend shows a continued decrease in the CO for both data sets, but the monthly average data for the most recent winters does not show this continued decrease. The 95% confidence interval on the monthly maximum 8-h averaged CO only includes 9 ppm for December, and the predictions through December 1993 do not include 9 ppm.

The results of the analysis of the Colorado Springs (I-25 and Uintah) data are a bit different than for the other sites. Figure 5 shows the monthly averaged concentrations for CO at this site. Prior to about 1985, the CO concentrations measured during the summer and fall were higher and much more variable than for the more recent years. The regression analysis suggests that monthly averaged CO at the Colorado Springs site increased by 12% due to the use of oxygenated fuels, but this result is not significant at the 5% level. For this data set, the probability of rejecting the null hypothesis when the alternative hypothesis, that CO decreased by 10%, is true is only 0.39. Both the $\overline{R^2}$ for the analyses and the power of the test are lower than for the other data sets. This is probably due to the large variability in the early years. Also shown in Figure 5 is the monthly maximum 8-h averaged CO concentration data for the Colorado Springs site, which is more consistent with the other sites. The oxygenated fuels parameter for these data suggest that the monthly maximum 8-h averaged CO increased by 1.5% as a result of oxygenated fuels use. As shown in Table 3, the $\overline{R^2}$, the minimum change due to oxygenated fuels that would be significant at the 5% level and the power of the test are

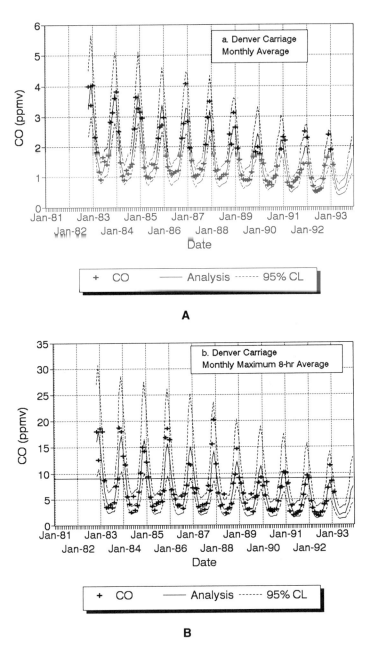

Figure 2 Monthly averaged (a) and monthly maximum 8-h averaged (b) concentration data and trend analysis results for carbon monoxide at Denver's Carriage air monitoring station.

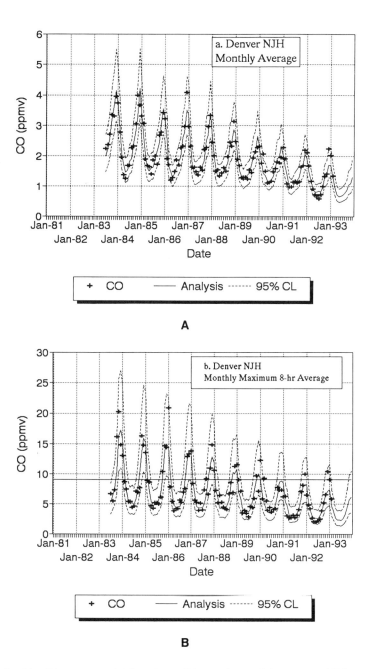

Figure 3 Monthly averaged (a) and monthly maximum 8-h averaged (b) concentration data and trend analysis results for carbon monoxide at Denver's National Jewish Hospital air monitoring station.

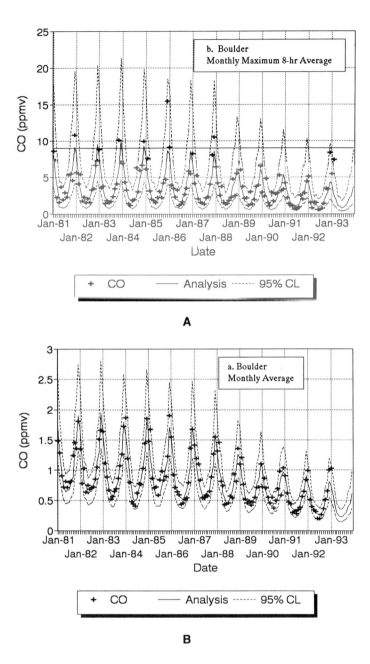

Figure 4 Monthly averaged (a) and monthly maximum 8-h averaged (b) concentration data and trend analysis results for carbon monoxide at the Boulder air monitoring station.

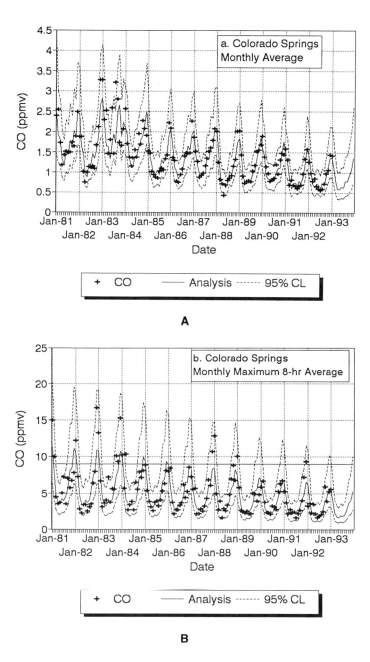

Figure 5 Monthly averaged (a) and monthly maximum 8-h averaged (b) concentration data and trend analysis results for carbon monoxide at Colorado Springs' I-25 and Uintah air monitoring station.

not inconsistent with the other sites. Both the winter concentration data and the trend analysis are consistent with a continued decrease in the concentrations of CO at this site. The 95% confidence limits continue to include 9 ppm for the monthly maximum 8-h averaged CO at this monitoring site.

CO data have been analyzed for two other Denver metropolitan area monitoring sites. CO data collection at the north urban Welby (78th and Steele) site began only 15 months prior to the start of oxygenated fuels use. This analysis includes data that extended through December 1991, and suggests that CO decreased by 2.2%, due to the use of oxygenated fuels. In this case, the oxygenated fuels parameter is significant above the 5% level. Caution must be used when interpreting the results of this analysis, since it is possible for some unusual winter weather to bias the results of this shorter, historical time series analysis. We believe that the time series data preceding the use of oxygenated fuels for Welby is too short to be useful for the intervention analysis. We have also analyzed south suburban Highland Reservoir (8100 S. University) CO data. The results of this analysis are consistent with those from the other sites, but the R^2 for this set of data is quite small, only 0.44. The monthly averaged CO concentrations measured at Highlands are between 0.2 and 1.4 ppm. The resolution of the analytical technique for CO is 0.1 ppm, hence much of the variability in the CO data is due to the low signal-to-noise ratio for this data set.

COMPARISONS WITH EMISSIONS INVENTORY

As we have seen, there has been a strong downward trend in the ambient CO concentration in Denver. Table 4 shows the monthly averaged and monthly maximum 8-h averaged CO concentrations from the time series analysis for CAMP, Carriage, NJH, Boulder, and Colorado Springs during the months of December 1982, December 1987, and December 1990. Also shown are the percentage changes in these concentrations from the base year, 1982. This table also includes emission inventory data for these same three years, along with the percentage changes from 1982. The changes in the analysis results for the monthly averaged CO concentration are quite similar to those for the monthly maximum 8-h averaged CO concentration for each site. The decreases that have been observed in the ambient concentrations of CO at the Denver CAMP site are much greater than the decreases predicted in the emission inventories.[9,11] The decreases observed in the CO concentrations for Carriage and Boulder are only slightly greater than the decreases in the emission inventories. The changes in the CO concentrations for Colorado Springs are quite consistent with the changes from the emissions inventory for CO. This suggests that overall the emission inventory may underestimate the reductions in the CO that have been achieved. The emission inventories are based on different emission models, which would make the results somewhat inconsistent. The inventory is for the Denver metropolitan area, which includes all of the monitoring sites except Colorado Springs. This table suggests that the CO decreases at the downtown Denver CAMP station between 1982 and 1990, where violations of the CO air quality standard have occurred most often, are greater than the decreases observed for the other sites included in this analysis.

AURARIA MONITORING STUDIES

Our research group has been monitoring aldehydes on the Auraria campus adjacent to downtown Denver since December 1987. Aldehyde sampling has been conducted by collecting 4-h average samples, 24 h a day, 7 days a week for most of 46 months since our sampling began in December 1987. In addition to monitoring aldehydes at this site, we are currently measuring carbon monoxide, ozone, nitrogen oxides, carbon dioxide, temperature, wind speed, wind direction, humidity, barometric pressure, and precipitation.

Table 4 Comparisons between the changes observed in the analysis results for the monthly averaged and the monthly maximum 8-h averaged CO concentrations at various Colorado air monitoring sites compared to 1982, and the changes expected from the CO emissions inventory data for the Denver metropolitan area

	Denver Camp		Denver Carriage		Denver NJH		Boulder		Colorado Springs		Denver Emissions Inventory	
	CO	% of 1982	CO	% of 1982	CO	% of 1982	CO	% of 1982	CO	% of 1982	Tons/day	% of 1982
Monthly Averaged CO Trend												
December 1982	6.01		4.10				1.95		2.65		2366	
December 1987	3.67	61.1%	3.19	77.8%	3.40		1.60	82.1%	2.09	78.9%	1966	83.1%
December 1990	2.29	38.1%	2.23	54.4%	2.32		0.96	49.2%	1.77	66.8%	1476	62.4%
Monthly Max 8-h Averaged CO Trend												
December 1982	25.42		18.27				9.56		10.94			
December 1987	16.62	65.4%	14.23	77.8%	12.79		8.56	89.5%	8.36	76.4%		
December 1990	11.94	47.0%	10.57	57.9%	8.80		5.43	56.8%	7.01	64.1%		

ALDEHYDE SAMPLING AND ANALYSIS TECHNIQUES

We use dinitrophenylhydrazine (DNPH)-coated silica packed cartridges, similar to EPA Method TO-11.[28] The cartridges used in these studies have a larger cross-sectional area, and hence have a smaller pressure drop than most commercially available cartridges. The pressure drop is typically less than 4 in. mercury at flow rates up to 3 l/min. Collection efficiency tests using these cartridges show an average of 97% efficiency, even at these high flow rates. Samples were collected using an automated low-volume sequential sampler (Anderson & Associates, Model SEQ-3). This sampler is equipped to handle six sample cartridges and a seventh field blank cartridge. All seven cartridges are connected to a common inlet manifold. Air is drawn through six successive cartridges for the required 4-h sampling period. These aldehyde-DNPH adducts are extracted into 5-ml volumetric flasks using acetonitrile. Each sample is analyzed by high performance liquid chromatography with UV absorption detection at 360 nm. The normal analysis run begins with the injection of three identical standards to condition and check the operation of the system, followed by injections of a set of six samples and a field blank. Each set of samples is followed by the injection of one of a series of five different standards. For every six samples analyzed, there is one standard and one field blank analyzed. The calibration curve for the analysis run is constructed from the standards analyzed throughout the run. All sample concentrations are calculated from field blank corrected data.

RESULTS OF ALDEHYDE CONCENTRATION STUDIES

Table 5 shows a summary of the monthly average formaldehyde and acetaldehyde concentrations, the standard deviation, the range of concentrations, and the number of samples analyzed for that month. Through February 1993, we have analyzed more than 7000 samples for formaldehyde and acetaldehyde in downtown Denver. These data are described in greater detail elsewhere.[29] In general, we have found that the concentration of formaldehyde tends to be higher and more variable during the late fall and winter than it is during the spring and summer. The formaldehyde concentrations reported in this work are as high or higher than the concentrations reported in many other areas, including Southern California.[15,16,30] Formaldehyde concentrations measured in the Denver metropolitan area are normally two to three times higher than the acetaldehyde concentrations.

DIURNAL BEHAVIOR OF FORMALDEHYDE

Currently, we have over 1 year of data with both formaldehyde and CO concentrations measured at the Auraria air monitoring site. Figure 6 shows plots of the diurnal average profiles for 4-h averaged concentrations of both formaldehyde and CO measured in downtown Denver during the months of November 1991 and June 1992. These were chosen to be representative of typical winter and summer months. The winter data show a morning and an evening peak for both formaldehyde and CO, with the morning peak for formaldehyde being more pronounced. During June 1992, the highest formaldehyde on average was observed during the 8 a.m. to noontime period, while the CO was highest during the 4 a.m. to 8 a.m. and 8 p.m. to midnight periods. During both of these time periods, the concentration of formaldehyde was about 3×10^{-3} times the CO concentration.

RELATIONSHIP BETWEEN FORMALDEHYDE AND CO

Since neither the formaldehyde nor the CO concentrations are normally distributed, we used Spearman rank correlations.[31] Table 6 shows the diurnal behavior of the Spearman rank correlation between the formaldehyde and CO concentrations measured at the Auraria site, for each month between October 1991 and January 1993. From this table it

Table 5 **Descriptive statistics for the formaldehyde and acetaldehyde sampling in Denver**

Month/Year	Formaldehyde (ppbv)				Acetaldehyde (ppbv)			
	Average	Standard Deviation	Range	Number	Average	Standard Deviation	Range	Number
Dec 1987	6.0	5.2	0.2–24.6	100	3.1	3.0	0.0–14.5	87
Jan 1988	4.4	3.7	0.2–20.0	111	2.8	2.3	0.0–10.4	59
Feb 1988	2.9	1.9	0.7–9.6	72	1.4	1.1	0.1–4.6	32
Mar 1988	2.2	1.3	0.3–6.7	103	1.3	0.6	0.2–3.6	59
Apr 1988	2.8	1.8	0.5–7.6	89	1.8	1.1	0.5–6.7	47
May 1988	3.2	1.4	1.3–7.2	23	2.0	0.8	0.7–3.9	23
Oct 1988	5.4	2.7	0.5–15.3	168	2.7	1.4	1.1–10.0	168
Nov 1988	3.7	2.6	0.6–13.4	150	1.6	1.2	0.4–6.9	151
Dec 1988	4.5	2.9	0.5–16.2	166	2.0	1.3	0.3–7.8	164
Jan 1989	3.7	2.7	0.5–14.3	124	1.5	1.1	0.2–7.0	118
Feb 1989	3.4	3.5	0.4–22.7	134	1.9	2.3	0.3–15.4	127
Mar 1989	2.8	2.0	0.4–10.8	118	1.5	1.0	0.1–5.7	111
Sept 1989	6.5	4.6	0.3–20.8	42	4.5	2.3	1.1–11.4	43
Oct 1989	6.7	3.9	0.1–18.2	168	3.2	1.9	0.1–8.9	171
Nov 1989	5.7	3.7	0.7–23.8	174	2.4	1.7	0.3–13.2	173
Dec 1989	5.7	3.6	0.2–31.9	182	2.5	2.3	0.1–17.5	182
Jan 1990	5.7	4.3	0.1–31.2	184	2.2	2.1	0.2–18.4	184
Feb 1990	6.3	4.8	0.2–32.2	141	2.5	2.0	0.1–12.7	143
Mar 1990	6.5	4.3	1.1–25.5	146	2.5	1.8	0.3–9.8	138
Apr 1990	5.4	2.9	0.8–20.9	138	2.3	1.4	0.1–10.7	137
Oct 1990	5.0	3.9	0.0–22.3	111	2.1	1.7	0.0–8.9	109
Nov 1990	4.0	3.2	0.0–17.7	133	1.6	1.3	0.0–6.8	128
Dec 1990	5.1	3.8	0.4–18.8	175	2.4	2.1	0.2–11.7	175
Jan 1991	6.4	4.0	0.2–25.6	171	2.7	2.0	0.1–15.1	170
Feb 1991	6.1	4.9	0.3–49.1	170	2.5	2.1	0.1–20.1	171
Mar 1991	3.4	2.1	0.6–14.2	171	1.9	1.3	0.1–11.1	168
Apr 1991	2.8	1.2	0.1–6.0	145	1.5	0.6	0.5–3.7	145
Jul 1991	5.7	7.2	0.0–36.6	164	5.1	3.0	1.0–17.2	164
Aug 1991	7.6	6.4	0.4–32.6	179	3.6	2.4	0.2–12.8	179
Sep 1991	4.0	2.1	0.0–10.7	180	1.7	0.8	0.0–5.0	180
Oct 1991	5.9	2.6	1.4–12.0	183	2.2	1.3	0.3–8.3	183
Nov 1991	6.0	3.3	0.3–19.1	180	2.3	1.6	0.3–9.8	180
Dec 1991	7.7	3.4	0.8–23.6	186	2.7	1.6	0.5–10.1	186
Jan 1992	4.8	2.9	0.1–19.0	185	1.6	1.1	0.1–7.4	185
Feb 1992	3.8	1.9	0.1–9.8	172	1.5	0.9	0.1–5.2	174
Mar 1992	3.3	1.7	0.2–9.0	174	1.1	0.6	0.0–3.9	172
Apr 1992	3.0	1.8	0.2–11.9	165	1.1	0.7	0.0–4.1	165
May 1992	2.4	1.3	0.1–6.4	173	1.1	0.7	0.0–4.4	167
Jun 1992	3.0	1.1	0.8–7.7	179	1.5	0.8	0.1–4.8	178
Jul 1992	3.1	1.2	0.4–7.5	186	1.7	0.7	0.1–4.2	186
Aug 1992	3.0	1.5	0.9–10.2	174	1.6	0.9	0.1–6.0	174
Sep 1992	3.4	1.7	0.6–9.0	176	1.8	1.0	0.1–4.9	176
Oct 1992	4.9	4.1	0.1–35.3	178	2.5	1.8	0.2–14.3	179
Nov 1992	3.3	2.3	0.2–11.5	159	1.6	1.3	0.0–6.3	128

Table 5   (Continued)

Month/ Year	Formaldehyde (ppbv)				Acetaldehyde (ppbv)			
	Average	Standard Deviation	Range	Number	Average	Standard Deviation	Range	Number
Dec 1992	5.1	3.5	0.8–18.0	185	2.9	2.5	0.0–10.3	170
Jan 1993	6.0	3.9	0.1–21.9	167	3.1	2.5	0.1–14.8	152
Feb 1993	4.6	2.8	0.2–22.4	150	2.3	1.6	0.1–14.0	111
Overall	4.6		0.0–49.1	7104	2.2		0.0–20.1	6772

is apparent that the time periods when formaldehyde is most strongly related to CO, with the most significant correlations are between 2000 and 1200. Between 1200 and 2000, the number of significant correlations decreases, as does the correlation coefficient. Since over 80% of the CO emitted in Denver comes from motor vehicles,[11] this strong correlation is believed to suggest that a large portion of the formaldehyde in the atmosphere during the periods of high correlation also comes from motor vehicles. The formaldehyde concentrations measured between 1200 and 2000 are believed to be more greatly affected by photochemical production. We have chosen the time period between 2000 and 0800 as that period when formaldehyde concentrations will be least impacted by photochemical processes. The average correlation between 2000 and 0800 for each month is higher than the correlation between 1200 and 2000. The correlations for all time periods during the fall and winter months are in general higher than the correlations for the spring and summer months, suggesting that a larger fraction of the formaldehyde comes from motor vehicle sources during the winter than during the summer. This is consistent with the belief that photochemical sources of formaldehyde are of greater importance during the summer than during the winter.

Another way to look at the seasonal dependence of the relationship between formaldehyde and CO is to look at a plot of the ratio of formaldehyde-to-CO throughout the data set. Figure 7 shows a plot of the daily average ratio of the formaldehyde-to-CO concentration between October 1991 and January 1993. For the first fall and winter, the ratio was relatively high, between 2 and 4×10^{-3}. Between February and April 1992, the ratio was near 1×10^{-3}; through May and June, the ratio was nearer 4×10^{-3}; and from July 1992 through January 1993, the ratio was near 2×10^{-3}. There is clearly a very strong seasonal dependence to this ratio, but the ratio during the fall and early winter of 1991–1992 appears to be quite different than that during the fall and early winter of 1992–1993.

If we compare the ratio of formaldehyde-to-CO during the day (from 1200 to 2000) with that at night (between 2000 and 0800), the daytime ratio is generally higher than the nighttime ratio. The daytime ratio is much higher during the late spring and summer than during the winter. Again, this suggests that there is a net daytime photochemical source for formaldehyde that is most important during the summer.

VEHICLE EMISSIONS OF FORMALDEHYDE

Since the major known source of CO emissions in Denver is motor vehicles,[11] and there are no known sinks for CO that are of significance in an urban area, we will treat CO as a conservative tracer of vehicle emissions. Formaldehyde has other sources and sinks. The most important sinks are probably deposition and photochemical destruction, either by direct photolysis or by reaction with hydroxyl radicals produced by photochemical processes.[30] The dry deposition of aldehydes is expected to be slow, except for formaldehyde over water;[32] hence, we will ignore these processes. During the nighttime periods,

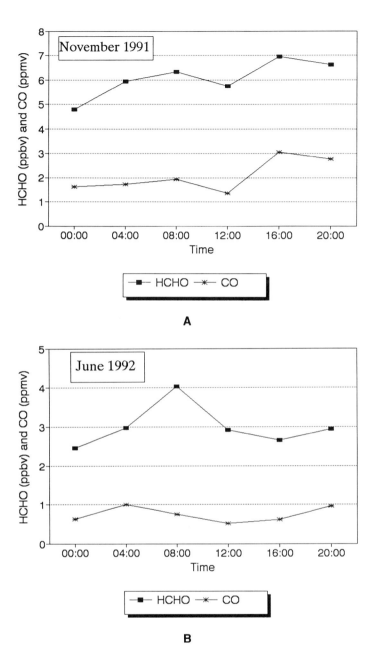

Figure 6 Diurnal averaged profiles for 4-h averaged formaldehyde and CO concentrations measured at the Auraria air monitoring site in Denver during a typical winter month (November 1991) and a typical summer month (June 1992).

Table 6 Spearman rank correlation between 4-h averages of formaldehyde and CO measured at the Auraria monitoring site for each month between October 1991 and January 1993

Month/Year	Time					
	0000–0400	0400–0800	0800–1200	1200–1600	1500–2000	2000–2400
Oct 1991	0.934**	0.911**	0.893**	0.186	0.312	0.842**
n	11	11	11	12	12	12
Nov 1991	0.843**	0.936**	0.823**	0.790**	0.870**	0.864**
n	26	26	25	27	27	26
Dec 1991	0.899**	0.778**	0.894**	0.871**	0.851**	0.935**
n	31	31	31	31	31	31
Jan 1992	0.864**	0.894**	0.855**	0.808**	0.941**	0.876**
n	25	23	25	24	23	24
Feb 1992	0.941**	0.880**	0.856**	0.786*	0.872**	0.922**
n	12	12	12	11	13	13
Mar 1992	0.571*	0.878**	0.226	0.456	0.317	0.896**
n	17	17	16	17	16	16
Apr 1992	0.615**	0.741**	0.746**	0.322	0.585*	0.609**
n	23	24	25	23	25	25
May 1992	0.535*	0.548*	0.683**	0.351	0.217	0.534*
n	22	22	21	23	20	22
Jun 1992	0.610**	0.880**	0.651**	0.557**	0.619**	0.823**
n	29	29	29	30	29	28
Jul 1992	0.478	0.806**	0.408	−0.100	0.299	0.834**
n	23	23	23	24	24	23
Aug 1992	0.590**	0.721**	0.554**	0.566**	0.403	0.900**
n	28	29	29	28	28	28
Sept 1992	0.866**	0.880**	0.893**	0.608**	0.829**	0.947**
n	30	30	29	29	29	29

Table 6 *(Continued)*

Month/Year	0000–0400	0400–0800	0800–1200	Time 1200–1600	1600–2000	2000–2400
Oct 1992	0.598**	0.851**	0.753**	0.234	0.619**	0.692**
n	27	27	29	29	28	28
Nov 1992	0.498	0.447	0.436	−0.386	0.639*	0.535
n	12	13	13	13	14	16
Dec 1992	0.942**	0.875**	0.859**	0.775**	0.900**	0.972**
n	13	13	12	13	14	14
Jan 1993	0.738**	0.729**	0.903**	0.730**	0.770**	0.843**
n	19	20	20	19	19	19

* 1-tailed significance 0.01.

** 1-tailed significance 0.001.

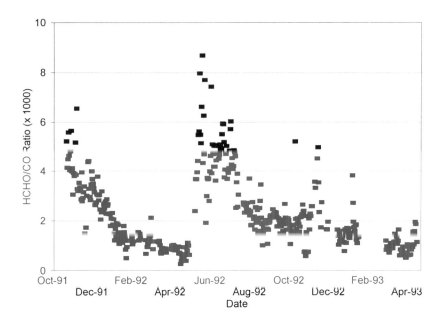

Figure 7 Plot of the daily average ratio of formaldehyde to CO concentrations measured at the Auraria air monitoring site in Denver between October 1991 and January 1993.

both these photochemical sources and sinks are negligible. The smallest formaldehyde-to-CO ratio should best represent the formaldehyde-to-CO ratio in direct vehicular emissions. Altshuller[32] discusses an additional secondary source for formaldehyde that can occur at night and just after sunrise in the early morning. This is due to the nighttime oxidation of alkenes by the nitrate radical and ozone, and the early morning oxidation of alkenes and alkanes by the hydroxyl radical. Altshuller believes that these can be a significant contribution to the early morning aldehyde concentrations. This additional source of aldehydes might correlate fairly well with CO, since much of the hydrocarbon precursors and NO_x required for production of the nitrate radical may come from vehicle exhaust. For the purpose of this analysis, we will assume that during the period when the formaldehyde to CO ratio is the smallest, and there is strong correlation between form-aldehyde and CO, this ratio is most representative of the ratio of formaldehyde-to-CO in vehicular exhaust.

Figure 7 shows that the formaldehyde to CO ratio is smallest from January through April 1992. The average nighttime (2000 to 0800) ratio of formaldehyde to CO for these four months is 1.10×10^{-3}, and the daytime (1200 to 2000) ratio is 0.92×10^{-3}. Since the daytime formaldehyde-to-CO ratio is lower than the nighttime ratio, this may suggest that the daytime photochemical loss of formaldehyde is greater than its production, or that the nighttime production of formaldehyde may be of significance. Even if we only consider the on-road vehicle emissions inventory for CO (1063 metric tons/day, December 1990), the emissions of formaldehyde from these on-road vehicles should be at least 1 metric ton/day (1.1 tons/day) in the Denver metropolitan area. It is also apparent from Figure 7 that there are substantial periods of time when the formaldehyde-to-CO ratio is near 4×10^{-3}, which suggests that the total formaldehyde source may be as large as about 4 metric tons/day in the Denver metropolitan area. This is much greater than the total known formaldehyde emissions from all identified stationary sources in the entire state of Colorado, which is only about 1 ton/year,[33] consistent with the current Toxics Release Inventory for the Denver metropolitan area.[34]

More complete analyses of the carbonyl data collected at the Auraria monitoring station in Denver is presented elsewhere.[35,36]

CONCLUSIONS

Contrary to the commonly held belief, ambient concentration data can be used to assess the effectiveness of recently implemented air quality control programs, such as the use of oxygenated fuels. We have examined ambient CO data graphically and with both descriptive and inferential statistical techniques. All lead to the same conclusions. (1) There was a significant downward trend in ambient concentrations of CO that began in about 1983, roughly corresponding to the most recent reduction in federal motor vehicle emission standards for CO (1981). This same downward trend can be observed in data nationwide.[37] (2) The effect of oxygenated fuels, while generally having the expected sign, does not reduce ambient concentrations of CO in Denver by the magnitude predicted by air quality modeling. The effects were smallest in the downtown Denver area that most commonly exceeds the air quality standard.

Carbon monoxide concentrations in urban areas are most significantly controlled by meteorological factors. The emissions of CO vary relatively little from one winter weekday to the next, but the atmospheric concentration of CO varies dramatically. This analysis has used the ambient data in two different forms, the monthly averaged concentration and the monthly maximum 8-h averaged concentration. The monthly averaged CO concentration data are expected to reflect more closely the changes in CO emissions, since the concentration data result from averaging over all meteorological (and other) conditions in the month. The monthly maximum 8-h averaged data set reflects the concentrations under a very unique set of meteorological conditions, those that allow the highest CO concentrations of the month to be attained. As seen in Table 3, the results of the analysis of this extreme data set are quite similar to those for the monthly averaged data. This result would be expected unless the meteorological conditions that allowed the unusually high CO concentrations to be reached also led to unusually high or low emissions of CO. Certainly, meteorological factors do affect CO emissions; large quantities of snowfall may result in less driving, thereby reducing CO emissions, and very cold evenings may encourage more wood burning, thus increasing CO emissions. The consistency of the results from the two sets of analyses suggests that the factors responsible for the accumulation of the highest CO concentrations in a month do not lead to substantial changes in CO emissions. The analysis of the monthly averaged CO concentration data provides essentially the same results as the monthly maximum data, but with narrower uncertainty bounds.

The results of this analysis are inconsistent with the results of the models used to predict the effects of oxygenated fuels on ambient CO concentrations. The structural time series analysis used in this project is capable of detecting the predicted CO reductions with statistical significance. However, our analysis of ambient CO data along Colorado's Front Range has found no significant effect of using oxygenated fuels on the concentration of CO during the recent winters. The structural time series approach applied in this paper provides a valuable technique for analyzing ambient concentration data, not just for CO, to determine trends and for intervention assessment.

Our current attempts to analyze ambient formaldehyde concentration data to determine the effects of oxygenated fuels use on the concentration of formaldehyde in the air have not led to statistically significant conclusions. The lack of historical data for formaldehyde concentrations prior to the use of oxygenated fuels makes it more difficult to assess the effects. We have learned a lot about the sources of formaldehyde from this work. Among them are that motor vehicles are a major source of formaldehyde, particularly during the winter. This observation, along with the vehicle emissions test data, suggest that the concentration of formaldehyde should increase when oxygenated fuels are used.

Our research has engendered questions that should help determine the direction of future research. Applications of the model we use for assessing the effects of the intervention in Denver to data from other oxygenated fuels cities and comparison of morning and evening data to assess cold-start effects are underway. We are also continuing our efforts to develop improved techniques for the analysis of ambient concentration data to assess the effectiveness of various air quality control programs. Our research has also underscored the importance of comparing model predictions to outcome measures. It is imperative that the analysis of ambient concentration data be included among the tools used to assess the effectiveness of air quality control programs.

ACKNOWLEDGMENTS

We would like to express our appreciation to Charles Machovec, who was responsible for much of the early work on this project. Many other students and faculty have been involved in various aspects of this project, most significantly Neil Anderson, Philip Anderson, Kerry Grant, Jim Koehler, John Koeppe, Robert Meglen, Joyce Miyagishima, and Heather Spurgeon.

REFERENCES

1. Colorado Department of Health, 1988 Oxygenated Fuels Program, Air Pollution Control Division Final Report to the Colorado Air Quality Control Commission, May 1988.
2. Colorado Department of Health, 1988–89 Oxygenated Fuels Program, Air Pollution Control Division Final Report to the Colorado Air Quality Control Commission, June 1989.
3. Colorado Department of Health, 1989–90 Oxygenated Fuels Program, Air Pollution Control Division Report to the Colorado Air Quality Control Commission, April 1990.
4. Colorado Department of Health, 1991 Oxygenated Fuels Program, Air Pollution Control Division Report to the Colorado Air Quality Control Commission, April 1991.
5. Colorado Department of Health, 1992 Oxygenated Gasoline Program, Air Pollution Control Division Report to the Colorado Air Quality Control Commission, May 1992.
6. Colorado Air Quality Control Commission, Regulation No. 13, The Reduction of Carbon Monoxide Emissions from Gasoline Powered Motor Vehicles Through the Use of Oxygenated Fuels, September 1992.
7. **Anderson, L. G., Leto, J. J., Rolniak, P., Swanson, J., Milliken, J. G., Pielke, R. A., Smith, C. S., Stedman, D. H., and Woodard, G. C.,** Benefits and Costs of Oxygenated Fuels in Colorado, Milliken Research Group, August 1990.
8. **Anderson, L. G., Machovec, C. M., Lanning, J. A., Aloi, M. J., and Meglen, R. R.,** The effects of using oxygenated fuels on the concentration of carbon monoxide in Denver. Paper 92–91.07. *Proc. 85th Annu. Meet. Air & Waste Management Assoc.,* Kansas City, MO, June 25–30, 1992.
9. **Rogers, F.,** Airshed Modeling of Denver for Carbon Monoxide: A Comprehensive Overview, Colorado Department of Health, February 1986.

10. PRC Environmental Management, Inc., Final Report of the Performance Audit of Colorado's Oxygenated Fuels Program, December 1992.

11. Colorado Department of Health, Air Pollution Control Division, 1990 Base-Year Carbon Monoxide Emission Inventory for the Denver, Colorado Nonattainment Area, November 1992.

12. Committee on Tropospheric Ozone Formation and Measurement, *Rethinking the Ozone Problem in Urban and Regional Air Pollution*, National Academy Press, Washington, D.C., 1991.

13. **Calvert, J. G., Heywood, J. B., Sawyer, R. F., and Seinfeld, J. H.,** Achieving acceptable air quality: Some reflections on controlling vehicle emissions, *Science*, 261, 37, 1993.

14. **Hoekman, S. K.,** Speciated measurements and calculated reactivities of vehicle exhaust emissions from conventional and reformulated gasolines, *Environ. Sci. Technol.*, 26, 1206, 1992.

15. **Grosjean, D.,** Ambient levels of formaldehyde, acetaldehyde, and formic acid in Southern California: Results of a one-year base-line study, *Environ. Sci. Technol.*, 25, 710, 1991.

16. **Fung, K.,** Carbonyl observations during SCAQS, Paper 89–152.3, *Proc. 82nd Annu. Meet. Air and Waste Management Assoc.*, Anaheim, CA, June 25–30, 1989.

17. **Grosjean, D. and Williams, III, E. L.,** Ambient levels of aldehydes in Phoenix: Baseline data prior to the implementation of an oxygenated fuel program, February 1989, Ambient levels of aldehydes in Phoenix, Winter 1989–1990, June 1990, and Ambient levels of aldehydes in Tucson, Winter 1989–1990, Prepared for Arizona Department of Environmental Quality, DGA, Inc., June 1990.

18. **Anderson, L. G.,** Tucson aldehyde study, 1990–91, Prepared for Arizona Department of Environmental Quality, Anderson & Associates, July 1991.

19. Alcohol Week's: New Fuels Report, Senators Call for MTBE Testing; Use of Alaska Oxy-Fuel Program Questioned, MTBE Health Ailments Reported in Montana; Officials may Switch to Ethanol, ARCO Chemical, OFA React Swiftly to Charges of MTBE Risks Discussed at Hearing, *New Fuels Rep.*, Vol. 14, No. 11, March 15, 1993.

20. U.S. Environmental Protection Agency, Motor Vehicle-Related Air Toxics Study, Office of Mobile Sources, EPA 420-R-93–005, Ann Arbor, MI, 1993.

21. IRIS, Integrated Risk Information System [data base]. Carcinogenicity Assessment for Lifetime Exposure to Formaldehyde. U.S. Environmental Protection Agency, Office of Health and Environmental Assessment, Environmental Criteria and Assessment Office, Cincinnati, OH.

22. U.S. Environmental Protection Agency, Assessment of Health Risks to Garment Workers and Certain Home Residents from Exposure to Formaldehyde. Office of Pesticides and Toxic Substances, Washington, D.C., 1987.

23. U.S. Department of Health and Human Services, NIOSH Pocket Guide to Chemical Hazards, Washington, D.C., 1990.

24. **Box, G. E. P. and Jenkins, G. M.,** *Time Series Analysis: Forecasting and Control*, Holden Day, San Francisco, 1976.

25. **Harvey, A. C.,** *Forecasting, Structural Time Series and the Kalman Filter*, Cambridge University Press, Cambridge, 1989, pp. 2–3, 10.

26. **Marija J. Norusis,** *SPSS/PC+: Trends*, SPSS Inc., Chicago, 1990.

27. **Effron, B.,** *The Jackknife, the Bootstrap, and Other Resampling Techniques*, SIAM, Philadelphia, 1982, chap. 5.

28. **Winberry, Jr., W. T., Murphy, N. T., and Riggin, R. M.,** Second Supplement to Compendium of Methods for the Determination of Toxic Organic Compounds in Ambient Air, Method TO-11. EPA-600/4–89–018, U.S. Environmental Protection Agency, Research Triangle Park, NC, 1990.

29. **Anderson, L. G., Machovec, C. M., Barrell, R. A., Wolfe, P., and Lanning, J. A.,** Ambient concentrations of formaldehyde and acetaldehyde in Denver, Colorado, manuscript in preparation.

30. **Carlier, P., Hannachi, H., and Mouvier, G.,** The chemistry of carbonyl compounds in the atmosphere — A review, *Atmos. Environ.*, 20, 2079, 1986.

31. **Snedcor, G. W. and Cochran, W. G.,** *Statistical Methods*, 8th ed., Iowa State University Press, Ames, 1989, p. 194.

32. **Altshuller, A. P.,** Production of aldehydes as primary emissions and from secondary atmospheric reactions of alkenes and alkanes during the night and early morning hours, *Atmos. Environ.*, 27A, 21, 1993.

33. Radian Corporation, Colorado Toxic Air Pollutant Emission Inventory and Prioritization for Further Study, DCN No. 84-240-016-60-5, October 1984.

34. Colorado Department of Health, Toxic Release Inventory, 1991.

35. **Anderson, L. G., Lanning, J. A., Barrell, R., Miyagishima, J., Jones, R. H., and Wolfe, P.,** Sources and sinks of formaldehyde and acetaldehyde: An analysis of Denver's ambient concentration data, *Atmos. Environ.*, submitted for publication.

36. **Anderson, L. G., Lanning, J. A., and Wolfe, P.,** Acetone in the urban atmosphere of Denver, Colorado, *Israel J. Chem.*, accepted for publication.

37. U.S. Environmental Protection Agency, National Ambient Air Quality and Emissions Trend Report, 1991, EPA 450-R-92–001, Research Triangle Park, NC, 1992.

38. **Draper, N. R. and Smith, H.,** *Applied Regression Analysis*, John Wiley & Sons, New York, 1966, pp. 266–278.

39. **Neter, J., Wasserman, W., and Kutner, M. H.,** *Applied Linear Models*, 3rd ed., Richard D. Irwin, Boston, 1990, p. 74.

APPENDIX

A. ORTHOGONALIZING THE POLYNOMIAL[38]

Visual inspection of the data suggests the trend is not strictly linear. We wish to include a cubic polynomial consisting of $t = 1, 2, 3, \ldots n$, t^2, and t^3 into the analysis; but these vectors are highly collinear. To avoid this problem, we center the linear component t to get $t_i = -65.5, -64.5, \ldots 65.5$, regress t^2 on t_1 saving the residuals as t_2, regress t^3 on t_2 and t_1 saving the residuals as t_3. The vectors t_1, t_2, t_3 are, by construction, orthogonal to each other and to the vector of 1s representing the constant term in the X matrix of the regression analysis.

B. JUSTIFICATION FOR AVERAGING

Inasmuch as the data are serially correlated, the amount of information lost by averaging over 30-day intervals will be less than if the data were independently distributed. We quantify the loss of information in terms of the reduction in the variance. The variance for averaged data can be written

$$\text{var}(\overline{X}) = \frac{\sigma^2}{n^2}\left(n + 2(n-1)\rho_1 + \ldots + 2\rho_{n-1}\right)$$

where ρ_i is the correlation between an observation at time t and an observation at time $t - i$. When $\rho_i = 1$, $var(\overline{X}) = $ var (X); and when $\rho_i = 0$, as for independently distributed data, the $var(\overline{X}) = \sigma^2/n$. For n = 30, the variance for averaged independent data would decrease to $0.033\ \sigma^2$. The estimated variance for daily averaged CO over the 11 years of our data is 3.193 ppm, while the estimated variance for the monthly averaged data is 1.334 ppm, which is 41.8% of the variance for the daily data, or $0.418\ \sigma^2$.

C. BOOTSTRAP PROCEDURE FOR ESTIMATING THE DISTRIBUTION OF $\hat{\beta}$[27]

Bootstrap estimates for the distribution of the β values are constructed as follows:

1. Estimate the least squares regression, save the residuals and the fitted values $\hat{Y}_i$.
2. Draw a random sample (with replacement) from the residuals, denoted $\hat{e}_i$, i = 1,2, ... N. Using $\tilde{Y}_i = \hat{Y}_i + \hat{e}_i$ re-estimate the β values and save them. Repeat 1000 times.
3. Check the resulting distributions. We compared the skewness and kurtosis to those of a normal distribution.

D. POWER OF THE TEST[38]

The regression coefficients and their variances reported in Table 3 are based on Equation 2; hence, a 10% reduction in CO requires $e^\beta = 0.90$, or $\beta = -0.105$. Because the oxygenated fuels program is intended to lower ambient CO, we apply a one-tailed test and consider H_o: $\hat{\beta}_{15} = 0$ against H_a: $\hat{\beta}_{15} = -0.105$. The power of the test is the probability of rejecting H_o when H_a is true; this can be stated more formally as the proportion of the distribution centered on -0.105 that lies to the left of -0.067, which in this case is the $\hat{\beta}$ that corresponds to the critical value of t with 115 degrees of freedom for $\alpha = 0.05$. Letting TCF be an operator that returns the cumulative t distribution for its argument, and noting that $(\beta - \hat{\beta})/s_{\hat{\beta}}$ is distributed as t, we conclude the power of the test is 0.834 since

$$TCF\left[(-0.067 - (-0.105))/0.040\right] = 0.834$$

Chapter 6

Modeling the Effects of Alternative Fuels on Ozone in Ontario, Canada

Robert McLaren, George Paraskevopoulos, and Donald L. Singleton

CONTENTS

INTRODUCTION

The photochemical formation of ozone (O_3) in the troposphere continues to be a concern in many regions of Canada due to its adverse effects on human health, crops, and forests. More than half of the Canadian population lives in areas where the 1-h air quality ozone objective of 82 ppb is routinely exceeded in the summer. Ozone formation is a complex phenomenon in that ozone is a secondary pollutant, not directly emitted to the troposphere. Ozone is formed from secondary reactions of nitrogen oxides (NO_x) and volatile organic compounds (VOC) in the presence of sunlight, under widely varying meteorological conditions. Mobile sources, especially automobiles, emit a significant fraction of the anthropogenic NO_x and VOC precursors in urban areas. Reductions in this sector can, and have been made by the development of catalysts, better engine designs, and improved emission control technology. Despite these improvements, emissions from motor vehicles using conventional fuels will be significant for some time to come due to the increasing numbers of vehicles. The use of alternative and reformulated fuels is being considered in Canada for their potential benefits to the environment, including reduced emissions of ozone precursors, toxics, and improvements in visibility. The alternative fuels being considered include methanol and ethanol blends, compressed natural gas (CNG), and liquified petroleum gas (LPG). The impact of emissions from alternatively fueled vehicles on tropospheric ozone is not yet known because significant penetrations into the conventional gasoline market are not currently a reality. Photochemical modeling is one of the tools we have to understand the ozone formation process. It can also be used to estimate the effects of large market penetrations in future years.

Previous reports on the benefits of alternative fuels with respect to ozone formation have been dominated by studies of methanol fuels, particularly M85, a mixture of 85% methanol and 15% gasoline, for use in flexible fueled vehicles (FFVs). Results from the studies have been mixed, showing both increases and decreases in the ozone-forming potential of emissions from FFVs operating on M85 compared to gasoline.[1-4] Some of this variability is due to the emissions data used for input. The levels of formaldehyde in the exhaust of methanol vehicles have been shown to be crucial for predicting the overall O_3-

forming potential of the emissions.[1,5] This sensitivity is the result of formaldehyde's high reactivity, including its ability to be a radical-generating species.[4] Higher levels of formaldehyde were typically present in emissions from older FFVs compared to the emissions from newer vehicles.[4] The NO_x emissions from these FFVs operating on M85 are generally lower than conventional vehicles operating on gasoline, although the same is not true for the comparison of gasoline and M85 emissions from the same FFV.[3]

The few studies that have considered the ozone potential of emissions from CNG and LPG vehicles have generally agreed that the reactivity of the organic emissions are less than that for gasoline.[6] The NO_x emissions are somewhat inconclusive, being very sensitive to the conversion technology and the tuning of the vehicle.[7]

While much of the modeling has been done for regions in the U.S., it is anticipated that Canada may have some unique environmental conditions for ozone formation different from those prevailing in more southern climates. These include generally lower temperatures and longer hours of red sunlight during summer ozone seasons. The lower temperatures will affect emission rates of biogenic and anthropogenic species as well as the rates of chemical reactions. The changes in the solar flux at northern latitudes can be anticipated to affect the reactivity of sensitive species such as formaldehyde, because their photolysis rates will decline relative to the NO_2 photolysis rate. Photochemical modeling can address these conditions to predict the relative effects of alternative fuel emissions on ozone.

We are currently modeling the effects that alternative and reformulated fuel use in light-duty vehicles (LDV) will have on tropospheric ozone in different regions of Canada, including the Windsor to Quebec City corridor (WQC) and the Lower Fraser Valley in British Columbia. In this chapter, results will be presented on our modeling of the impacts of alternative fuels on tropospheric ozone in southern Ontario, the region of Canada which most frequently exceeds the ozone air quality objective. The alternative fuels that we have investigated include CNG, LPG, and M85 in FFVs. Comparisons are made to industry-average gasoline of a fuel volatility (9.0 psi Reid vapor pressure) that is proposed for summer conditions in areas of Canada most affected by ground-level ozone.[8]

In this study, the effects of NO_x and VOC emissions from alternatively and conventionally fueled vehicles on tropospheric ozone during episodic conditions are compared. A measure of the ozone-forming potential of organic emissions from these vehicles relative to emissions from conventional gasoline emissions is presented. The effect of NO_x availability on this reactivity is also investigated.

PHOTOCHEMICAL MODELING

Photochemical modeling was performed with OZIPR,[9] a two-level box model that simulates complex chemical and physical processes of the lower atmosphere. The chemistry of the lower troposphere can be handled by optional mechanisms in this model. The mechanism used in this study is based upon the Statewide Air Pollution Research Center/ Environmental Research and Technology, Inc. (SAPRC/ERT) OZIPM chemical mechanism.[10] The mechanism was modified to include explicit chemistry for benzene, isobutene, and propane, as exists in the corresponding detailed mechanism,[10] and explicit chemistry for methane, methanol, ethanol, isoprene, methylvinyl ketone, and methacrolein. Twenty primary organic emission species are represented in the mechanism as indicated in Table 1. The mechanism includes explicit chemistry for all the major chemical species in alternative fuels (methane, propane, methanol, ethanol) and their first oxidation species (formaldehyde, acetone, propionaldehyde, acetaldehyde).

Modeling was performed for a 127×127-km^2 area of southern Ontario, including the city of Toronto in the southern half of the box. The box area is coincident with grid 16:18 of the Acid Deposition and Oxidant Model (ADOM),[11] a regional Eulerian grid model.

Table 1 **Primary emitted organic species in the chemical mechanism**

Species Name	Surrogate	Representation
ALK4	C4 and C5 normal and isoalkanes	C4 and C5 alkanes
ALK7	C6-C8 alkanes	>C5 alkanes
ETHE	Ethylene	Ethylene
PRPE	Propene	All terminal alkenes
TBUT	*trans*-2-butene	Internal and cyclic alkenes
IBUT	Isobutene	2-substituted terminal alkenes
ISOP	Isoprene	Isoprene and 1,3-butadiene
BENZ	Benzene	Benzene
TOLU	Toluene	Monoalkyl benzenes
XYLE	*m*-xylene	Dialkyl benzenes
TMBZ	Mesistylene	Tri- and tetraalkyl benzenes
HCHO	Formaldehyde	Formaldehyde
ALD2	Acetaldehyde	Acetaldehyde and benzaldehyde
RCHO	Propionaldehyde	>C2 aldehydes
ACET	Acetone	Acetone
MEK	Methylethylketone	>C3 ketones
METH	Methane	Methane
MEOH	Methanol	Methanol
ETOH	Ethanol	Ethanol
ALK3	Propane	Propane

Episodic emissions and meteorological data for this area were available for a period during the Eulerian Model Evaluation Field Study (EMEFS). The specific period of the base case included 4 days, August 01–04, 1988. An ozone episode occurred over Ontario during this time period. Day-specific hourly emissions (point, area, and mobile) and meteorological input for ADOM were obtained from the Ontario Ministry of the Environment and transformed appropriately for use in the OZIPR model. The meteorological data used in the box model included hourly temperature, humidity, and mixed layer heights. During the episode, winds were light and predominantly from the south to southwest. The model was initialized each day at 7:00 a.m. with NO_x, non-methane organic gases (NMOG), methane, and ozone, derived from day-specific monitoring station data. An average of initial conditions from multiple stations was used as input to the box model. NMOG was derived from total hydrocarbon and methane measurements and included an assumed 5% aldehyde composition.[12] The speciation of NMOGs for the surface and aloft layers were taken from Jeffries all-city average, as reported by Carter.[13]

Alternative fuel use in light-duty gasoline vehicles (LDGVs) was modeled for a full penetration effect in the year 2005, a target year for attainment of the Canadian ozone objective of 82 ppb in all regions in Canada. Adjustments to the non-LDGV emissions for 2005 were made based on an emission forecast for Ontario,[8] including emission reductions anticipated in Phase I of the Canadian Council of Ministers of the Environment Plan for NO_x and VOCs. The concentration of NMOG, NO_x, and CO in the initial conditions were reduced for the future scenario in the same proportion as the anticipated total emission reductions. Methane was increased at a rate of 0.9% per year to account for the increasing global background concentrations.[14]

A mobile emissions factor model adapted for Canadian conditions, MOBILE4.1C,[15] was used to predict Total Organic Gases (TOG), NO_x, and CO emission factors for the Ontario LDGV fleet in the year 2005 assuming the temperatures of the episode. The Reid vapor pressure of gasoline for these base conditions was 9.0 psi. Increases in the LDGV fleet over 1988 were made based upon population projections.[16] Emission factors for each alternative fuel fleet were estimated by applying correction ratios to the MOBILE4.1C exhaust and evaporative mode speciated emissions factors for the base fuel,

$$E_{f,p,m} = \frac{e_{f,p,m}}{e_{G,p,m}} \times E_{M41C,p,m} \qquad (1)$$

where $E_{f,p,m}$ is the fleet emission in the future year for alternative fuel f, pollutant p, and emissions mode m. The base fuel fleet emissions are represented by $E_{M41C,p,m}$. The average emission factors, $e_{f,p,m}$ and $e_{G,p,m}$ (grams per mile or grams per test), were derived from a database of vehicle emissions, to be discussed below, for each alternative fuel and the base gasoline fuel, respectively.

The adjustments to the LDGV fleet emissions during the simulation day for each alternative fuel account for the effect that will be experienced in the modeling region during the day. One of the uncertainties in this modeling approach is with respect to the speciation and concentrations of VOCs and NO_x for the initial conditions. The base case initial conditions were derived from day-specific measurements from a number of stations in the modeling region in which gasoline was used. Failure to account for the effect of alternative fuel use on initial conditions could result in an underestimation of the fuel's impact on tropospheric ozone. To account for this, corrections were made to the initial surface and aloft concentrations and chemical speciation for each fuel. This adjustment to initial VOCs, NO_x, and CO for the alternative fuels was made by subtracting from the initial mass, a LDGV inventory fraction of the base fuel emissions with appropriate chemical speciation, followed by replacement with a corresponding mass and speciation for the alternative fuel. The approximation made in this type of simulation is that all the mass in the surface layer comes from an airshed in which a fuel switch is made. Calculations were done for a 100% replacement of LDGVs with each alternative fleet. While 100% replacement is not anticipated, interpolation is possible for small changes in ozone.

VEHICLE EMISSIONS

Speciated emissions from alternatively fueled vehicles and conventional fueled vehicles were collected from a variety of sources: for CNG;[17-20] for LPG;[17,20-22] for M85;[17,19,23] and for industry-average gasoline.[23] The technology represented in the database includes predominantly dual-fuel vehicles with conventional mechanical-type carburetion for CNG and LPG, prototype flexible fueled vehicles operating on M85, and "current" technology gasoline vehicles with closed loop computer-controlled fuel injection. The database was used for two purposes. The first purpose was to estimate fleet average exhaust emissions of total organic gases (TOG, including methane), NO_x and CO, and evaporative emissions for each vehicle/fuel combination. These averages are represented by $e_{f,p,m}$ and $e_{G,p,m}$ in Equation 1. The second purpose of the database was to calculate average chemical profiles of organic gases for each emission mode. The latter averages account for changes in the reactivity of organic emissions when introduced into the model. The requirement for inclusion of the data in the averages used for this study include: 1) vehicle mileage less than 30,000 miles; 2) vehicle model year greater than 1988; 3) presence of a catalytic convertor; 4) maximum inclusion of two identical vehicles in the average; 5) testing under a consistent set of conditions, the Urban Dynamometer Driving Schedule (UDDS) of the Federal Test Procedure (FTP); and 6) chemical speciation of organic emissions.

Table 2 **Carbon fractions of total vehicle emissions**

Species Type	Gasoline	Natural Gas	Propane	M85
Alkanes (>C3)	0.478	0.023	0.051	0.241
Alkenes	0.088	0.008	0.072	0.028
Aromatics	0.310	0.009	0.029	0.146
Formaldehyde	0.003	0.004	0.008	0.020
Aldehydes and ketones (>C1)	0.009	0.002	0.010	0.003
Methane	0.093	0.913	0.206	0.061
Methanol	0.000	0.000	0.000	0.495
Propane	0.019	0.041	0.624	0.006

Evaporative emissions for conventional vehicles and FFVs were taken from the Auto/ Oil Air Quality Improvement Research Program working data set.[23] Fleet averages were calculated for diurnal and hot soak emissions. In the absence of better data, evaporative emissions for CNG and LPG vehicles were assumed to be zero in this study. This is not a bad approximation because of the closed nature of the fuel delivery systems in these vehicles. Leakage and refueling emissions are not accounted for in this study.

The fleet average organic chemical profiles were lumped for the chemical mechanism. A summary of the organic profiles for total vehicle emissions are presented in Table 2, indicating general chemical categories. The chemical profiles for CNG and LPG reflect exhaust emissions only, while those for gasoline and M85 reflect a mixture of exhaust, hot soak, and diurnal emissions. The particular weighting applied to each mode is derived from the output of the mobile emissions model. Of the other mobile model outputs, resting loss emissions were treated as being chemically equivalent to diurnal emissions while running loss emissions were treated as being equivalent to a 50:50 mixture of hot soak and diurnal emissions.

The emissions and impacts of the alternatively fueled vehicles while operating on gasoline are not considered in this study. In general, these dual fuel vehicles have higher emissions while operating on gasoline than conventional vehicles.[3,7] While this is a fact that should be considered, it may represent a short-term impact on the way to OEM (original equipment manufacturer) technology for CNG, LPG, and methanol vehicles in the future. The attempt here is to estimate the effect that alternative fuels may have on ozone in the future, as estimated by the data and technology present here today. Because of evolving technology, this approach represents a snapshot in time.

RESULTS AND DISCUSSION

Figure 1 shows the 4-day base case simulation of ozone for 1988. The solid line represents the simulated ozone concentration, while the symbols represent monitoring data from a number of stations in the model area. The simulation matches the severity of the episode reasonably well in terms of peak ozone except for the first day. On this day, the model overestimates the peak ozone concentration in the afternoon. There is also an overesti- mate of ozone on the second evening when the actual observations show a sharper drop in ozone than the simulation. On the first day, a thundershower moved through the area in the afternoon. This event would suppress the peak ozone concentration because of the drop in photolysis rates and through rainout of ozone and precursors. Rain was also recorded on the second night. The model does not account for such events; it is intended for "clear sky" conditions. Clear sky conditions were assumed for the future year

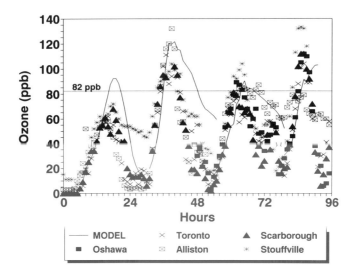

Figure 1 Base case simulation for the Aug. 01–04, 1988 ozone episode. The solid line is the simulated ozone concentration, while the symbols represent data from monitoring stations in the domain. Hour zero starts on Aug. 01 at midnight.

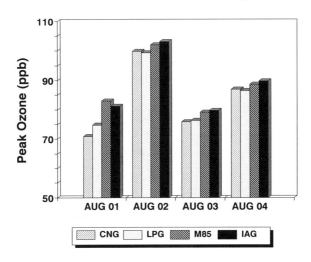

Figure 2 Calculated peak hourly-average ozone in the year 2005 with LDGV fleets operating on industry average gasoline (IAG), compressed natural gas (CNG), liquified petroleum gas (LPG), and 85/15 methanol/gasoline mixture (M85).

simulations. A convenient measure of ozone to compare alternative fuel effects in these simulations is the peak hourly average ozone. This peak will usually occur late in the afternoon, as seen in Figure 1.

Figure 2 shows the calculated peak hourly average ozone concentrations in the future year for each alternative fleet and each day of the episode. The CNG and LPG vehicles show the most reduction in peak ozone for the 4 days, although the absolute ordering changes with the day. The FFVs operating on M85 also show reduced ozone compared to gasoline vehicles on all days except for day 1, when it shows the greatest ozone. Days 1 and 3, which have the lowest initial VOC/NO_x ratio of the 4 days, show the greatest range of ozone concentrations for the different fuels.

The different impacts of fuels on different days is a result of the nature of the chemistry of ozone formation. A simplified reaction scheme for ozone formation includes the following reactions:

$$NO_2 + h\nu \rightarrow NO + O$$

$$O + O_2 + M \rightarrow O_3 + M$$

$$O_3 + NO \rightarrow NO_2 + O_2$$

$$RO_2 + NO \rightarrow NO_2 + RO$$

Formation of ozone is favored by high light intensities, $h\nu$, and large values of the ratio [NO_2]/[NO]. Peroxy radicals, RO_2, formed by the oxidation of organic compounds, facilitate the formation of ozone by conversion of NO to NO_2. The two extreme conditions for ozone formation involve the situation where the system is limited by VOCs or limited by NO_x, corresponding to low and high values of the VOC/NO_x ratio. In the VOC limiting regime, ozone formation is expected to be very sensitive to changes in emissions of organic species, and differences in their reactivity are expected to be most pronounced. Also under these conditions, further additions of NO could lead to reductions of ozone through the direct reaction with NO.

In the NO_x limiting regime, at high VOC/NO_x ratios, ozone formation is expected to be relatively less sensitive to changes in either the mass or speciation of VOC emissions. Additions of NO_x are expected to lead to increased ozone formation. In general, low VOC/NO_x ratios are characteristic of regions near emission sources, and high ratios occur further downwind due to the more rapid consumption of NO_x relative to VOCs.

Because of the nonlinearity of ozone formation discussed above, differences in NO_x emissions from alternatively fueled vehicles will impact ozone in different ways depending on the VOC/NO_x ratio. At a low ratio, the vehicle that emits more NO_x will reduce ozone locally, albeit only temporally, and at the expense of other components of the physical environment. The control of ozone through an increase in NO_x is largely thought to be unacceptable as a control strategy.[24] Higher NO_x emissions can contribute to acid deposition, visibility impairment, higher NO_2 concentrations (NO_2 is a respiratory irritant), and eventually, higher ozone downwind.

Two of the days in the modeling scenario (Aug. 01 and 03) were actually VOC limited (low VOC/NO_x ratio). On these days, fuels with higher emissions of VOCs and lower emissions of NO_x give the greatest peak ozone, while fuels with the lowest VOC emissions and highest NO_x emissions give the lowest ozone. Thus, the simulations with the CNG fleet gave the lowest peak ozone concentrations, while those with M85 gave the highest or second highest ozone on those days.

With this short discussion, the problem of comparing the impact of alternative fuels on ozone formation becomes readily apparent. The possibility exists to have a tradeoff between NO_x emissions and VOC emissions at low VOC/NO_x ratios. It is thus useful to separate the effects of VOC and NO_x emissions in the modeling study. This can be done by considering the impact of the organic emissions from alternatively fueled vehicles alone by holding NO_x emissions constant in the simulation. This was done for the future scenario by holding the initial NO_x concentrations and daytime light-duty vehicle NO_x emissions equivalent to that used for the base fuel. CO was treated as an organic species.

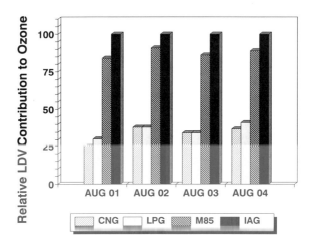

Figure 3 Relative LDGV contributions to peak ozone in the year 2005 for each alternative fuel when NO$_x$ emissions are held constant.

To present these results, it is useful to calculate relative LDGV contributions to peak ozone, where all fuels are compared relative to the base fuel, industry-average gasoline,

$$R^{VOC}{}_f = \frac{O_{f,NO_x} - O_{0,NO_x}}{O_{IA,NO_x} - O_{0,NO_x}} \times 100 \qquad (2)$$

where $R^{VOC}{}_f$ is the relative light duty vehicle contribution to ozone for the organic emissions of fleet f, O_{f,NO_x} is the peak ozone simulated with fleet f, O_{IA,NO_x} is the peak ozone simulated with the base industry average gasoline fleet, and O_{0,NO_x} is the peak ozone simulated with no light-duty vehicle organic emissions. The NO$_x$ subscripts in Equation 2 are there to remind us of constant NO$_x$ emissions in the simulations. By definition, the value of $R^{VOC}{}_{IA}$ is 100% for industry-average gasoline. The $R^{VOC}{}_f$ values are presented in Figure 3 for simulation days 1 to 4 of the episode in the future year scenario. Each fuel has a range of values for the 4 days reflecting different meteorological and emissions characteristics of the day. The emissions from the natural gas and propane fleets have $R^{VOC}{}_f$ values ranging from 25 to 40%, depending on the day. Those for M85 range from 80 to 90%. It will be shown later that the range of $R^{VOC}{}_f$ values for one fuel can be correlated with the NO$_x$ availability. The differences in the $R^{VOC}{}_f$ for each fuel reflect the mass of organic emissions from the vehicles as well as the reactivity of the organic emissions.

UNCERTAINTY ANALYSIS

A number of uncertainty/sensitivity analyses have been performed for the current modeling. Uncertainties exist with all the inputs to the model. The uncertainties will be manifest in the peak ozone values as well as in the answers that we hope to obtain from the modeling. The tests chosen reflect our uncertainty with the model inputs in some cases, and in other cases were based on recommendations in an oxidant model evaluation protocol.[25] In this modeling study, our attempt is to compare the impact of different fuels on ozone formation. Thus, the effect of the uncertainties on our relative measure, $R^{VOC}{}_f$, will be investigated along with their impact on peak hourly ozone.

The results of the uncertainty analysis are indicated in Tables 3 and 4. For each sensitivity test, simulations were run for 4 days and three fuels (IA, M85, and CNG).

Table 3   **Sensitivity of peak ozone to modeling conditions**

Sensitivity Test	Average % change in Peak Ozone[1]			
	Day 1	**Day 2**	**Day 3**	**Day 4**
1. Change NMOG speciation[2]	−5.1	−0.8	−0.2	0.2
2. Double ozone aloft	2.2	7.7	16.9	8.9
3. Decrease photolysis rates by 30%	−34.1	−17.3	−23.3	−13.1
4. Halve the deposition rates	10.3	5.5	4.4	4.4
5. Mixing heights increased 50%	3.9	−1.0	−0.4	−0.9
6. Mixing heights decreased 50%	−5.7	1.3	0.4	1.4
7. HONO added (2% of NO_x)	2.9	0.7	0.7	0.4
8. Reduce initial NO_x by 25%	28.0	−2.8	1.2	−2.9

[1] Average values for three fuels.

[2] The speciation of the initial conditions were changed from those of Jeffries (Reference 13) to the defaults recommended for the SAPRC mechanism (Reference 10).

Table 4   **Relative light-duty vehicle contributions of organic emissions (R^{VOC}_f) to ozone**

Fuel	Day 1	Day 2	Day 3	Day 4
CNG	25.5 ± 0.6	38.2 ± 0.8	34.5 ± 1.4	36.9 ± 1.5
LPG	30.3 ± 0.6	40.0 ± 1.0	36.2 ± 1.3	41.3 ± 1.2
M85	84.1 ± 0.6	90.9 ± 0.9	86.2 ± 1.5	89.1 ± 1.2

Sensitivity runs were done for variable NO_x and for constant NO_x. For the variable NO_x simulations, the percent change in ozone between each pair of simulations, the base simulation vs the sensitivity simulation, was calculated for the matrix of 4 days and three fuels. The average of these values for the three fuels is given for each day in Table 3. Some of the sensitivity runs can lead to large differences in peak ozone concentrations. For example, doubling the ozone aloft can increase the peak ozone by 17% on day 3, the day with the largest dilution factor. A photolysis rate reduction of 30% decreases the peak ozone by 13 to 34%, and halving the deposition rates increases the peak ozone by 4 to 10%. To project these uncertainties through to our relative measures, values of R^{VOC}_f were calculated for all sensitivity simulations with constant NO_x. The appropriate simulations with no vehicle VOC emissions were run for each sensitivity test in order to calculate these values. For a given day and fuel, the R^{VOC}_f values were found to have very little variance associated with the tests listed in Table 3, apart from test 8. In test 8, the initial NO_x was reduced by 25%, thus increasing the VOC/NO_x ratio by 33%. The values of R^{VOC}_f increased with the VOC/NO_x ratio. The values of R^{VOC}_f were averaged for each day and each fuel over all sensitivity tests except test 8. The average deviation from the base values of R^{VOC}_f were calculated for each day and fuel and are presented in Table 4 along with the R^{VOC}_f values. The average deviations listed in the Table 4 reflect variance associated with the numerical calculations and variance associated with the sensitivity tests listed in Table 3. The average deviations do not reflect any uncertainty associated with the vehicle emissions used for the simulation.

REACTIVITY OF ORGANIC EMISSIONS

As mentioned previously, the values of R^{VOC}_f reflect the mass and reactivity of organic emissions from the vehicles, Table 2. There is considerable uncertainty in the mass of emissions from alternative vehicles, both today and in the future. The technology in alternatively fueled vehicles is changing rapidly and is expected to have a large impact on the emission rates. New regulations are also expected to have future impacts on emission rates, as they will dictate the level of emissions control that is necessary to achieve the limits. It would thus be desirable to compare the values of R^{VOC}_f on a mass basis in order to compare the reactivity of organic emissions from different fuels. In doing so, we assume that the speciation of the vehicle emissions will not change as the emission rate of the vehicle changes. This may not be totally true, as new technology may affect the speciation in the future, but we will assume the speciation in Table 2 as being typical. Extra simulations were run to test the linearity of R^{VOC}_f with the mass of organic emissions. Linearity was found on all days for LDGV inventory fractions up to 20%. It is thus possible to scale these values by the ratio of emissions from gasoline and the alternative fuel simulation to give a measure of the relative reactivity of the organic mixtures,

$$ RF_f = \frac{O_{f,NO_x} - O_{0,NO_x}}{M_f} \times \frac{M_{IA}}{O_{IA,NO_u} - O_{0,NO_u}} \tag{3} $$

where RF_f is the reactivity of the organic mixture with respect to ozone formation relative to the reactivity of emissions from current gasoline vehicles and M_f is the mass of emissions. The units chosen for mass must be specified as this will affect the values of RF_f, especially for those fuels that contain significant oxygen content. For comparison to other values, the mass of non-methane carbon was used for scaling. Methane has been excluded in the mass although the small effects of its reactivity have been included in the simulation. The RF_f value for gasoline is 1, by definition. For equivalent carbon NMOG emissions and equivalent NO_x emissions, a reactivity value greater than 1 indicates higher peak ozone values, while the reverse is true for values less than 1.

The values of RF_f are presented in Figure 4 for each simulation day. The average deviations from the base case in Table 4 have been propagated through to the values of RF_f. The days have been ordered in terms of increasing initial VOC/NO_x ratio for the base fuel simulation, which are indicated on the bottom axis. As can be seen, the relative reactivities of the M85, LPG, and CNG emission mixtures are all less than that for gasoline on a carbon basis. A trend is also seen from left to right; the relative reactivity values increase for the alternative fuel mixtures as the VOC/NO_x ratio increases. The relative reactivity values vary from 0.65 to 0.97, 0.67 to 0.91, and 0.79 to 0.86, for CNG, LPG, and M85, respectively for the 4 days. The increase in the relative reactivities of the fuel emissions for simulation days ordered according to increasing initial VOC/NO_x ratio is consistent with the observation that R^{VOC}_f values increased when the initial VOC/NO_x ratio was increased by 33% in sensitivity test # 8, Table 3. In effect, we see that a leveling effect occurs; the relative reactivity of the organic mixtures become more similar as the NO_x availability is reduced. This is consistent with the earlier work of Dodge.[26] The modeling study by Dodge showed that the reactivity of individual hydrocarbons decreases as the VOC/NO_x ratio increases, but the reactivity of very reactive hydrocarbons decreased more with the VOC/NO_x ratio compared to less-reactive species. This leveling between the reactivities of several VOCs at high VOC/NO_x ratio can probably be attributed to reactions of olefins with ozone and to some aromatic compounds being efficient sinks for NO_x under NO_x limiting conditions.[26]

It is interesting to note that the relative reactivity of M85 emissions in Figure 3 shows less of a trend with the VOC/NO_x ratio than CNG or LPG emissions. To account for this, we may look to methanol or formaldehyde, constituents of the M85 emissions that are

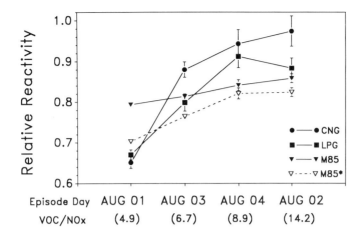

Figure 4 Relative reactivity of organic emissions from alternatively fueled vehicles compared to emissions from conventional vehicles on an equal carbon basis. The days of the episode are ordered according to the initial VOC/NO$_x$ ratio for the base gasoline simulation.

vastly different from the other alternative fuels. Methanol is less reactive than most species and the proportionate change in its reactivity with VOC/NO$_x$ ratio is not as significant as constituents of gasoline.[26] Based upon this, we would expect a more extreme leveling of the M85 emissions reactivity as the VOC/NO$_x$ is increased, contrary to what is observed. Formaldehyde, on the other hand, is a very reactive species. It will photolyze to produce peroxy radicals and would thus be even more reactive under VOC limited conditions in which peroxy radicals are at a premium. It's reactivity changes more rapidly with the VOC/NO$_x$ ratio than most species, similar to the internal olefins and xylene constituents of gasoline.

To test this, a simulation was done in which the formaldehyde was removed from the organic profile in Table 2, scaling the other constituents in order to keep equal carbon mass. The results of this simulation are indicated in Figure 4 by the dashed line. The relative reactivity of the M85 mixture decreases when the chemical profile is changed to exclude formaldehyde. But the change in relative reactivity is more pronounced at low VOC/NO$_x$ ratios, and is minimal at high VOC/NO$_x$ ratios. The formaldehyde emissions from FFVs are thus most detrimental in terms of ozone formation at low VOC/NO$_x$ ratios.

The variation of the relative reactivities with VOC/NO$_x$ ratio makes it difficult to predict the effect of organic emissions from alternatively fueled vehicles on tropospheric ozone formation. The greatest impact is expected at low VOC/NO$_x$ ratios, where VOCs contribute more to ozone formation and where the relative reactivity of emissions are more differentiated from those of gasoline emissions. At higher ratios, the VOCs are less important to the ozone formation but in addition, the residual reactivity that does exist is less differentiated from that of gasoline vehicle emissions. Thus, organic emissions from alternatively fueled vehicles will make less of an impact on ozone formation in the high VOC/NO$_x$ regime.

Comparison of the results in this study can be made with other similar studies. Using trajectory box modeling, Chang et al.[1] calculated a per vehicle ozone reduction potential for the total emissions from a FFV on M85. The formaldehyde component of the total vehicle emissions was 3% by carbon, slightly greater than what was used here. Simulations were done for 20 urban areas in the U.S. for current and future year scenarios. The average ozone reduction potential for the M85 emissions was 19%. This value reflects equivalent carbon emissions from the M85 and gasoline vehicles while NO_x emissions were held constant. The value, 0.81 (1.00 − 0.19), is comparable to the relative reactivity for total M85 vehicle emissions obtained here, ranging from 0.79 to 0.86 for the 4 simulation days.

Jeffries[4] recently reported simulations using a trajectory model with a modified Carbon Bond IV chemical mechanism. The simulations were for conditions representative for North Carolina. On an equal carbon basis, the M85 simulations showed more ozone than industry average gasoline in some cases. Admittedly, the data were from older FFVs with a total formaldehyde composition of 7%. Newer data from the Auto/Oil AQIRP indicate a 2% formaldehyde composition. Jeffries then performed outdoor chamber studies using the new lower formaldehyde levels. The result was that the M85 emission mixtures produced slightly less ozone than gasoline emission mixtures on an equal carbon basis. This implies a relative reactivity of the M85 emissions that is less than 1.0 on a carbon basis.

Recently, Russell et al.[2] reported simulations for the year 2000 using a three-dimensional Eulerian grid model. The M85 emissions were estimated in that simulation, as opposed to using emissions data from vehicles, but they were expected to be reasonable. M85 emissions showed slightly less ozone compared to gasoline in those simulations, again indicating a relative reactivity less than 1.0.

The impacts of alternative fuels on tropospheric ozone presented here are based upon the results of 1-day simulations and thus estimate benefits within a 1-day transport from the source region. For areas where a regional build-up of precursors are expected to lead to ozone formation over multiple days, the benefits of alternative fuels may be overestimated due to low kinetic reactivity of the mixture. Multiple-day episodes allow greater time for less reactive species to react, thus increasing their contribution to ozone formation. The effect of alternative fuels during oxidant episodes affected by transport over large distances and multiple days is an issue that needs to be addressed.

SUMMARY

Air quality simulations for the southern Ontario region have been performed for a future year scenario that indicate that emissions from alternatively fueled vehicles can lead to lower peak ozone concentrations than emissions from vehicles operating on industry-average gasoline. The results are based upon specific technology that is expected to evolve, both for alternative and gasoline fleets. Due to uncertainty in present and future emission rates, the ozone-forming potential of vehicle emissions mixtures have been compared on an equal carbon scale. The relative reactivity of the organic mixtures from CNG, LPG, and M85 vehicles are less than that for industry-average gasoline on an equivalent carbon basis. The difference in this reactivity compared to gasoline decreases rapidly as the VOC/NO_x ratio increases for CNG and LPG mixtures, much less so for M85 mixtures. The removal of formaldehyde from the emissions of flexible fueled vehicles operating on M85 is seen to decrease the reactivity of the vehicle mixture relative to gasoline. This decrease is more pronounced at low VOC/NO_x ratios, and minimal at higher VOC/NO_x ratios.

ACKNOWLEDGMENTS

Several individuals provided data or software that were invaluable in this modeling study: J. Bottenheim (Environment Canada); W. P. L. Carter (University of California, Riverside); Coordinating Research Council of the Auto/Oil Air Quality Improvement Research

Program; M. Dodge (U.S. EPA); C. Fung (Ontario Ministry of the Environment); P. Gabelle (U.S. EPA); P. Kiely (Ontario Ministry of the Environment); G. Rideout, M. Kirshenblatt, and C. Prakash (Environment Canada). This work was supported in part by Energy, Mines and Resources Canada through the Panel on Energy, Research and Development, by Transport Canada and by the National Research Council of Canada.

REFERENCES

1. **Chang, T. Y., Rudy, S. J., Kuntasal, G., and Gorse, R. A.,** Impact of methanol fuels on ozone air quality, *Atmos. Environ.*, 23, 1629, 1989.

2. **Russell, A. G., St. Pierre, D., and Milford, J. B.,** Ozone control and methanol fuel use, *Science*, 247, 201, 1990.

3. Emissions and Air Quality Modeling Results from Methanol/Gasoline Blends in Prototype Flexible/Variable Fuel Vehicles, Technical Bulletin No. 7, Auto/Oil Air Quality Improvement Research Program, Atlanta, GA, 1992.

4. **Jeffries, H.,** A different view of ozone formation: Fuel reformulation alone may not reduce urban ozone, *Fuel Reformulation*, 3, 58, 1993.

5. **Seinfeld, J. H.,** Urban air pollution: State of the science, *Science*, 243, 745, 1989.

6. **Carter, W. P. L.,** Ozone Reactivity Analysis of Emissions from Motor Vehicles, prepared for the Western Gas Association, U.S.A., 1989.

7. **Wilson, B.,** Evaluation of Aftermarket Fuel Delivery Systems for Natural Gas and LPG Vehicles, NREL/TP-420-4892, National Renewable Energy Laboratory, Golden, CO, 1992.

8. Canadian Council of Ministers of the Environment Management Plan for Nitrogen Oxides and Volatile Organic Compounds — Phase I, CCME-EPC/TRE-31E, Ottawa, Ontario, 1990.

9. **Gery, M. W. and Crouse, R. R.,** User's Guide for Executing OZIPR, 9D2196NASA, U.S. Environmental Protection Agency, Research Triangle Park, NC, 1989.

10. **Lurmann, F. W., Carter, W. P., and Coyner, L. A.,** A Surrogate Species Chemical Mechanism for Urban-Scale Air Quality Simulation Models, Volume I - Adaptation of the Chemical Mechanism, EPA/600/3-87/014a, U.S. Environmental Protection Agency, Research Triangle Park, NC, 1987.

11. **Venkatram, A., Karamchandani, P., Kuntasal, G., Misra, P. K., and Davies, D. L.,** The development of the acid deposition and oxidant model (ADOM), *Environ. Pollut.*, 13, 189 1992.

12. **Baugues, K.,** Procedures for Applying City-Specific EKMA, EPA-450/4-89-012, U.S. Environmental Protection Agency, Research Triangle Park, NC, 1989.

13. **Carter, W. P. L.,** Development of Ozone Reactivity Scales for Volatile Organic Compounds, EPA/600/3-91/050, U.S. Environmental Protection Agency, Research Triangle Park, NC, 1991.

14. **Glooschenko, W. A., Roulet, N., Barrie, L. A., and Schiff, H. I.,** Northern Wetlands Study Report, Canadian Institute for Research in Atmospheric Chemistry, Toronto, Ontario, Canada, 1991.

15. **Terrillon, F.,** Adapting MOBILE4.1 to Model the Canadian Motor Vehicle Fleet, Environment Canada, Ottawa, Ontario, Canada, 1991.

16. Population Projections for Canada, Provinces and Territories: 1989-2011, 91-520, Statistics Canada, Ottawa, Ontario, Canada, 1991.

17. Proposed Reactivity Adjustment Factors for Transitional Low-Emission Vehicles, Technical Support Document, State of California Air Resources Board, Sacramento, CA, 1991.

18. **Gabelle, P., Knapp, K. T., Ray, W. D., Snow, R., Crews, W., Perry, N., and Lanning, J.,** Ambient Temperature and Driving Cycle Effects on CNG Motor Vehicle Emission, SAE Technical Paper Series-90206, presented at International Fuels and Lubricants Meeting and Exposition, Tulsa, OK, Oct. 22-25, 1990.

19. **Gabelle, P.,** Mobile Source Emissions Research Branch, U.S. Environmental Protection Agency, Research Triangle Park, NC, unpublished data, 1991.

20. **Prakash, C.,** Emission performance of four identical passenger cars running on gasoline, M85, CNG and LPG, 91-106.9, presented at the 84th Annual Meeting and Exhibition of the Air and Waste Management Association, Vancouver, British Columbia, Canada, June 16-21, 1991.

21. **Hilden, D. L., Mulawa, P. A., and Cadle, S. H.,** Liquified Petroleum Gas as an Automotive Fuel: Exhaust Emissions and Their Atmospheric Reactivity, GMR-7442, General Motors Research Laboratories, Warren, MI, 1991.

22. **Rideout, G.,** Comparison of LPG & Gasoline Fueled Emissions from a Light-Duty Truck, 91-06, Environment Canada, Ottawa, Ontario, Canada, 1991.

23. **Pollack, A. K., Cohen, J. P., and Noda, A. M.,** Auto/Oil Air Quality Improvement Research Program: Description of Phase I Working Data Set, SYSAPP-91/100, Systems Applications International, San Rafael, CA, 1991.

24. **Finlayson Pitts, B. J. and Pitts, J. N.,** *Atmospheric Chemistry: Fundamentals and Experimental Techniques*, John Wiley & Sons, New York, 1986, 615.

25. Model Evaluation Protocol, in *Proceedings of the Workshop on Oxidant Model Evaluation and Application 1. Development of Protocol*, Montreal, Quebec, Canada, November 9-10, 1992, 163.

26. **Dodge, M. A.,** Combined effects of organic reactivity and NMHC/NO$_x$ ratio on photochemical oxidant formation — a modeling study, *Atmos. Environ.*, 18, 1657, 1984.

Chapter 7

Solar Photochemistry

John S. Connolly

CONTENTS

PROLOGUE

"The ability to synthesize fuels and chemicals from abundant, renewable resources could have a significant impact on the energy budget of the entire world. Thus, transformation of solar energy to usable forms by photochemical mechanisms is one of the great social as well as scientific challenges of this century. Our dependence on, and profligate use of, fossil fuels exacerbated by the energy crises of the 1970s (gave rise) to severe inflation (that) ... affected ... developing and developed nations alike. Nevertheless, the use of our most abundant resource—the sun—has increased only marginally over the last several decades.

"In addition to dwindling supplies of nonrenewable energy sources, other important items ... include the closely related problems of population, pollution, the global increase of CO_2, and the possibility that (the supply of) fresh water may prove to be one of the limiting resources of the planet. To increase worldwide utilization of solar energy and, hence, to decrease our dependence on fossil fuels will require that the resources of all nations be focused on the problem. This will entail not only implementation of existing solar technologies but also support for the fundamental and applied research needed to develop new technologies for the 21st century."

These words, written over 10 years ago,[1] are as applicable now as then.

INTRODUCTION

The proven worldwide reserves of fossil and nuclear fuels (i.e., uranium) are only about 3×10^{22} J, whereas the earth at its surface receives from the sun nearly 100 times that amount every year.[2] If only 0.1% of the earth's surface were covered with solar collectors operating at a net efficiency of 10%, all of the world's current annual energy needs could be supplied by the sun. However, only about 0.3% of the earth's insolation is converted to and stored as chemical energy by green-plant photosynthesis; the remainder maintains the hydrological cycle and heats the land masses and the oceans. While solar energy can, in principle, supply our foreseeable energy needs, much research and development will be needed to make this a reality.

Although the sun delivers more than enough energy to supply society's needs for as long as anyone can foresee, its output is extremely diffuse at the earth's surface. Therefore, large areas are required to receive and convert the low power densities available

119

(~100 to 300 W/m^2, depending on geography) to more useful forms. The intermittency of solar radiation is another drawback, which requires some type of storage so that the energy collected can be made available on demand. These two factors are primarily responsible for the considerable economic barriers that stand in the way of large-scale utilization of the sun's energy. Thus, while the source is "free", the means to harness it effectively are not.

BACKGROUND

Over the last 3 billion years or so, biological photosynthesis has evolved elegant supramolecular assemblies that carry out highly specific reactions at efficiencies optimal for the survival of each organism in its own particular environment. Purely synthetic systems, on the other hand, offer the potential advantages of higher solar efficiencies and much greater flexibility in terms of the overall chemistry, coupled with the ability to "tailor" the molecular constituents to carry out specific processes, from water splitting to chemical reduction of CO_2.

Direct quantum processes offer the best hope for discovering new molecular systems to harness the sun's energy cleanly and efficiently. The prospect of using solar photons to initiate useful chemical reactions has attracted the interest of photochemists for many years.[3] Nearly 2 decades ago, Balzani and co-workers[4] listed the minimal requirements for a practical photochemical solar-conversion system, which defined the boundaries within which the search for photochemical devices must be carried out.

Photochemistry involves alteration of chemical bonds by quantized absorption of light in the ultraviolet (UV) to near-infrared (NIR) region of the electromagnetic spectrum. In the context of solar-driven processes, the the practical wavelength limits are roughly 300 nm (ca. 4.1 eV ~ 396 kJ/mol ~ 95 kcal/mol) to 1000 nm (ca. 1.2 eV ~ 116 kJ/mol ~ 28 kcal/mol). Since these energies are in the same range as those of chemical bonds, photochemical reactions offer the prospect of high theoretical conversion efficiencies coupled with essentially limitless versatility. However, of all proposed solar-energy conversion schemes, photochemical processes remain the farthest from commercial development. Consequently, a concise outline in terms of specific starting materials, techniques, and end-products is not possible.

SOLAR PHOTOCHEMISTRY

A general outline of solar photochemical processes is presented in Table 1,[1] where some useful reactions are categorized according to system (homogeneous or heterogeneous) and type (direct or photosensitized). In this scheme, "homogeneous" generally refers to the liquid phase, and "heterogeneous" is taken to mean a suspension or immersion of one or more insoluble constituents in a liquid. For reasons associated with environmental considerations,[4] water would probably be the preferred solvent in a practical device of either kind.

Direct and sensitized photoreactions are represented schematically as follows:

Direct:
$$A \rightarrow A^*$$
$$A^* \rightarrow \text{product(s) (unimolecular)}$$
$$\text{or} \quad A^* + B \rightarrow \text{product(s) (bimolecular)}$$

Sensitized:
$$S \rightarrow S^*$$
$$S^* + A \rightarrow S + A^*$$
$$A^* \rightarrow \text{product(s) (unimolecular)}$$
$$\text{or} \quad A^* + B \rightarrow \text{product(s) (bimolecular)}$$

Table 1 **Solar photochemical processes**

A. Homogeneous Systems (usually liquid phase)
   1. Photoredox reactions
      a. Water splitting
      b. N_2 fixation
      c. CO_2 reduction
      d. Formation of specific products
      e. Production of electrical energy (photogalvanic cells)
   2. Molecular energy storage
      a. Isomerization
      b. Dissociation
      c. Synthesis
B. Heterogeneous Systems (mainly liquid and solid phases)
   1. As above, but with one or more components embedded in a solid matrix or sequestered, e.g., in a micelle, vesicle, membrane, or polymer film
      a. Artificial photosynthesis
   2. Photoelectrochemical reactions (i.e., involving one or more photoactive components with semiconducting properties)
      a. Electricity production
      b. Synthesis of fuels and chemicals
      c. Destruction of undesirable molecules (e.g., pollutants)

From Connolly, J. S. and Turner, J. A., in *Photochemical Conversions*, Braun, A. M., Ed., Presses Polytechnique Romandes, Lausanne, Switzerland, 1983, chap. 3, and reference cited therein. (With permission.)

where A and B are reactants, S is a sensitizer, and A^* and S^* represent excited electronic states. In sensitized reactions, it is possible to utilize wavelengths that are not absorbed by the reactant(s).

Twenty or so years ago, photogalvanic action was thought to be a viable mechanism for solar generation of electricity,[5-7] but such processes were later shown[8] to have inherent limitations that preclude practical uses. Similarly, molecular energy storage by means of, e.g., photoisomerization reactions (whether direct or sensitized) also has inherent problems that appear to be insurmountable, even for reactions with quantum yields on the order of unity.[1]

Photoredox processes seem to be the most promising type of reaction in terms of solar applications. More specifically, the need for vectorial electron transfer and/or separation of the product(s) from the reactant(s) indicate that heterogeneous systems are much more likely than homogeneous systems to lend themselves to assembly of a practical device. Thus, commercial solar photochemical production of useful chemicals (e.g., fuels) will probably be accomplished by some means of artificial photosynthesis, including photoelectrochemical (PEC) reactions. In fact, the majority of recent papers presented at the biennial International Conferences on Photochemical Conversion and Storage of Solar Energy (IPS)[6-14] and at the annual Solar Photochemistry Research Conferences of the U.S. Department of Energy[15] have been devoted to these two areas.

In principle, an advantage of PEC-based systems is that the requirement for exceedingly low photodegradation efficiencies[1] need not be as stringent as for molecular systems. This is because degradation of a semiconductor may involve nothing more than etching, which exposes a new layer of the photoactive surface. In contrast, photodegradation of molecular devices involves destruction of the photoactive sites (molecules, complexes, or assemblies). Unless a regenerative mechanism can be incorporated (as in biological photosynthesis), the acceptable quantum yield of degradation would be on the order of 10^{-7} to 10^{-9} for a device with a $1/e$ lifetime of only 1 year.[1]

Two chapters in this book, "Photoelectrochemical Production of Hydrogen," by A. J. Nozik (Chapter 12) and "Molecular Approaches to Artificial Photosynthesis," by D. Gust, T. A. Moore, and A. L. Moore (Chapter 8), discuss recent progress in understanding photoredox processes of these respective types.

RECENT DEVELOPMENTS

Most people who were working in solar photochemistry 15 years ago envisioned primarily the potential to manufacture useful substances from inexpensive and abundant, renewable resources. To use a baseball analogy, the "home run" in this respect is depicted by the (photosensitized) reaction:

$$CO_2 + 2H_2O \xrightarrow{h\nu} CH_4 + 2O_2$$

If achievable, such a process could not only supply a high-energy fuel, but could also diminish (or at least not contribute to the further growth of) atmospheric CO_2.

What was not foreseen by most of us was the potential to use photochemical reactions to destroy undesirable compounds (e.g., pollutants) by degrading them into environmentally benign end-products. In this application, solar energy can be used to remediate certain environmental problems at the source. Another advantage of this approach is that the restrictions on allowable degradation quantum yields (see above) can be relaxed, depending on the costs and properties of the specific constituents.

Chapters 10 and 11 in this book discuss somewhat different approaches to "solar detoxification." The contribution by D. M. Blake, "Solar Processes for the Destruction of Hazardous Chemicals," entails direct, heterogeneous photoelectrochemical reactions using suspensions of the anatase form of the semiconductor, TiO_2. "The Detoxification of Waste Water Streams Using Solar and Artificial Light Sources," by J. R. Bolton and co-workers, presents some comparisons of this approach with an alternative (proprietary) process.[16] Both types of systems might well find commercial applications in the near future.

EPILOGUE

Evidently, a complete description of solar photochemical processes is beyond the scope of this introductory paper. Readers who wish to learn more about the details of the wide variety of possible photochemical systems are referred to the supplementary reading list.[17-46]

REFERENCES

1. **Connolly, J. S. and Turner, J. A.,** Status and prospects for solar photochemistry, in *Photochemical Conversions*, Braun, A. M., Ed., Presses Polytechniques Romandes, Lausanne, Switzerland, 1983, chap. 3, and references cited.
2. **Bolton, J. R. and Hall, D. O.,** Photochemical conversion and storage of solar energy, *Annu. Rev. Energy*, 4, 353, 1979.
3. "The reactions caused by light are so many that it should not be difficult to find some which are of practical value." Ciamician, G., The photochemistry of the future, *Science*, 36, 385, 1912.
4. **Balzani, V., Moggi, L., Manfrin, M. F., Bolletta, F., and Gleria, M.,** Solar energy conversion by water photodissociation, *Science*, 189, 852, 1975.

5. **Lichtin, N. N., Ed.,** The current state of knowledge of photochemical formation of fuel, *Report of a Workshop Held at Boston University's Osgood Hill Conference Center, North Andover, Massachusetts, September, 1973.* National Technical Information Service, Springfield, VA, NTIS Document PB-246229, 1974.

6. **Bolton, J. R., Ed.,** *Solar Power and Fuels,* Academic Press, New York, 1977; Proc. First Int'l. Conf. Photochem. Conv. Storage Solar Energy, London, Ontario, Canada, August 1976.

7. Proc. Second Int'l. Conf. Photochem. Conv. Storage Solar Energy, Cambridge, England, August 1978, *J. Photochem.,* 10(1), Special Issue, 1979.

8. **Archer, M. D. and Ferreira, M. I. C.,** in *Photochemical Conversion and Storage of Solar Energy,* Connolly, J. S., Ed., Academic Press, New York, 1981; Proc. Third Int'l. Conf. Photochem. Conv. Storage Solar Energy, Boulder, CO, August 1980.

9. **Rabani, J., Ed.,** *Photochemical Conversion and Storage of Solar Energy, 1982,* Weizmann Sci. Press, Jerusalem, 1982; Proc. Fourth Int'l. Conf. Photochem. Conv. Storage Solar Energy, Jerusalem, August 1982.

10. Proc. Fifth Int'l. Conf. Photochem. Conv. Storage Solar Energy, Osaka, August 1984, *J. Photochem.,* 29(1,2), Special Issue, 1985.

11. Proc. Sixth Int'l. Conf. Photochem. Conv. Storage Solar Energy, Paris, August 1986, *New J. Chem.,* 11, 1987.

12. **Norris, J. R., Jr. and Meisel, D., Eds.,** *Photochemical Energy Conversion,* Elsevier, New York, 1989; Proc. Seventh Int'l. Conf. Photochem. Conv. Storage Solar Energy, Evanston, IL, July 31–August 5, 1988.

13. **Pelizzetti, E. and Schiavello, M., Eds.,** *Photochemical Conversion and Storage of Solar Energy,* Kluwer, Dordrecht, 1991; Proc. Eighth Int'l. Conf. Photochem. Conv. Storage Solar Energy, Palermo, July 1990.

14. **Tian, Z. W. and Cao, Y., Eds.,** *Photochemical and Photoelectrochemical Conversion and Storage of Solar Energy,* International Academic Publishers, Beijing, 1993; Proc. Ninth Int'l. Conf. Photochem. Conv. Storage Solar Energy, Beijing, August 1992.

15. Proceedings of the First (1977) through Seventeenth (1993) DOE Solar Photochemistry Research Conferences, National Technical Information Service, Springfield, VA.

16. **Safarzadeh-Amiri, A.,** Photocatalytic method for treatment of contaminated water, U.S. Patent Application #996198 (filed 22 Dec. 1992; Notice of Allowance, August 1993).

17. **Bolton, J. R.,** Solar fuels, *Science,* 202, 705, 1978.

18. **Hautala, R. R., King, R. B., and Kutal, C., Eds.,** *Solar Energy, Chemical Conversion and Storage,* Humana Press, Clifton, NJ, 1979.

19. **Silverman, J., Ed.,** *Energy Storage,* Pergamon Press, Oxford, 1980; *Trans. First Int'l. Assem. Energy Storage,* Dubrovnik, May 27–June 1, 1979.

20. **Lichtin, N. N.,** Fixing sunshine abiotically, *Chemtech.,* 10, 252, 1980.

21. **Cardon, F., Gomes, W. P., and Dekeyser, W., Eds.,** *Photovoltaic and Photoelectrochemical Solar Energy Conversion,* Plenum Press, New York, 1981.

22. **Harriman, A. and West, M. A., Eds.,** *Photogeneration of Hydrogen,* Academic Press, New York, 1982.

23. **Hall, D. O., Palz, W., and Pirrwitz, D., Eds.,** *Photochemical, Photoelectrical, and Photobiological Processes,* Reidel, Dordrecht, 1983.

24. **Grätzel, M., Ed.,** *Energy Resources through Photochemistry and Catalysis,* Academic Press, New York, 1983.

25. **Kutal, C.,** Photochemical conversion and storage of solar energy, *J. Chem. Ed.,* 60, 882, 1983.

26. **Newton, M. D. and Sutin, N.,** Electron-transfer reactions in condensed phases, *Annu. Rev. Phys. Chem.*, 35, 437, 1984.

27. **Bolton, J. R., Strickler, S. J., and Connolly, J. S.,** Limiting and realizable efficiencies of solar photolysis of water, *Nature (London)*, 316, 495, 1985.

28. **Fendler, J. H.,** Photochemical solar energy conversion. An assessment of scientific accomplishments, *J. Phys. Chem.*, 89, 2730, 1985.

29. **Marcus, R. A. and Sutin, N.,** Electron transfers in chemistry and biology, *Biochim. Biophys. Acta*, 811, 265, 1985.

30. **Bilgen, E. and Hollands, K. G. T., Eds.,** *INTERSOL 85, Proc. 9th Bienn. Congr. Int. Sol. Energy Soc.*, 1985, Pergamon, New York, 1986.

31. **Balzani, V., Ed.,** *Supramolecular Photochemistry*, Series C: Mathematical and Physical Sciences, Vol. 214, Reidel, Dordrecht, 1987.

32. **Fox, M. A. and Chanon, M., Eds.,** *Photoinduced Electron Transfer. Part A: Conceptual Basis; Part B: Experimental Techniques and Medium Effects; Part C: Organic Substrates; Part D: Inorganic Substrates and Applications*, Elsevier, New York, 1988.

33. **Grassi, G. and Hall, D. O., Eds.,** *Photocatalytic Production of Energy-Rich Compounds*, Elsevier Appl. Sci., New York, 1988.

34. **Lehn, J.-M.,** Supramolecular chemistry. Scope and perspectives: molecules, supermolecules, and molecular devices (Nobel lecture), *Angew. Chem., Int. Ed. Engl.*, 27, 89, 1988.

35. **Grätzel, M.,** *Heterogeneous Photochemical Electron Transfer*, CRC Press, Boca Raton, FL, 1989.

36. **Meyer, T. J.,** Chemical approaches to artificial photosynthesis, *Acc. Chem. Res.*, 22, 163, 1989.

37. **Serpone, N. and Pelizzetti, E., Eds.,** *Photocatalysis. Fundamentals and Applications*, John Wiley & Sons, New York, 1989.

38. **Boxer, S. G.,** Mechanisms of long-distance electron transfer in proteins: Lessons from photosynthetic reaction centers, *Annu. Rev. Biophys. Biophys. Chem.*, 19, 267, 1990.

39. **Norris, J. R. and Schiffer, M.,** Photosynthetic reaction centers in bacteria, *Chem. Eng. News*, 68, 22, 1990.

40. **Bolton, J. R., Mataga, N., and McLendon, G. L., Eds.,** *Electron Transfer in Inorganic, Organic and Biological Systems*, Adv. Chem. Ser., Vol. 228, American Chemical Society, Washington, D.C., 1991.

41. **Cox, A.,** Photochemical aspects of solar energy conversion, (Chem. Soc., specialist periodical report), *Photochemistry*, 22, 505, 1991. (See also, previous annual reviews in this series.)

42. **Grätzel, M.,** The artificial leaf: molecular photovoltaics achieve efficient generation of electricity from sunlight, *Comments Inorg. Chem.*, 12, 93, 1991.

43. **Grätzel, M. and Kalyanasundaram, K., Eds.,** *Kinetics and Catalysis in Microheterogeneous Systems*, Marcel Dekker, New York, 1991.

44. **Honda, K., Ed.,** *Photochemical Processes in Organized Molecular Systems*, North-Holland, Amsterdam, 1991.

45. **Schneider, H.-J. and Duerr, H., Eds.,** *Frontiers in Supramolecular Organic Chemistry and Photochemistry*, Verlag Chemie, Weinheim, Germany, 1991.

46. Photochemistry, *Chem. Rev.*, 92(3), 1992.

Chapter 8

Molecular Approaches to Artificial Photosynthesis

Devens Gust, Thomas A. Moore, and Ana L. Moore

CONTENTS

INTRODUCTION

Natural solar energy conversion currently fills most of humanity's energy needs. The power of sunlight harvested by ancient photosynthetic organisms is stored in coal, oil, natural gas, and other fossil fuels. Firewood is a major energy source in some parts of the world, and many of our food, fiber, and building-material requirements are also met, directly or indirectly, by modern photosynthesis. The importance of photosynthesis has prompted scientists not only to try to understand the natural process, but also to attempt to mimic it in the laboratory using synthetic or partially synthetic assemblages. Indeed, the impetus for the development of modern photochemistry came in part from a desire to understand and copy the photosynthetic process. At the turn of the century, one of the fathers of the discipline, Giacomo Ciamician, envisioned a future in which huge photochemical reactors fill many of society's needs for energy and materials processing.[1] This is the future still, but our current knowledge of photosynthesis, coupled with advances in the physical sciences, has allowed significant progress. This chapter will review this progress in one of several major approaches to artificial photosynthesis being investigated today.

NATURAL PHOTOSYNTHESIS

As the design of artificial photosynthesis requires a knowledge of the workings of the natural process, we will first briefly describe the highlights of this phenomenon. Although several types of natural photosystems occur, as far as is known the major features of all of them are more or less the same. The photosystems of certain bacteria are presently best understood,[2-4] and photosynthesis in these organisms will be discussed.

To be most useful, light must be converted to other forms of energy that can be stored and accessed as needed. The energy conversion process in photosynthetic organisms begins with the absorption of light by pigments such as chlorophylls, other tetrapyrroles, and carotenoid polyenes. The result is the excited singlet state of the pigment. Typically, photosynthetic organisms construct large arrays of such pigments that act as antennae to collect light and funnel excitation energy to an entity known as the reaction center, where excitation is converted to other forms of energy. These antenna systems are optimized for the range of wavelengths and photon fluxes encountered by the organism in question. For example, organisms living in the deep shade of tropical forests or beneath the surface of the sea have significantly different antenna requirements from those living in desert areas. The antenna system is designed to provide excitation to the reaction centers at a rate which maximizes their energy production, while preventing damage to the cellular components at excessively high light levels. Organisms use the photophysical processes of singlet and triplet energy transfer to transport excitation energy to reaction centers and excess energy to sites for disposal as waste heat.

The heart of photosynthesis is the reaction center, where excitation energy is converted to chemical potential. In the initial stages of this process, the reaction center functions as a photovoltaic device. It uses light energy to transfer an electron across the thickness of the lipid bilayer in which it resides, and thereby generates a relatively long-lived, high-energy charge-separated state. The energy stored in this state is subsequently converted to various forms of chemical energy which are of use to the organism. The reaction center consists mainly of protein material that is embedded in and spans the lipid bilayer membrane. Implanted in turn within the protein and held there by noncovalent interactions is a suite of relatively small cofactors including chlorophylls and pheophytins, carotenoid polyenes, and quinones. In the bacteria mentioned above, the primary excited singlet state is that of a bacteriochlorophyll dimer in which the two bacteriochlorophyll molecules interact rather strongly. Within about 3 ps of excitation, this "special pair", which resides near one side of the membrane, transfers an electron to a bacteriopheophytin that lies deeper within the membrane. The resulting charge-separated state, which stores within it some of the energy of the initial photon, has a strong thermodynamic driving force for charge recombination, wherein the electron jumps back to the special pair to regenerate the ground state of the system and liberate the stored energy as heat. This wasteful recombination is prevented by an additional electron transfer from the bacteriopheophytin radical anion to a nearby quinone molecule, which occurs within about 200 ps. From there, an electron is transferred to a second quinone within the membrane. The positive charge localized on the special pair is then neutralized by electron transfer from a porphyrin moiety of a cytochrome molecule on the surface of the membrane. At this point, the charge separation effectively spans the membrane. As the rate constants of electron transfer reactions depend strongly (approximately exponentially) on donor-acceptor separation, the transmembrane charge separation precludes significant charge recombination and makes the stored energy available to the other energy-transducing elements of the organism. A key feature of natural photosynthesis is the multistep electron transfer strategy by which a high yield of high-energy charge separation is achieved and at the same time rapid charge recombination is prevented.

ARTIFICIAL PHOTOSYNTHESIS

The brief description of bacterial photosynthesis presented above reveals that although the basic structure of the reaction center, and therefore the crucial environment for electron and energy transfer, is provided by the protein, the actual photochemistry is carried out by the organic cofactors. This suggests one approach to the mimicry of photosynthesis: construct an artificial reaction center using synthetic chlorophyll, quinone, and carotenoid moieties related to those that occur naturally, but replace at least the structural role of the protein with covalent bonds. Properly designed covalent linkages might be able to control satisfactorily the electronic interactions of the various chro-

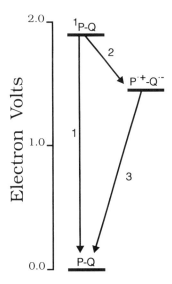

1

Figure 1 Transient species and relevant interconversion pathways for porphyrin-quinone dyad **1** in
benzonitrile solution. The energies of the various states
have been estimated from spectroscopic (excited singlet state) and cyclic voltametric (charge-separated
state) data. Step 1 includes decay of the first excited
singlet state of the porphyrin by intersystem crossing,
internal conversion, and fluorescence. Steps 2 and 3
signify photoinduced charge separation and charge
recombination, respectively.

mophores, donors, and acceptors so that photochemistry reminiscent of that of the natural
reaction centers would be observable. This strategy has been pursued by many laboratories and will be exemplified below with work from Arizona State University. It must be
remembered, however, that there are many ways to attack the problem of artificial
photosynthesis, of which this is only one.

THE SIMPLEST ARTIFICIAL REACTION CENTERS

Clearly, the most basic artificial reaction center must contain a chromophore whose
excited state can act effectively as an electron donor (or acceptor) and a nearby electron
acceptor (or donor) moiety. In the photosynthesis area, the first successful molecules of
this type were two-part "dyads" consisting of porphyrins (chlorophyll relatives) covalently linked to quinones. The initial studies were reported in the late 1970s by Kong
and Loach[5] and by Tabushi and co-workers.[6] Dyad **1** is an example from our laboratories.[7]
The porphyrin (P) absorbs strongly in the visible region of the spectrum, and from a
thermodynamic point of view its first excited singlet state is a good electron donor. The
quinone (Q) is an electron acceptor related to those found in natural reaction centers. The
two moieties are held together by the amide linkage, which constrains the separations and

relative orientations available to them. The photochemistry of dyad **1** is diagrammed in Figure 1. Excitation with visible light yields the first excited singlet state of the porphyrin, ^{1}P-Q, that lies 1.90 eV above the ground state in energy. In an isolated porphyrin, this state decays rapidly (~10 ns) by internal conversion, intersystem crossing to the triplet, and fluorescence (step 1). In the dyad, photoinduced electron transfer to the quinone is faster than these processes and yields the P$^{\bullet+}$-Q$^{\bullet-}$ charge-separated state. In benzonitrile solution, for example, the rate constant for step 2 in Figure 1 is 2.0×10^9 s^{-1} and the quantum yield of charge separation is 0.94. The energy of the P$^{\bullet+}$-Q$^{\bullet-}$ state is about 1.45 eV above that of the uncharged ground state, as estimated from cyclic voltametric measurements [7]

Dyad **1** is a reasonable mimic of the initial steps of photosynthesis in that excitation produces the charge-separated state in high quantum yield, and this state preserves about 1.45 of the 1.90 eV inherent in the porphyrin first excited singlet state. Investigations of porphyrin-quinone dyads are legion, and have led to a great deal of new knowledge concerning photoinduced electron transfer and its dependence on thermodynamic driving force, solvent, temperature, and the electronic interactions between the donor and acceptor.[8-11] However, studies of the lifetimes of the P$^{\bullet+}$-Q$^{\bullet-}$ charge-separated states reveal a major drawback in the use of these molecules as mimics of the natural photosynthesis process: charge recombination is very rapid. With dyad **1** in benzonitrile, for example, transient absorption measurements[7] are consistent with a lifetime of only about 1.7 ps ($k_3 = $ ~6×10^{11} s^{-1}). This is much too short to allow efficient, convenient harvesting of the stored energy.

MOLECULAR TRIADS

Natural photosynthesis also faces the challenge of rapid charge recombination. As described above, it employs a multistep electron transfer strategy to help surmount this problem. A long-lived charge-separated state is achieved by separating the positive and negative ions across the thickness of a lipid bilayer membrane. Charge recombination through this relatively large distance is slow. Photoinduced charge separation across the membrane in a single step would also be slow, and unable to compete with relaxation of the first excited singlet state of the special pair by other mechanisms. The reaction center achieves a high quantum yield of charge separation by using a series of short-range, rapid, and efficient electron transfer steps to attain the final state.

In 1983, we reported results for triad **2**, which adapts the multistep electron transfer strategy to artificial photosynthetic systems.[12,13] The triad consists of a porphyrin-quinone similar to **1** but bearing a carotenoid polyene (C). The photochemistry of the triad is shown in Figure 2. Excitation of the porphyrin moiety yields the first excited singlet state, C-^{1}P-Q, which decays by electron transfer to the adjacent quinone to give C-P$^{\bullet+}$-Q$^{\bullet-}$. In benzonitrile, time-resolved fluorescence and absorption studies[7] yield a rate constant for step 2 in the figure of 2.4×10^9 s^{-1}, which is similar to that observed for the analogous step in **1**. The quantum yield of C-P$^{\bullet+}$-Q$^{\bullet-}$ is 0.85. As is the case for **1**, this charge-separated state has a large rate constant for charge recombination to the ground state (k_3 ~6×10^{11} s^{-1}). However, electron transfer from the carotenoid polyene to the porphyrin radical cation (step 4, k_4 ~1×10^{11} s^{-1}) competes with charge recombination to yield a final C$^{\bullet+}$-P-Q$^{\bullet-}$ charge-separated state. This final state can be readily detected by monitoring the transient absorbance of the carotenoid radical cation in the 955-nm region. It is formed with a quantum yield of 0.13 and has a lifetime of 370 ns.

The triad, by employing a sequential, two-step electron transfer, achieves an energetic (1.15 eV) final charge-separated state in reasonable quantum yield whose lifetime is over 100,000 times longer than that of the related state in the porphyrin-quinone dyad. Thus, it illustrates well the applicability of the multistep electron transfer strategy to control charge recombination in artificial photosynthetic constructs. A variety of triad systems

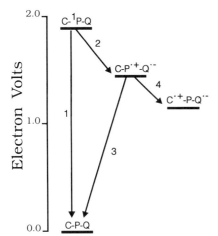

Figure 2 Transient species and relevant interconversion pathways for carotenoporphyrin-quinone triad **2** in benzonitrile solution.

featuring organic or inorganic chromophores have now been prepared, and the results for these molecules demonstrate the generality of the multistep electron transfer concept.[9,11,14-17]

MORE COMPLEX ARTIFICIAL REACTION CENTERS

The usefulness of the multistep electron transfer approach in the triads suggests that more complex assemblies of donors and acceptors, if properly designed, might display useful multistep electron or energy transfer functions that are precluded in simpler systems. As illustrated below, this has proven to be the case.

Molecular Tetrads

The absorption spectrum of tetrad **3** in the visible and near UV regions of the spectrum features readily visible porphyrin bands at 420, 550, 592, and 652 nm and carotenoid maxima at 478 and 512 nm in dichloromethane. The porphyrin fluorescence, with maxima at 655 and 720 nm, is strongly quenched relative to that of a carotenoporphyrin model compound. This quenching signals electron transfer to the naphthoquinone moiety. Time-resolved fluorescence measurements give quantitative information concerning the quenching process. The $C^{-1}P-Q_A-Q_B$ state has a lifetime of 0.38 ns, whereas a model carotenoporphyrin lacking the diquinone moiety decays with a time constant of 3.4 ns in dichloromethane. Assuming that the curtailment of the lifetime of $C^{-1}P-Q_A-Q_B$ is attributable to photoinduced electron transfer by step 2 in Figure 3, the rate constant for this step may be calculated from Equation 1,

$$k_2 = \frac{1}{\tau} - \frac{1}{\tau_o} \qquad (1)$$

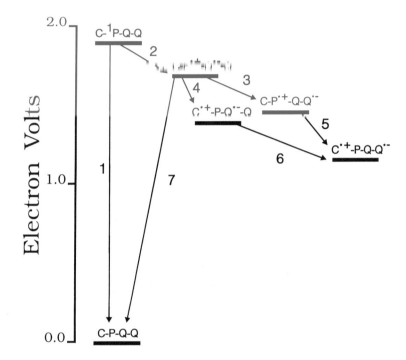

Figure 3 Transient species and relevant interconversion pathways for carotenoporphyrin-diquinone tetrad **3** in dichloromethane solution.

where τ is the porphyrin excited singlet state lifetime in tetrad **3** and τ_o is the corresponding lifetime in the model carotenoporphyrin. Thus, k_2 equals 2.3×10^9 s^{-1}. The quantum yield of C-P$^{\bullet+}$-Q$_A$$^{\bullet-}$-Q$_B$, Φ_2, is given by

$$\Phi_2 = k_2 \tau \tag{2}$$

and equals 0.87.

As shown in Figure 3, the C-P$^{\bullet+}$-Q$_A$$^{\bullet-}$-Q$_B$ state in **3** can either undergo charge recombination to the ground state by step 7 or evolve via steps 3 to 6 to yield a final charge-separated state of the form C$^{\bullet+}$-P-Q$_A$-Q$_B$$^{\bullet-}$. These possibilities were investigated using transient absorption spectroscopy on the nanosecond time scale. Excitation of a dichloromethane solution of the tetrad with a 590 nm, ~15-ns laser pulse resulted in the formation of a new transient species with a strong absorption in the 970-nm region. This species is the C$^{\bullet+}$-P-Q$_A$-Q$_B$$^{\bullet-}$ charge-separated state, and the absorption is characteristic

4

of the carotenoid radical cation. The state has a lifetime of 7.4 µs, and is formed with an overall quantum yield of 0.49.

In **3**, the existence of the final $C^{\bullet+}$-P-Q_A-$Q_B{}^{\bullet-}$ state is inferred from the observation of the carotenoid radical cation absorption and the photochemistry of related model systems. The quinone radical anion was not spectroscopically observed. The closely related tetrad **4** shows photochemical behavior similar to that described for **3**, but the quantum yield of $C^{\bullet+}$-P-Q_A-$Q_B{}^{\bullet-}$ is 0.23 in dichloromethane and the lifetime is only 460 ns.[19,20] An investigation of photoinduced electron transfer in **4** was undertaken using time-resolved electron paramagnetic resonance (EPR) techniques.[21,22] A major advantage of the time-resolved Fourier transform (FT) EPR technique is that, when combined with pulsed laser excitation, it allows complete EPR spectral information to be determined for relatively short-lived species such as the charge-separated states of the tetrad. The FT-EPR spectrum obtained from tetrad **4** following excitation of the porphyrin moiety with a 591-nm laser pulse is similar to that of the radical anion of a model benzoquinone bearing a bicyclic bridge such as that in **4**, and thus confirms the presence of the quinone radical anion. Observation of the carotenoid radical cation was precluded in the FT-EPR experiment because of its large line width. However, a two-pulse electron spin-echo experiment allowed simultaneous detection of both the carotenoid radical cation and the benzoquinone radical anion. Additional experiments demonstrated that the observed electron spin polarization arises from a small degree of electron spin-spin interaction between the radical ions, and that the electron transfer reaction does indeed proceed from a singlet excited state precursor.

Tetrads **3** and **4** are more difficult to prepare than are triads such as **2**, and their photochemistry is considerably more complex. On the other hand, the availability of additional electron transfer pathways results in a higher overall quantum yield for charge separation. In fact, the tetrads use two different types of multistep electron transfer strategies. A sequential multistep electron transfer process, exemplified by steps 2, 3, and 5 in Figure 3, moves the positive and negative charges apart in a series of stages to achieve the final state. Each step occurs over a relatively short distance so that the overall quantum yield is reasonably high. In the final state, the radical ions are well separated spatially so that charge recombination is slow. This sequential process is similar in concept to that employed by natural reaction centers. The tetrads also exhibit a parallel multistep electron transfer chemistry. For example, step 7 in Figure 3 is likely the most rapid of the charge-recombination processes, as the radical ions are close together in C-$P^{\bullet+}$-$Q_A{}^{\bullet-}$-Q_B. Electron transfer steps 3 and 4, occurring in parallel, compete with charge recombination and produce a higher yield of the final $C^{\bullet+}$-P-Q_A-$Q_B{}^{\bullet-}$ state than could be achieved by either one operating alone.

Pentad Artificial Reaction Centers

The most complex photosynthesis mimics thus far prepared in our laboratories are a series of molecular pentads including structures **5** to **7**.[14,17,23,24] These consist of a diporphyrin moiety covalently linked to a diquinone reminiscent of that in **3** and a carotenoid polyene.

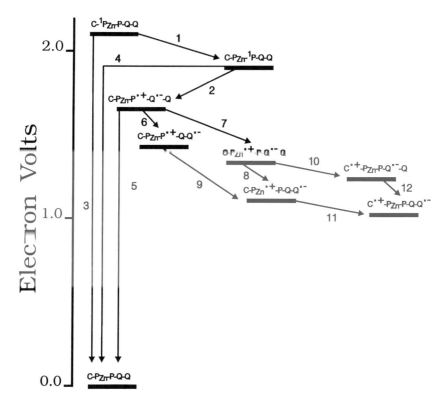

Figure 4 Transient species and relevant interconversion pathways for pentad **5**, in which the porphyrin bearing the carotenoid moiety contains a zinc ion.

The amide linkages, with their partial double-bond character, assure that the molecules exist in extended conformations with an overall length of about 80 Å, as estimated from molecular mechanics calculations.

The photochemistry of the pentads will be illustrated with the results for C-P_{Zn}-P-Q_A-Q_B pentad **5**. The estimated energies of the various transient states and the relevant interconversion pathways are shown in Figure 4. The absorption spectrum of **5** is essentially a linear combination of those of the various component chromophores. Fluorescence studies were carried out in chloroform solution. The emission spectra display maxima at 655 and 720 nm that are characteristic of the free-base porphyrin. Excitation at wavelengths where absorption is due mainly to the zinc porphyrin gives rise to an additional very weak emission band at 611 nm, which is ascribed to that of porphyrin. The emission from the free-base porphyrin is quenched, relative to that from a model porphyrin, but the emission from the zinc porphyrin is quenched much more strongly. This finding is consistent with rapid singlet-singlet energy transfer from the zinc porphyrin to the free base, as has been observed in related porphyrin dyads.[25]

The possibility of singlet-singlet energy transfer was investigated by obtaining the fluorescence excitation spectrum of **5** in chloroform solution. The corrected excitation spectrum for emission in the 720-nm region, where fluorescence is due essentially entirely to the free-base porphyrin, was normalized to the absorption spectrum at 650 nm, where absorption is due only to the free base. The resulting corrected and normalized excitation spectrum is nearly coincident with the absorption spectrum throughout the region where only the two porphyrin pigments absorb. These results show that the quantum yield for singlet-singlet energy transfer from the zinc porphyrin to the free base (step 1 in Figure 4) is very high.

The behavior of the porphyrin excited singlet states was further probed using time-resolved fluorescence spectroscopy. The sample was excited at 590 nm, where both porphyrin moieties absorb, and the decay of the porphyrin fluorescence was measured at 14 wavelengths in the 610 to 730-nm region. The data were analyzed globally[26] as four exponential components. The two most significant components had lifetimes of 0.039 and 1.2 ns. Excitation at 650 nm, where only the free-base porphyrin absorbs, yielded spectra featuring the 1.2-ns decay, but lacking the 39-ps constituent.

The 39-ps component is associated with the decay of C-$^1P_{Zn}$-P-Q_A-Q_B, mainly by singlet-singlet energy transfer to the free-base porphyrin to yield C-P_{Zn}-^1P-Q_A-Q_B. The rate constant for singlet-singlet transfer (k_1 in Figure 4) may be estimated from Equation 3, where τ is the 0.039-ns lifetime of C-$^1P_{Zn}$-P-Q_A-Q_B and k_3 is estimated as the reciprocal of the 0.37-ns fluorescence lifetime of a zinc carotenoporphyrin model compound. Thus, k_1 equals 2.3×10^{10} s-1.

$$1/\tau = k_1 + k_3 \tag{3}$$

The quantum yield for singlet transfer, given by $k_1 \times 0.039 \times 10^{-9}$, is 0.90, which is consistent with the steady-state fluorescence excitation results discussed above, within experimental error.

The spectrum of the other major component of the fluorescence decay of pentad **5** (1.2 ns) shows that it is due to the free-base porphyrin moiety. The short lifetime relative to that of a model monomeric porphyrin (7.8 ns) is consistent with the quenching of the steady-state fluorescence intensity mentioned earlier. The decreased lifetime is attributed to electron transfer via step 2 in Figure 4, which is similar to that observed in **1** to **4**. The rate constant for this photoinduced electron transfer (k_2) may be estimated using Equation 1 as 7.1×10^8 s^{-1}. The corresponding quantum yield of C-P_{Zn}-P$^{\bullet+}$-$Q_A^{\bullet-}$-Q_B, calculated from Equation 2 and based on C-P_{Zn}-^1P-Q_A-Q_B, is 0.85.

Figure 4 shows that C-P_{Zn}-P$^{\bullet+}$-$Q_A^{\bullet-}$-Q_B can recombine via step 5 or evolve through steps 6, 7, and subsequent electron transfers to yield C$^{\bullet+}$-P_{Zn}-P-Q_A-$Q_B^{\bullet-}$. The fate of this initial charge-separated state was studied using transient absorption spectroscopy. Excitation of **5** in chloroform solution with a 650-nm, ~15-ns laser pulse led to the formation of a long-lived transient absorption with a maximum at 970 nm that was assigned to the carotenoid radical cation of the C$^{\bullet+}$-P_{Zn}-P-Q_A-$Q_B^{\bullet-}$ charge-separated state. A lifetime of 55 μs was obtained from an exponential fit of the decay with a floating baseline. The quantum yield of C$^{\bullet+}$-P_{Zn}-P-Q_A-$Q_B^{\bullet-}$ is 0.83.

In dichloromethane, similar results were obtained. The rate constant for singlet-singlet energy transfer step 1 is 2.5×10^{10} s^{-1} in this solvent, and the quantum yield is 0.89. The rate constant for electron transfer from C-P_{Zn}-^1P-Q_A-Q_B to yield C-P_{Zn}-P$^{\bullet+}$-$Q_A^{\bullet-}$-Q_B (step 2 in Figure 4) is 2.9×10^8 s^{-1}, and the corresponding quantum yield is 0.71. Transient absorption experiments revealed the formation of a C$^{\bullet+}$-P_{Zn}-P-Q_A-$Q_B^{\bullet-}$ final charge-

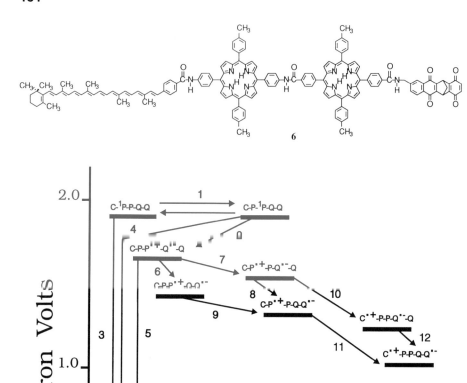

Figure 5 Transient species and relevant interconversion pathways for pentad **6**, in which both porphyrins are in the free-base form.

separated state following 650-nm excitation with a quantum yield of 0.60. The lifetime of the final state is ~200 μs.

Pentad **5** makes effective use of the sequential and parallel electron transfer processes indicated in Figure 4 to produce a very long-lived charge-separated state with a quantum yield approaching unity, which is characteristic of natural reaction centers. This could only be achieved by careful adjustment of the redox properties of the donor and acceptor moieties and the electronic coupling between them in order to maximize the rates of the desired electron transfer processes and retard the others. Particularly noteworthy is the fact that in chloroform, the quantum yield of the final $C^{\bullet+}$-P_{Zn}-P-Q_A-$Q_B^{\bullet-}$ state is virtually identical with that of the initial C-P_{Zn}-$P^{\bullet+}$-$Q_A^{\bullet-}$-Q_B species. This means that the desirable forward electron transfer by steps 6 through 12 in Figure 4 competes very favorably with the charge recombination processes, including step 5.

The delicacy of the balance of rates of the various processes in pentad **5** is shown by the fact that minor changes in conditions or molecular structure can have large effects on the yield and lifetime of the final charge-separated state. For example, when the solvent is changed from chloroform to dichloromethane, the quantum yield of $C^{\bullet+}$-P_{Zn}-P-Q_A-$Q_B^{\bullet-}$ in **5** decreases from 0.83 to 0.60.[23,24] Two factors contribute to this change. In the first place, the rate constant for photoinduced electron transfer step 2 in dichloromethane is only about 0.4 times that in chloroform, and this reduces the quantum yield of the initial step from 0.85 to 0.71. Similar reductions in the rate of photoinduced electron transfer upon changing the solvent from chloroform to dichloromethane have been observed in porphyrin-quinone dyads.[27] Secondly, it is evident that the change in solvent affects the rates of dark reactions so that steps 6 to 12 in Figure 4 are unable to compete as efficiently with charge recombination of the various intermediates.

Pentad **6** is identical to **5**, with the exception that both porphyrins are present in the free-base form.[23,24] The relevant transient species and their interconversion routes are shown in Figure 5. Because the two free-base porphyrin moieties have very similar absorption spectra, excitation of the pentad produces nearly equimolar quantities of C-^{1}P-P-Q_A-Q_B and C-P-^{1}P-Q_A-Q_B. These two species exchange singlet energy very rapidly and decay by various photochemical processes (steps 3 and 4 in Figure 5) and by electron transfer step 2 to give C-P-$P^{\bullet+}$-$Q_A^{\bullet-}$-Q_B. Fluorescence decay studies of **6** and appropriate model compounds yield a value for k_2 of 2.3×10^8 s^{-1} in dichloromethane solution. The quantum yield of C-P-$P^{\bullet+}$-$Q_A^{\bullet-}$-Q_B is 0.36. Transient absorption studies show that this initial charge-separated state evolves via steps 6 to 12 in Figure 5 to yield a final $C^{\bullet+}$-P-P-Q_A-$Q_B^{\bullet-}$ species with an overall quantum yield of 0.15 and a lifetime of ~340 μs.

Removal of the zinc from pentad **5** does not significantly affect the rate of photoinduced electron transfer step 2, but it does result in a substantially lower quantum yield of the initial C-P-$P^{\bullet+}$-$Q_A^{\bullet-}$-Q_B charge-separated state in **6**. This is due to the fact that removal of the zinc renders C-^{1}P-P-Q_A-Q_B and C-P-^{1}P-Q_A-Q_B essentially isoenergetic and leads to rapid singlet-singlet energy transfer between the two porphyrins. Although the decay pathways for C-P-^{1}P-Q_A-Q_B are not directly affected by the carotenoid, the lifetime of C-^{1}P-P-Q_A-Q_B is somewhat reduced through quenching by the attached carotene. Thus, when the two porphyrin singlet states are rapidly exchanging excitation, the quantum yield of C-P-$P^{\bullet+}$-$Q_A^{\bullet-}$-Q_B is reduced relative to that in **5**.

In addition, the yield of the final $C^{\bullet+}$-P-P-Q_A-$Q_B^{\bullet-}$ state in **6** is reduced from 0.36 to 0.15 due to inefficient competition of steps 6 to 12 in Figure 5 with charge recombination. As a first approximation, it is reasonable to postulate that step 5 is the most rapid charge-recombination reaction, as the π-electron systems bearing the positive and negative charges are most strongly coupled electronically in the C-P_{Zn}-$P^{\bullet+}$-$Q_A^{\bullet-}$-Q_B state. Although the thermodynamic driving forces for steps 5 and 6 in Figure 5 are expected to be essentially identical to those for the comparable steps in pentad **5** (Figure 4), the driving force for step 7 in free-base pentad **6** (0.11 eV) is reduced from that for step 7 in zinc pentad **5** (0.32 eV). The Marcus theory for electron transfer predicts that step 7 will be substantially slower in **6** than in **5**. This undoubtedly accounts for the decrease in efficiency of the dark reactions in **6**.

Metallation of both porphyrin moieties of the pentad yields **7**.[24] The absorption spectrum of this molecule in dichloromethane includes carotenoid bands as observed for **5** and **6** and zinc porphyrin absorptions at 424, 548, and 590 nm. The emission spectrum is typical of zinc porphyrins, with maxima at 602 and 650 nm. Time-resolved fluorescence emission studies show that the two zinc porphyrin moieties, whose first excited singlet states are essentially isoenergetic at 2.07 eV, exchange singlet excitation energy rapidly (Figure 6). In addition, C-$^1P_{Zn}$-P_{Zn}-Q_A-Q_B decays by various photochemical pathways included in step 3 with a rate constant of 3.2×10^9 s^{-1}, based on results for a

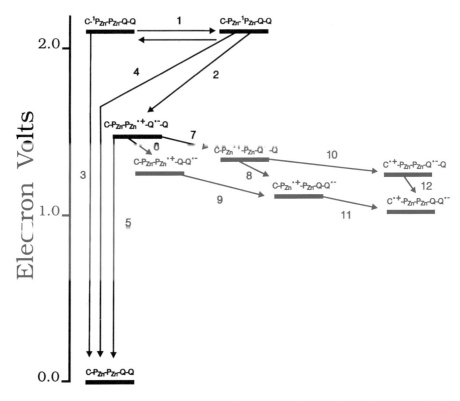

Figure 6 Transient species and relevant interconversion pathways for pentad **7**, in which both porphyrin moieties are metallated with zinc.

model zinc carotenoporphyrin. The other porphyrin first excited singlet state, C-P_{Zn}-$^1P_{Zn}$-Q_A-Q_B, decays by step 4; k_4 is estimated to be 6.8×10^8 s^{-1} from results for a model zinc porphyrin. The C-P_{Zn}-$^1P_{Zn}$-Q_A-Q_B state also relaxes by photoinduced electron transfer to the naphthoquinone with a rate constant equal to 4.8×10^9 s^{-1}. This produces the C-P_{Zn}-$P_{Zn}{}^{\bullet+}$-$Q_A{}^{\bullet-}$-Q_B charge-separated state with a quantum yield of 0.55. Subsequent electron transfer by steps 6 to 12 gives the final C$^{\bullet+}$-P_{Zn}-P_{Zn}-Q_A-$Q_B{}^{\bullet-}$ species with a quantum yield of 0.15 and a lifetime of 120 μs.

As was the case for free-base pentad **6**, the dimetallated compound **7** also has a reduced quantum yield for the final charge-separated species relative to **5**. Two factors conspire to produce this result. The rate of photoinduced electron transfer step 2 for **7**, where the metalloporphyrin is the donor, is about 17 times greater than that for **5**, where the free-base porphyrin donates the electron. This results from the larger driving force for step 2 in the dimetallated pentad (~0.60 eV, Figure 6), which arises in turn from a higher-energy first excited singlet state and a more stable porphyrin radical cation. This enhancement might be expected to result in a higher quantum yield for the initial $C\text{-}P_{Zn}\text{-}P_{Zn}^{\bullet+}\text{-}Q_A^{\bullet-}\text{-}Q_B$ charge-separated state. This is not the case because the zinc porphyrin first excited singlet state $C\text{-}P_{Zn}\text{-}^1P_{Zn}\text{-}Q_A\text{-}Q_B$ has a substantially shorter lifetime in the absence of electron transfer than does $C\text{-}P_{Zn}\text{-}^1P\text{-}Q_A\text{-}Q_B$ and because this state rapidly exchanges singlet excitation with the adjacent porphyrin to yield $C\text{-}^1P_{Zn}\text{-}P_{Zn}\text{-}Q_A\text{-}Q_B$, which is also short-lived.[24] As a result, the quantum yield of $C\text{-}P_{Zn}\text{-}P_{Zn}^{\bullet+}\text{-}Q_A^{\bullet-}\text{-}Q_B$ in **7** is only 0.55 in dichloromethane, whereas that of $C\text{-}P_{Zn}\text{-}P^{\bullet+}\text{-}Q_A^{\bullet-}\text{-}Q_B$ in **5** is 0.71.

The yield of $C^{\bullet+}\text{-}P_{Zn}\text{-}P_{Zn}\text{-}Q_A\text{-}Q_B^{\bullet-}$ is also reduced by inefficient competition of steps 6 to 12 with charge recombination. This is likely due, for the most part, to the large decrease in driving force, and therefore rate, for step 7 in **7** relative to **5** that results from metallation of both porphyrin moieties.

CONCLUSIONS

The examples discussed above demonstrate that it is possible to synthesize artificial reaction centers in the laboratory that mimic the basic features of natural photosynthetic electron transfer. Excitation of properly designed donor-acceptor dyads with visible light can lead to high quantum yields of energetic charge-separated states. The triad molecular architecture provides a mechanism for temporal stabilization of charge separation so that it can in principle be accessed through diffusion-controlled chemical reactions or other means. More complex structures such as tetrads and pentads provide additional design flexibility that can be used to maximize overall efficiencies and lifetimes. Pentad **5**, for example, rivals natural photosynthetic energy conversion in several respects. Other multicomponent systems have now been reported that mimic the ultra-fast electron transfer at low temperatures found in natural bacterial reaction centers and probe the role of superexchange interactions in mediating photoinduced electron transfer.[9-11]

A next stage in the evolution of artificial reaction centers involves the development of methods for harvesting the energy stored in the long-lived charge-separated states. Chemical and electrochemical reactions in a variety of phase-separated or micro-compartmentalized media are being investigated. Progress in understanding the basic science underlying natural and biomimetic photochemical solar energy conversion is rapid today, although practical applications are still goals for the future.

ACKNOWLEDGMENTS

We gratefully acknowledge the contributions of the graduate students, postdoctoral associates, and colleagues who have contributed to the work described here, and who are noted in the references. This work was supported by grants from the Division of Chemical Sciences, Office of Basic Energy Sciences, Office of Energy Research, U.S. Department of Energy (DE-FG0287ER13791), and the National Science Foundation (CHE-8903216). This is publication 163 from the Arizona State University Center for the Study of Early Events in Photosynthesis. The Center is funded by U.S. Department of Energy grant DE-FG02–88ER13969 as part of the USDA/DOE/NSF Plant Science Center program.

REFERENCES

1. **Ciamician, G.,** The photochemistry of the future, *Science*, 36, 385, 1912.
2. **Norris, J. R. and Schiffer, M.,** Photosynthetic reaction centers in bacteria, *Chem. Eng. News*, 68, 22, 1990.
3. **Feher, G., Allen, J. P., Okamura, M. Y., and Rees, D. C.,** Structure and function of bacterial photosynthetic reaction centres, *Nature (London)*, 339, 111, 1989.

4. **Kirmaier, C. and Holten, D.,** Primary photochemistry of reaction centers from the photosynthetic purple bacteria, *Photosynth. Res.*, 13, 225, 1987.

5. **Kong, J. L. Y. and Loach, P. A.,** Covalently linked porphyrin quinone complexes as RC models, in *Frontiers of Biological Energetics: From Electrons to Tissues*, Vol. 1, Dutton, P. L., Leigh, J. S., and Scarpa, H., Eds., Academic Press, New York, 1978, 73.

6. **Tabushi, I., Koga, N., and Yanagita, M.,** Efficient intramolecular quenching and electron transfer in tetraphenylporphyrin attached with benzoquinone of hydroquinone as a photosystem model, *Tetrahedron Lett.*, 257, 1979.

7. **Hung, S.-C., Lin, S., Macpherson, A. N., DeGraziano, J. M., Kerrigan, P. K., Liddell, P. A., Moore, A. L., Moore, T. A., and Gust, D.,** Kinetics of multistep photoinitiated electron transfer reactions in a molecular triad, *J. Photochem. Photobiol. A: Chem.*, 77, 207, 1994.

8. **Connolly, J. S. and Bolton, J. R.,** Intramolecular electron transfer: History and some implications for artificial photosynthesis, in *Photoinduced Electron Transfer, Part D*, Fox, M. A. and Channon, M., Eds., Elsevier, Amsterdam, 1988, 303.

9. **Wasielewski, M. R.,** Photoinduced electron transfer in supramolecular systems for artificial photosynthesis, *Chem. Rev.*, 92, 435, 1992.

10. **Asahi, T., Ohkohchi, M., Matsusaka, R., Mataga, N., Zhang, R. P., Osuka, A., and Maruyama, K.,** Intramolecular photoinduced charge separation and charge recombination of the product ion pair states of a series of fixed-distance dyads of porphyrins and quinones: Energy gap and temperature dependences of the rate constants, *J. Am. Chem. Soc.*, 115, 5665, 1993.

11. **Bixon, M., Fajer, J., Feher, G., Freed, J. H., Gamliel, D., Hoff, A. J., Levanon, H., Mobius, K., Nechushtai, R., Norris, J. R., Scherz, A., Sessler, J. L., and Stehlik, D.,** Primary events in photosynthesis: Problems, speculations, controversies, and future trends, *Israel J. Chem.*, 32, 449, 1992.

12. **Gust, D., Mathis, P., Moore, A. L., Liddell, P. A., Nemeth, G. A., Lehman, W. R., Moore, T. A., Bensasson, R. V., Land, E. J., and Chachaty, C.,** Energy transfer and charge separation in carotenoporphyrins, *Photochem. Photobiol.*, 37S, S46, 1983.

13. **Moore, T. A., Gust, D., Mathis, P., Mialocq, J.-C., Chachaty, C., Bensasson, R. V., Land, E. J., Doizi, D., Liddell, P. A., Lehman, W. R., Nemeth, G. A., and Moore, A. L.,** Photodriven charge separation in a carotenoporphyrin-quinone triad, *Nature (London)*, 307, 630, 1984.

14. **Gust, D., Moore, T. A., and Moore, A. L.,** Molecular mimicry of photosynthetic energy and electron transfer, *Acc. Chem. Res*, 26, 198, 1993.

15. **Gust, D. and Moore, T. A.,** Mimicking photosynthetic electron and energy transfer, *Adv. Photochem.*, 16, 1, 1991.

16. **Gust, D. and Moore, T. A.,** Photosynthetic model systems, *Topics in Curr. Chem.*, 159, 103, 1991.

17. **Gust, D. and Moore, T. A.,** Multistep electron and energy transfer in artificial photosynthesis, in *The Photosynthetic Reaction Center*, Vol. II, Norris, J. R. and Deisenhofer, J., Eds., Academic Press, New York, 1993, 419.

18. **Lee, S.-J., DeGraziano, J. M., Macpherson, A. N., Shin, E.-J., Seely, G. R., Kerrigan, P. K., Moore, A. L., Moore, T. A., and Gust, D.,** Photoinduced charge separation in a carotenoid-porphyrin-diquinone tetrad: Enhancement of quantum yields via control of electronic coupling, *Chem. Phys.*, 176, 321, 1993.

19. **Gust, D., Moore, T. A., Moore, A. L., Barrett, D., Harding, L. O., Makings, L. R., Liddell, P. A., De Schryver, F. C., Van der Auweraer, M., Bensasson, R. V., and Rougée, M.,** Photoinitiated charge separation in a carotenoid-porphyrin-diquinone tetrad: Enhanced quantum yields via multistep electron transfers, *J. Am. Chem. Soc.*, 110, 321, 1988.

20. **Gust, D., Moore, T. A., Moore, A. L., Seely, G., Liddell, P., Barrett, D., Harding, L. O., Ma, X. C., Lee, S.-J., and Gao, F.,** A carotenoid-porphyrin-diquinone tetrad: Synthesis, electrochemistry and photoinitiated electron transfer, *Tetrahedron*, 45, 4867, 1989.

21. **Hasharoni, K., Levanon, H., Tang, J., Bowman, M. K., Norris, J. R., Gust, D., Moore, T. A., and Moore, A. L.,** Singlet photochemistry in model photosynthesis: Identification of charge separated intermediates by Fourier transform and CW-EPR spectroscopies, *J. Am. Chem. Soc.*, 112, 6477, 1990.

22. **Hasharoni, K., Levanon, H., Bowman, M. K., Norris, J. R., Gust, D., Moore, T. A., and Moore, A. L.,** Analysis of time-resolved CW-EPR spectra of short-lived radicals at different times after laser excitation, *Appl. Magnetic Resonance*, 1, 357, 1990.

23. **Gust, D., Moore, T. A., Moore, A. L., Lee, S.-J., Bittersmann, E., Luttrull, D. K., Rehms, A. A., DeGraziano, J. M., Ma, X. C., Gao, F., Belford, R. E., and Trier, T. T.,** Efficient multistep photoinduced electron transfer in a molecular pentad, *Science*, 248, 199, 1990.

24. **Gust, D., Moore, T. A., Moore, A. L., Macpherson, A. N., Lopez, A., DeGraziano, J. M., Gouni, I., Bittersmann, E., Seely, G. R., Gao, F., Nieman, R. A., Ma, X. C., Demanche, L., Luttrull, D. K., Lee, S.-J., and Kerrigan, P. K.,** Photoinduced electron and energy transfer in molecular pentads, *J. Am. Chem. Soc.*, 115, 11141, 1993.

25. **Gust, D., Moore, T. A., Moore, A. L., Gao, F., Luttrull, D., DeGraziano, J. M., Ma, X. C., Makings, L. R., Lee, S.-J., Trier, T. T., Bittersmann, E., Seely, G. R., Woodward, S., Bensasson, R. V., Rougée, M., De Schryver, F. C., and Van der Auweraer, M.,** Long-lived photoinitiated charge separation in carotene-diporphyrin triad molecules, *J. Am. Chem. Soc.*, 113, 3638, 1991.

26. **Wendler, J. and Holzwarth, A.,** State transitions in the green alga *Scenedesmus Obliquus* probed by time-resolved chlorophyll fluorescence spectroscopy and global data analysis, *Biophys. J.*, 52, 717, 1987.

27. **Schmidt, J. A., Siemiarczuk, A., Weedon, A. C., and Bolton, J. R.,** Intramolecular photochemical electron transfer. 3. Solvent dependence of fluorescence quenching and electron transfer rates in a porphyrin-amide-quinone molecule, *J. Am. Chem. Soc.*, 107, 6112, 1985.

Chapter 9

Photovoltaic Technologies: Thin-Film and High-Efficiency Devices

David Ginley, Ken Zweibel, and John Benner

CONTENTS

Pho.to.vol.ta.ic \ *fot-o-val-'ta-ik, -,vol-\ adj*: of, relating to, or utilizing the generation of an electromotive force when radiant energy falls on the boundary between dissimilar substances.

INTRODUCTION

Though the PV effect was discovered by Edmund Becquerel in 1839, it was not until 1954 that scientists at Bell Laboratories demonstrated photovoltaic (PV) devices that promised to be a practical means for the direct conversion of the sun's radiant energy into electricity. In fact, it was the development of high-quality crystalline Si for the infant microelectronics industry that enabled this first PV technology, and the need for space power that drove it initially. Even into the 1970s, the promise of PV remained principally the purview of space power.

The energy crisis of the 1970s, the burgeoning Si microelectronics industry, and an increasing federal commitment to the technology led to an increasing investment in the development of terrestrial PV. The "holy grail" for PV was and is the production of stable modules which produce electricity at prices competitive with or superior to alternative technologies. This competitive price will depend on the application: i.e., remote power vs generators, and residential/utility power vs present generating plants (coal, gas, nuclear).

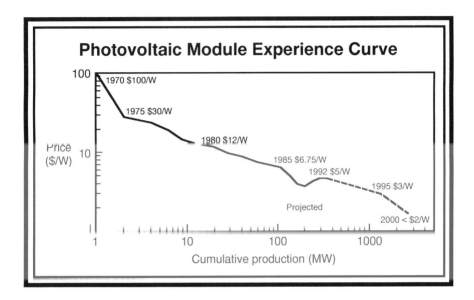

Figure 1 Price/watt vs cumulative production curve for photovoltaic modules for the period from 1970 to approximately 2000. (From Maycock, P. D., "Photovoltaic Technology, Performance Cost and Market Forecast 1990–2010," Photovoltaic Energy Systems, Inc., Casanova, VA, 22017. With permission.)

Increasingly, the value added because PV is an environmentally clean technology is playing a part in price competitiveness.

Figure 1 illustrates the price in dollars per watt of electricity produced vs module production in megawatts for the period from 1970 to approximately 2000. It is estimated that at $2.00 per watt for installed *systems* (as below), costs will begin to approach those for utility-generated power. At the current $5 to $10 per watt system price, some market penetration in remote power and developing countries exists. This drives an increasing production capacity and begins to let "economies of scale" affect the price. It is estimated that around the year 2000, the price per watt of some technologies will hit the $4 mark, and by 2005 to 2010 several may be near $2 per watt.

Coupled with the development of an increasing market is the continuous improvement of PV technology. Continuing progress in the semiconductor and related optoelectronic areas has developed new tools and approaches that have directly benefited the PV area. Tied to this is the development of new analytic tools that have allowed a deeper understanding of the materials on nearly the atomic level. These two forces have allowed significant progress in PV technology over the last few years, resulting in significant efficiency and reliability improvements. Such progress will be the primary focus of this chapter.

While the PV device is the direct converter of the sun's energy to electricity, a PV system consists of a variety of components. These are illustrated schematically for today's conventional Si solar cell in Figure 2. Si is grown from a high-purity feedstock, typically by the Czocralski or other melt processing technique, into single crystal or large-grain polycrystalline material with the correct doping level. The boule is wafered and polished. The slices are converted to individual PV devices by diffusion of a junction, texturing and the application of back and front contacts, and the application of an antireflection coating. Typically, the active PV devices are then interconnected, with the desired current and voltage properties, and encapsulated with appropriate windows/lenses

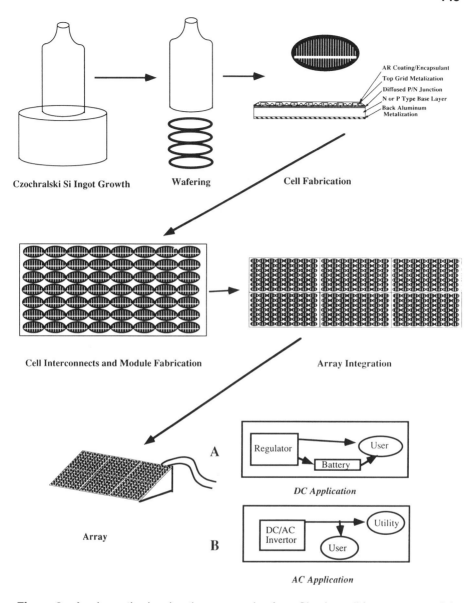

Figure 2 A schematic showing the progression for a Si solar cell from raw materials to an array integrated in an electrical system.

(if concentration of the light is desirable) and a rigid support in a maintainable, environmentally safe package called a module. Modules are mounted on support structures to make an array. The power from a module/array must be multiplexed and fed into a power conditioner to stabilize the DC power output or produce line-compatible AC power. In some cases, batteries are employed for storage. In the realization of a viable PV technology, the "balance of systems" cost for components other than the PV device can be as large as the module cost. Nonetheless, it is clearly critical to develop the most efficient, reliable PV cells and modules possible. This can be approached in two basic ways: *first*, the use of relatively expensive single crystal or epitaxially grown devices designed for optimum efficiency under concentrated sunlight; or *second*, to use polycrystalline or

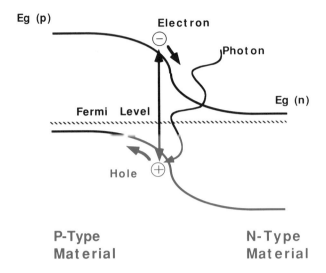

Eg (p)   Electron   Photon   Eg (n)   Fermi Level   Hole

P-Type
Material

N-Type
Material

Simple Si P/N Solar Cell/Diode

Figure 3 Typical band structure for a simple p/n Si solar cell. Light absorption generates an electron-hole pair, which is separated by the junction.

amorphous thin films to achieve moderate efficiency with very low device costs. It is not clear that either of these approaches is the most valid; in all likelihood, both will share the market. In the former case are single-crystal silicon (Si) (largest current commercial product) and compound III-V materials; in the latter, are amorphous-Si (amorphous silicon), cadmium telluride (CdTe), thin-film crystalline silicon, and copper indium diselenide (CuInSe$_2$, CIS) thin films. In this chapter, we will review recent advances in device technology. *Recent advances have resulted in new record efficiencies for thin films and high efficiency devices.*

PV — BASIC DEVICES
As Figure 3 shows, the basic concept of a PV device is very simple. Typically, two semiconductor materials with an electron affinity mismatch are joined to form a junction. This junction can be a homojunction (as illustrated), created by doping the same materials p- and n-type to make a p/n junction, a heterojunction formed from two different semiconductors, a semiconductor with a metal (Schottky barrier), or a thin insulator and a metal (MIS structure). The electron affinity mismatch between materials creates a built-in potential. If above-band-gap light is absorbed, carriers (holes and electrons) are created which are separated by the field and can be collected and employed in an external circuit. Many factors influence the efficiency of this process. From the simplest perspective, some of these are: 1) the band gap of the materials must be optimized to the incident spectrum, 2) the carriers must be efficiently separated and transported to the front and back contacts, and 3) the contacts need to be ohmic, low resistance, and not obscure the incident irradiation.

In reality, each interface in a cell presents a substantial materials science challenge. This is exacerbated for many of the heterostructure devices where unconventional materials are employed and oxides are combined with nonoxides. This can create lattice mismatch problems with the consequent formation of defects and electronic states that

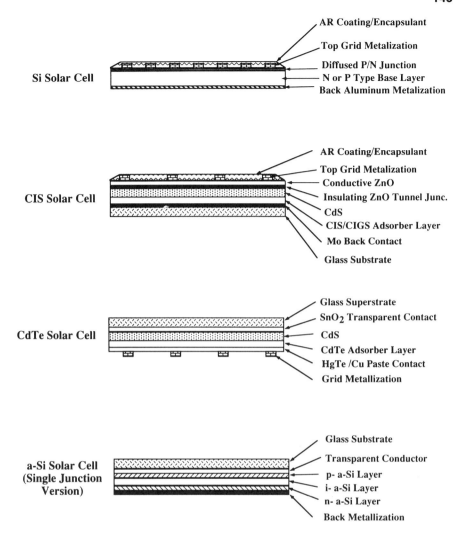

Figure 4 A number of typical device structures for Si, CIS, CdTe, and amorphous silicon solar cells, respectively.

force the premature recombination of carriers. Figure 4 illustrates a few typical device structures, including those for a single-crystal device, a back wall thin-film cell, and two front-wall thin-film cells.

The single crystal device is representative of a simple single-crystal Si PV device. As in Figure 2, the initial device is fabricated from a p-type slice of polished Si through the diffusion of an n-type junction region, the application of front and rear contacts, and the application of an antireflection coating on the top surface. The front grid must be optimized to minimize both electrical and optical losses. Texturing is sometimes employed to improve light gathering. Polycrystalline material can also be employed in this configuration. The same structure also applies to the III-V materials, such as GaAs or InP. The III-V materials are typically grown as epitaxial layers by techniques such as organo-metallic chemical vapor deposition (OMCVD) and molecular beam epitaxy (MBE). This allows the growth of integrated structures with binaries and ternaries of different band gaps. Through the introduction of a tunnel junction, integrated tandem devices can be grown with a wide band gap window cell and a smaller-gap underlying cell. It is important that the current in both cells be matched and the intervening tunnel junction not create any significant loss.

The first back-wall device is the typical structure for a CIS or CIGS solar cell and employs a Mo-coated soda-lime-glass substrate. The CIGS absorber layer is deposited on this substrate by a variety of techniques, and it is typically recrystallized during subsequent processing. Then a CdS layer is deposited. Originally, this was thought to make the junction, but recent results have indicated that the CIS devices may be a junction between stoichiometric CIS and an ordered vacancy layer that forms at the CIS surface. In this scenario, the CdS would serve primarily as a passivating layer for the CIS surface. A thin insulating ZnO layer serves as a tunnel junction, and a thick conductive ZnO layer is employed for contact. The structure is finished with a metal grid and an antireflection coating. Currently, a gallium-containing CIS layer with a gallium gradient from front to back is employed to get the higher gap from CIGS and to promote better adherence to the Mo.

In the case of the CdTe, the glass, soda-lime, or 7059 borosilicate glass is employed as a superstrate; i.e., top surface of the cell. Tin oxide is employed as a transparent conductive top contact, and then CdS again serves in some role both to form the junction and to passivate the interfaces. Highly resistive CdTe is then employed as the absorber layer. In this schematic, copper mercury telluride graphite paste is employed as an alloying back contact with a subsequent grid metallization.

The final structure is for an amorphous silicon single junction solar cell in a p-i-n configuration. This is a superstrate cell and requires a transparent conductive oxide, usually SnO_2, followed by a p^+ layer, an intrinsic layer, an n^+ layer, and a reflective metal back contact, typically aluminum. A p-i-n structure is favored because the transport in the intrinsic material is better than either of the doped materials. The field is established across the intrinsic region due to the n and p layers similar to a p/n junction. Many of the new cell designs actually incorporate three separate junctions to collect more of the solar spectrum and to reduce degradation from the well-known light-induced degradation.

Although, in concept, a PV device is a very simple one, in reality (as can be seen from the devices above) they are much more complex. This arises from the need to integrate the active device with contacts, transparent conductors, and antireflection coatings. This creates a variety of complex interfaces with materials and electrical properties that need to be optimized. In fact, it is not even clear where the junction lies in some of the thin-film cells. Nonetheless, all of the above structures have achieved greater than 10% efficiency, and some have achieved from 15 to 20%. In this chapter, we will discuss a variety of device concepts and will discuss in considerably more detail the devices and the prospects for their being integrated into economically viable systems.

CURRENT DEVICE TECHNOLOGY

HIGH EFFICIENCY
Silicon

The standard bearer for high efficiency is the silicon solar cell. This technology is relatively mature and exists as a commercial product. Low-cost crystalline and polycrystalline silicon substrates, obtained from ingots or grown in sheet (ribbon) form, are now extensively used for the commercial production of solar cells. The efficiencies of these cells typically range between 12 and 14% under 1-sun illumination. Research has shown that the commercial solar cell performance is limited by high concentrations of impurities and crystal defects introduced as a result of growing crystals under conditions that keep the material costs low.[1,2] Specifically, high thermal gradients required for faster growth

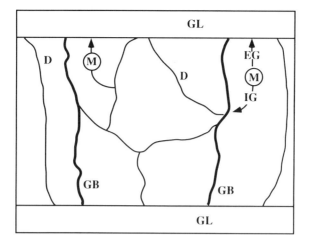

D = Dislocations
GL = Gettering Layer
GB = Grain Boundary
M = Metal
IG = Internal Gettering

Figure 5 A schematic illustration of gettering mechanisms.

are known to introduce stresses. Thermal stresses exceeding the yield stress produce defects in the substrate. Impurities may also be present in the lower quality polysilicon feedstock or may be introduced during the material growth. Although the techniques to minimize defects and impurities are known, the quality of the solar cell substrate is driven by economic considerations. Advancement of silicon PV technology can take two paths: improved growth processes or better methods for post-growth quality enhancement.

Post-growth quality enhancement relies on what is called *defect engineering*. This covers processing techniques that work with impurities and crystallographic defects to minimize adverse electronic interactions with the device by either removing the defect, electrically passivating it, or redistributing it to a less harmful physical or chemical configuration. A considerable effort in fundamental research is necessary to understand these complex processes to a degree where they can be included in a commercial cell processing schedule. Results to date have shown that a variety of phenomena can be exploited to improve the quality of low-cost substrates and yield cell efficiencies of 16 to 19%.[3,4]

Gettering is the process of removing impurities from the active region of a device to a predetermined site where they will be relatively benign. Figure 5 illustrates two approaches of gettering that can be applied to polycrystalline silicon substrates. In one case, impurities diffuse and pile up at an external gettering region applied to the surface(s). In this external gettering scheme, the impurities are driven to the surface(s) by means of a suitable heat treatment. In the second case, suitable thermal treatments cause impurities to be absorbed by the crystal defects in the material, leaving the majority of the bulk with a lower impurity concentration. Impurity gettering by phosphorus diffusions or aluminum alloying has been proposed to improve the performance of polycrystalline silicon solar cells.[5,6] Phosphorous gettering has been found to be very effective for all heavy-metal impurities (such as Fe, Ni, Cu, and Cr) and is well suited for solar cells because nearly all silicon solar cells use phosphorus diffusions for junction formation. The effectiveness of phosphorus gettering is due to the fact that the solubility of many 3d-elements is quite

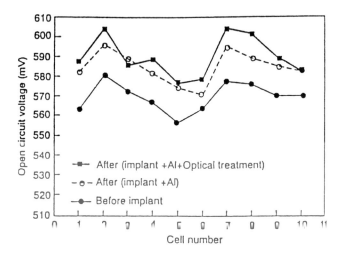

Figure 6 Changes in V_{oc} of several cells due to various steps in the backside hydrogenation process.

high in the heavily doped surface layer that is formed. Recently, aluminum gettering has also been proven to be quite useful for solar cells.[7]

Passivation of crystal defects and impurities by hydrogen is now recognized as a potential process for improving solar cell efficiency. Several techniques are used for hydrogenation which include low-energy ion implantation, plasma processing, plasma-enhanced chemical vapor deposition (PECVD), electron cyclotron resonance (ECR), and annealing in a hydrogen ambient. Typically, hydrogen is introduced from the junction side of the cell. However, it is known that high concentrations of hydrogen introduced from the junction side of a solar cell can have a detrimental effect on the cell performance due to the formation of defects near the junction. A backside hydrogen passivation technique has recently been developed that circumvents this difficulty. The passivation effect is more pronounced for lower-performance cells, typically when the cell efficiency is less than 12 to 13%.[8] Figure 6 shows the open-circuit voltage of a row of devices after each process step, showing that a major improvement in the cell performance can be obtained by hydrogen passivation. Although the basic approaches of post-growth treatments are known, the details of such complex mechanisms are only now becoming clear. It is recognized that these processes can be strongly influenced by point-defect phenomena.[9-11]

The other path focuses on improving the original growth of the silicon to design the material for a better cell. Most commercial silicon solar cells are limited by recombination in the base region (bulk of the device other than the diffused junction) of the device. In order to remove this limit, the material must either be grown with high crystallographic quality and purity to increase minority-carrier lifetime or grown in a thin layer to minimize the volume of the base region. Solar cell and module results using silicon wafers grown by the float-zone method show that improved materials quality creates a new regime of operation with silicon, namely 20 to 25% conversion efficiencies.[12,13] The device structures needed to achieve these efficiencies are quite complex, such as the passivated emitter rear locally diffused (PERL) cell shown in Figure 7. With the limit of base recombination removed, the design systematically addresses the other now-dominant losses due to recombination at the surfaces and contacts. In some respects, the cost of growing silicon by the float-zone method can actually be cheaper than for the Czochralski method.[14] However, the addition of magnetic fields to the Czochralski method for control

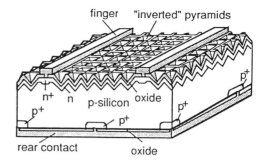

Figure 7 PERL high-efficiency Si solar cell.

of currents in the melt appears to produce silicon that can be processed into cells with efficiencies as high as float-zoned silicon.[15,16] It has also been shown that if the volume of the solar cell is reduced, while maintaining full optical absorption (difficult in Si which has an indirect gap), the base recombination limit can be lifted.[17] This approach appears to now be attracting commercial interest, with initial cell results in the range of 13 to 15% efficiency.[18,19]

Other Higher Efficiency Materials

Efficiencies of 30% are now a reality under both 1-sun[20] and concentrated sunlight.[21-23] These cells all use materials from the III-V system. These are compound semiconductors formed from elements in columns III and V of the Periodic Table. These are ideally suited for opto-electronic devices, including solar cells, because they can have a direct transition in the band structure and can be grown at a quality level needed to retain good electronic characteristics. For PV interests, this means that thin-film structures can be used. Inasmuch as production experience will ultimately drive product costs down toward the raw materials cost, minimizing materials consumption is critically important for all PV technologies. Thin-film cells can also be structured into optical stacks consisting of more than one cell, in which light unused by one cell can be transmitted to another for conversion. These structures can further increase the maximum achievable efficiency. An added benefit of this materials system is that, like silicon, there exists a substantial industry infrastructure marketing a wide range of products from comparable materials technology. This defrays a large portion of the research and development costs and provides alternative markets for advances in materials technology while the PV markets develop.

The highest efficiency solar cells demonstrated to date consist of two cells optically cascaded in a tandem structure. Figure 8 shows an example of one structure that recently achieved an efficiency of 29.5% under 1-sun, standard condition. The GaInP$_2$ material absorbs most of the light with energies higher than 1.85 eV. The lower energies pass through to the GaAs, with a band gap of 1.42 eV. Models of this type of structure[24] show that conversion efficiencies greater than 30% can be achieved at 1-sun. Under concentrated sunlight, the ideal efficiencies increase to about 40%. In practice, the increased current densities and adjustments of the semiconductor properties related to high illumination make it difficult to capture all of the theoretical benefits. A design of this structure with modifications for operation under concentrated sunlight achieved an efficiency of 29.7% under an intensity of 100 suns.

Future Approaches: GaAs on Silicon and Single-Crystal Ge

The major challenge today for development of high-efficiency technologies is to minimize the cost of the substrate. Owing to the abundance of the material, only silicon will be cheap enough to be used in thick (200 μm) wafers for flat-plate terrestrial conversion systems. For concentrator systems operating at more than 100 suns intensity, the cost of

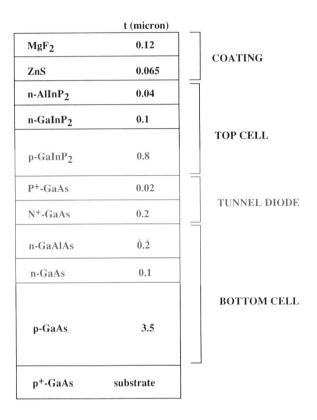

Figure 8 A schematic cross-section of the GaInP$_2$/GaAs tandem cell. A GaAs contacting layer is not shown.

the GaAs substrate is acceptable, but it adds a substantial thermal resistance to the cooling path. For today's high-value market for GaAs solar cells, power for extraterrestrial applications, the industry manufactures single-junction devices on germanium substrates with no loss in performance relative to GaAs substrates.[25] The added strength of the germanium reduces breakage in production and permits thinning for a desired weight reduction in the product. The move to silicon substrates presents greater difficulty. Silicon is not as well matched to GaAs as germanium is. The lattice constants differ by 4%. The thermal expansion coefficients differ by a factor of 2. Cell results for GaAs on silicon are about 25% reduced from the best devices on GaAs substrates.[26,27] However, recent improvements in the material may soon close this gap.[28,29]

The success in heteroepitaxy of GaAs on germanium opens an opportunity that crystallized thin films of germanium might serve as substrates for a high-efficiency, thin-film structure on glass. An initial test in this direction produced GaAs cells on polycrystalline germanium with an efficiency of 15.8%.[30] Throughout the literature, there is rapidly expanding interest and confidence that thin, highly crystalline films can be manufactured on amorphous or polycrystalline substrates.[30] One technique, currently used for display technology, lifts the film off an epitaxy substrate and transfers the film to glass. In the laboratory, this technique produced solar cells of 23.3% efficiency.[31] Regardless of the method used to form the crystalline film on glass, after the film formation the economies of thin-film PV processing apply for production in the remaining steps needed to complete the module.

TODAY'S THIN-FILM ALTERNATIVES: MATERIALS AND CELLS
Thin-Film Amorphous Si

Amorphous silicon modules are commercially available. They are the first truly commercial thin-film PV devices. However, like other pioneering technologies, amorphous silicon is not without its problems: efficiencies of present commercial modules are low (near 5%); and on exposure to light, modules suffer a well-known degradation, called the Staebler-Wronski effect (SWE) of about 15 to 30%, after which they are essentially stabilized. Due to these severe problems, amorphous silicon has not yet had a "breakthrough" impact on PV sales. Worldwide, amorphous silicon accounts for 25% of all PV sales in terms of power output.

In the U.S., several companies are developing the amorphous silicon technology. The key companies are United Solar Systems Corporation (USSC), Energy Conversion Devices Inc. (ECD), Solarex Corporation, Advanced PV Systems (APS), Utility Power Group, and Iowa Thin-film Technologies. The leading prototype amorphous silicon modules are given in Table 1. The efficiencies in Table 1 are for modules that have been exposed to sunlight for a period of time (days) to allow them to reach their "stable" output. (Note that this procedure is still relatively new and nonstandard. There is no certainty that the values are truly stable.)

Table 1 **Best stabilized module efficiencies**
(NREL solar simulator data after 600–2000 h light soaking)

Manufacturer	Type	Efficiency (%)	Area (cm^2)	Power (Watts)
SX	Si/Si/SiGe	8.8†	863	7.6
USSC	Si/Si	6.2	3,676	22.8
	Si/SiGe/SiGe	10.2*	903	9.18
	Si/Si/SiGe	9.4†	903	8.5
ECD	Si/Si/SiGe	7.8	3,906	30.6
FUJI	Si/Si	8.9	1,200	10.7
APS	Si	4.4‡	11,634	51.2
	Si/Si	6.0†	905	5.4
	Si/Si	4.6‡	11,522	53.0

† Not verified by measurements at NREL.

‡ Outdoor exposure and measurement.

* Light-soaked by USSC; measured at NREL.

Amorphous silicon PV has various strengths: 1) substantial investment in scale-up and manufacturing has resolved many manufacturing-related issues, allowing for greater confidence that laboratory progress will translate into commercial modules; 2) silicon materials are abundant and are not regarded as environmentally dangerous (at the product level); and 3) substantial knowledge of amorphous silicon is being developed outside of PV for non-PV amorphous silicon applications such as display technology.

Issues that drive the future of amorphous silicon can be regarded as follows:

1. The Staebler-Wronski effect
2. Relatively low device efficiencies
3. Manufacturability of innovations

These issues all revolve around the main problem of amorphous silicon devices, the light-induced degradation. This effect is unique to amorphous silicon, and is due to its amorphous structure. No other thin-film material displays the same problem. Until a few years ago, it was feared that the SWE would reduce amorphous silicon efficiencies to zero. However, it was found that a competing effect, thermal annealing during outdoor operation, tended to counter the degradation. This allowed an equilibrium to result after lengthy exposure, with the losses held to about 15 to 30%. Procedures are being developed to standardize measurements of this reduction. However, there are as yet no fully accepted methods. The current approach is to expose modules to sunlight (or simulated sunlight) for about 1000 h. This is believed to account for over 95% of the degradation. To counteract the problem of SWE degradation, manufacturers warrant the output of modules for 10 to 12 years at some fraction (80 to 90%) of the nameplate rating. This assures the buyer that they will receive a known minimum power output.

Much of the work in designing amorphous silicon devices is a result of trying to compensate for the SWE. For example, it was found in the early 1980s that thinner amorphous silicon layers resulted in either less or slower degradation. But cells made thin (i.e., with larger hundreds of nanometers instead of several micrometers thick) allow too much light to pass through unused. To capture more of the light, amorphous silicon devices use one of two strategies or a combination of both: they use a back reflector (such as silver coated zinc oxide [ZnO]) or they modify the device and make it a multijunction consisting of many layers having different band gaps. The latter approach has become the mainstream strategy for both more stable cells and higher efficiencies. In the earliest versions, multijunction amorphous silicon cells were used (without altering the band gap) merely to absorb more light. But to take advantage of the potential of multijunctions for higher efficiency, researchers sought to develop different band gap materials to incorporate in the multijunction design. Two materials evolved: a higher band gap alloy made with carbon; and a lower band gap alloy made with germanium. Today, the mainstream device design for high-efficiency cells is based on multijunctions using one or more of these alloys to supplement an amorphous silicon cell, either in a two- or three-junction design. These various alternatives can be seen in column 2 of Table 1, where the stacked layers are called out. As can be seen, fully functional triple-stacked alloy cells (using C and Ge for top and bottom cells, respectively) are only now emerging from the laboratory. They provide an important avenue for future improvements. Various secondary improvements (new conductive oxides, new back reflectors, microcrystalline silicon-doped layers) are also being developed to add to device efficiency.

The advent of multijunctions raises new manufacturing issues for amorphous silicon: multiple, very thin layers are a challenge to even a manufacturing-friendly technology. True understanding of these issues awaits more substantial amorphous silicon manufacturing volume. Current thinking suggests that manufacturability of thin layers is not by itself a particularly serious challenge. Rather, potential shunting is a problem because of textured ZnO or SnO_2 used for light trapping.

Greater or expanded amorphous silicon production is expected within the next 2 years (by 1995), as two companies (USSC and APS) carry out their announced plans to build 10 MW/year capacity amorphous silicon factories. The USSC plant is expected to be based on an advanced multijunction design.

It will be important to amorphous silicon that companies establish improved commercial module efficiencies with their next generation production. The economics of PV systems, and the market competition from traditional PV, require stabilized thin-film efficiencies in the 8% range, or higher, to begin serious market expansion. Similarly, other thin films (CIS and CdTe) will be maturing during this period and adding competition. The issue of SWE degradation remains critical. Even today's best amorphous

Polycrystalline Thin Film PV Efficiencies
(Best Laboratory Cells; Standard Conditions)

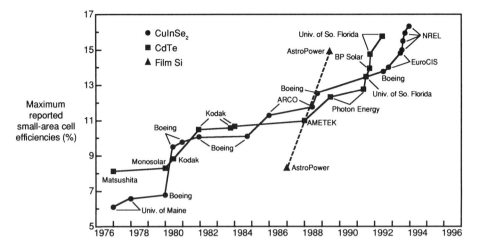

Figure 9 The best reported cell efficiencies in CIS and CdTe are now about 16% efficiency (standard conditions).

silicon multijunction device designs do not at present allow stabilized small-area amorphous silicon devices to approach the efficiencies of other thin films (11% stabilized, vs 16% for other thin-film cells). In addition, fundamental choices about manufacturing (such as the need for multiple, very thin layers) are being driven by the SWE. A solution to the SWE problem would change the nature of the amorphous silicon technology.

CdTe and CIS
Materials and Devices
Progress during the last 2 years in CIS and CdTe small-area cell efficiencies has been outstanding. Figure 9 shows efficiencies for these technologies (standard conditions, total-area basis). Summary tables on the performance of these devices have been published.[32] The recent improvement in polycrystalline thin-film (PTF) efficiencies has come from the special contributions of several universities: e.g., the EuroCIS consortium (University of Stuttgart, Royal Institute of Technology [Sweden], Swedish Institute of Microelectronics, Ecole Nationale Superieure de Paris) in CIS; and University of South Florida (USF) in CdTe. The most recent progress in CIS-based devices has also come from National Renewable Energy Laboratory (NREL) internal R&D. At this juncture, both CIS and CdTe have demonstrated about 16% efficiencies.

The EuroCIS group has found new ways to optimize CIS-based cells made by multiple-source evaporation. Besides the obvious value of their increased efficiencies (now reported to be 16.9% on an active-area basis[33]), their major contribution may have been to bring the use of alloying elements (Ga for In, S for Se) into the mainstream of CIS technology. Their work builds on previous innovative work in Ga alloying done by Boeing.[34,35] Most of the best CIS cells now being made are actually alloy cells with Ga or S. These include cells made by Boeing, EuroCIS, Siemens Solar Industries, and NREL. Figure 10 shows two NREL cells, both made with Ga. NREL has made the highest-efficiency CIS-based cell (16.4%) on a total-area basis. The NREL staff has investigated

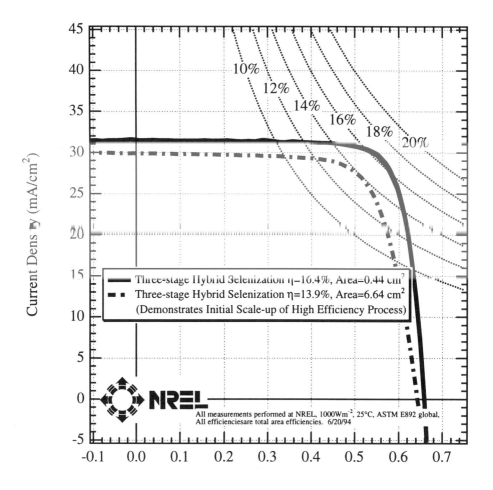

Figure 10 NREL has fabricated the most efficient CIS-based cell (15.9% on a total-area basis) and a larger version (13.9% at 6.6 cm²) that demonstrates initial scale-up. The NREL work is aimed at developing new, potentially more manufacturable methods of fabricating efficient CIS devices.

some important models of CIS film formation.[36-40] The fabrication effort at NREL focuses on two issues:

1. Fabrication of device structures based on $Cu(In,Ga)Se_2$ and $CuIn(S,Se)_2$ in order to achieve efficiencies over 17%
2. Validation of different reaction pathways needed to make the alloys/layers in ways that are potentially more manufacturable

Using a variety of processes and approaches, the NREL researchers have already gained significant insight into the formation mechanisms of the thin-film multinaries. Their aim is to understand the fundamental mechanisms controlling the formation of device-quality films, independent of the reaction pathways and processes. Knowledge gained from this approach is likely to aid the effort to resolve manufacturing problems associated with yield, adhesion, compositional variations, and manufacturing tolerances. It should also allow for higher cell/module efficiencies. The initial scale-up of NREL methods to larger size (6.6 cm^2) was done using reactors designed for small-area (1 cm^2) cells. Despite this, an excellent 13.9% efficiency was reached. This strong result supports the idea that thin films will eventually reach ambitious efficiency goals (15% for commercial modules).

The use of S or Ga alloys has some important implications. Using S or Ga to replace In or Se raises the band gap of CIS-based materials. By doing so, several aspects of module manufacturing can be simplified. For example, devices with a higher band gap absorber produce lower current densities (but higher voltages and fill factors), allowing module cells to be wider without incurring greater resistance losses. Having fewer, wider cells simplifies module scribing, reduces the number of sensitive scribe areas, and increases module active area. In addition, a higher band gap (and higher voltage) reduces the sensitivity of the device to temperature losses and provides a better match with the solar spectrum. All cells/modules lose efficiency with higher temperature, but the loss is inversely proportional to the voltage. Higher voltages will allow alloyed CIS devices to perform better at outdoor operating temperatures. Finally, the use of a higher-band-gap absorber reduces another major problem: it relaxes the thickness and doping constraints on the deposition of a transparant conducting oxide (TCO) such as ZnO. Finally, free-carrier absorption losses in the TCO are reduced when higher band gaps for the CIS are used.

The development of alloys as a major part of the CIS technology is an important contribution to the technical art. Another value of alloys is that they can take pressure off the need for large quantities of In and Se. Reducing the amounts of these relatively rare materials could be of value if the technology becomes very successful (multiple GW per year production).

Progress in CdTe cell efficiencies emerged from work by Chu et al.[41-44] at the University of South Florida (USF).[25-28] The Chu's retired in 1992, after having produced a world-record 14.6%-efficient CdTe cell. Their students, Chris Ferekides and Jeff Britt, carried on their work. The transition turned out quite well, with the fabrication of several record CdTe cells (see Figure 9).[45] In June 1992, their cells became the first non-single-crystal thin-film cells in the world to surpass 15%. They made many cells over 15% efficiency (all verified at NREL) and two with 15.8% efficiency. The achievement of these efficiencies was an important demonstration of the potential of CdTe. Combined with the EuroCIS and NREL results, it suggests that our thoughts about the performance limits of polycrystalline thin-film (PTF) materials need to be revised upward. For instance, the achievement of 18 to 20%, single-junction thin-film cells, heretofore deemed unrealistic by mainstream PV scientists, now appears to be possible. The attainment of ambitious cost goals, which depend on achieving very high **commercial module** efficiencies (over 13% and possibly over 15%), seem plausible.

The USF CdTe cells address a number of technical problems: CdTe contacting, thin CdS, and possible degradation barriers. The achievement of very high fill factors (many over 72%, some as high as 75%) was done with a new contacting approach.[45] Similarly, various degradation barriers have been investigated by USF. Figure 11 shows an analysis of two USF high-efficiency cells performed at Colorado State University (CSU). The analysis compares various device parameters (e.g., series and shunt resistance) over a period of 3 months. In the first case, an unencapsulated cell degraded from 13.4 to 12.5%, with most of the loss in the fill factor. No performance loss occurred in a more recent, 15%-efficient protected cell (also note the very high fill factor, 75.5%). Working with a similar set of six high-performance cells, CSU[46] found no change outside experimental error in the six cells tracked for a summer. Three were

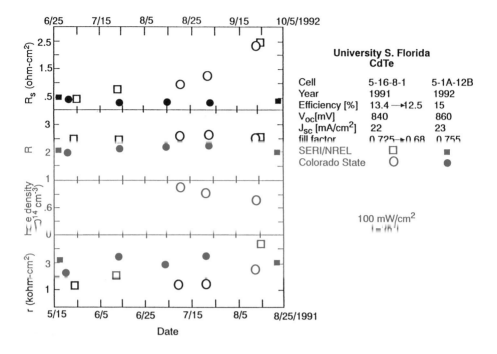

Figure 11 CSU examined a number of unencapsulated CdS/CdTe cells made by USF. Improvements made between 1991 and 1992 resulted in increased robustness to air exposure.

kept in desiccant and three in air. In contrast, the spring 1991 record efficiency cell showed a significant increase in series resistance which was responsible for decreases in fill factor and efficiency. Progress toward stable contacts and protective layers to improve the robustness of CdTe cells are valuable USF contributions.

CSU has observed a hysteresis in the measurement of CdTe device performance that may be causing some added uncertainty in the stability data of some CdTe cells and modules. An early study of the effect can be found in papers by Sasala and Sites[47] and Sites.[48,49] This transient effect is quite complex and is most likely due to an electronic property of the CdS/CdTe interface. It may be due to trap filling and relaxation time constants; for the purpose of this chapter, we are assuming that hypothesis. In any case, the observed effect is that efficiencies can vary up to 10% with the circumstances of the test. For example, some cells/modules are somewhat more efficient at maximum power point (under load) than they are if measured with a simple I-V measurement in which the load is tracked from short circuit to open circuit. During operation, or when exposed to light, efficiency improves from its initial value (presumably as traps are filled). When the module is in the dark, the traps slowly empty. When at short circuit, they empty faster. Thus, if a module is short-circuited before measurement, it measures below its real, outdoor operational efficiency. On the other hand, if it is light soaked at open-circuit, it measures somewhat above its real efficiency. Fortunately, the efficiency is nearly at its highest at maximum power (i.e., during actual operation outdoors), except for the expected effect of a small performance loss at elevated, outdoor operational temperature.

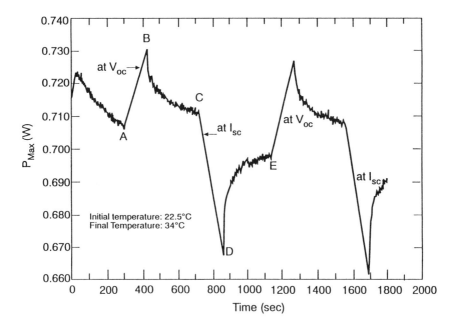

Figure 12 CdS/CdTe devices display a transient change in performance when exposed to light or bias. Efficiency increases with light exposure at open-circuit and decreases with no exposure in the dark. These are reversible effects, and actual outdoor performance at maximum power is unaffected. (The power output in the figure trends down slightly because the final temperature is 11.5 °C higher than the initial temperature.)

The result of this undesirable transient effect is that it causes some CdTe module efficiency measurements to be much more inaccurate than usual for PV. For example, if initial baseline testing is done with light-soaked modules at open-circuit, but subsequent tests are done after a short-circuited condition (emptying filled traps), an apparent (but unreal) degradation will be observed. The opposite is also true: base-lining with the traps empty and then measuring later with them filled will show a false increase in efficiency. Indeed, this effect varies with CdTe module supplier. Some cells/modules show it strongly; others show it very little. Finally, the effect is most obvious when a pulsed simulator is used to measure efficiencies because the pulses cause irreproducible trap filling/emptying. The only known way to truly measure CdTe module performance is to test modules at maximum power under continuous simulation and at a controlled temperature. Fortunately, this closely mimics outdoor use. However, for cases in which small changes need to be documented (such as for stability and qualification tests), the "transient effect" to some extent confounds the data.

Figure 12 (after K. Emery, NREL) shows the effect rather graphically. From point A to point B, the sample is exposed for about 200 s to light while being kept at open-circuit. After soaking, the efficiency is 24 mW higher (about 3%). Between points B and C, the sample is connected to a load at maximum power. The overfilled traps are emptied, and power drops to a mid-level of about 700 mW. Between points C and D, the module is short-circuited in the dark, and the traps are fully emptied (worst case). The sample is reconnected and loaded at maximum power between D and E, and power rises again towards 700 mW as the empty traps are partially filled. After point E, the entire process is repeated. There is a slight negative slope to the figure because sample temperature is rising slightly from 22.5 °C to a final level of 34 °C. One can see from Figure 12 that the worst-case efficiency is almost 10% lower than best-case efficiency. The "real" efficiency is the one at maximum power when traps are at a steady state. Not surprisingly, real efficiency is about in the middle of the range.

It is important to realize that: 1) the effect is in the positive direction (maximum power measurements are higher than those taken on unexposed samples), and 2) it is reversible and electronics driven. However, the significance of the effect is that it increases the uncertainty of CdTe efficiency measurements. For instance, NREL is careful not to introduce these potential artifacts during measurement of record cells. Although some of these issues have been discussed earlier, we feel that the research and technical issues for CIS and CdTe warrant some further discussion.

There are some issues that are generic to the CIS technology, e.g., the addition of Ga to the surface of the small aperture. There are a number of issues within the solar spectrum, and some issues that are process dependent. For example, we know that grading the CIS absorber layer with Ga by evaporation is possible. Increases in the open-circuit voltage, V_{oc}, have been observed (the recent NREL results, $V_{oc} = 0.687$ V) by this approach. Can similar grading be done by a selenization method? In yet another example, adhesion of CIS to the Mo electrode does not appear to be a problem in evaporation; however, we do see peeling of the CIS film during selenization processes, mainly due to the volume expansion and stress in the films. The addition of Te and Ga interlayers improves the adhesion of CIS to Mo. Achieving large-area stoichiometric and uniform films in manufacturing is still a formidable challenge and needs to be addressed with a great deal of attention. Also, work is being directed to replace the H_2Se gas with an Se vapor source for module fabrication, and to improve throughput, reproducibility, and yields. Much success has been achieved in replacing hydrogen selenide with selenium (for example, NREL has made 12.5% cells using selenium), but more effort is needed to reach state of the art and to transfer the approach to production.

In the case of CdTe, too, there are generic as well as process-dependent issues. Although 10 processes have yielded cell efficiencies of 10 to 16%, only four techniques have been used for module fabrication: electrodeposition, modified close-spaced sublimation, spraying, and screen-printing/sintering. CdTe is a self-compensating material. Thus, obtaining good, reproducible, and optimum doping of CdTe is still a research issue for the various deposition techniques. Achieving stable contacts for both cells and modules receives considerable attention. Contacts such as graphite (Cu and Hg-doped), Ni/Al, Cu/Ni have demonstrated fairly robust performance. Chemical and heat treatments have resulted in type conversion, junction formation, and enhanced grain growth (especially for CdTe grown at low temperatures, e.g., by electrodeposition). Although large-area modules with power outputs greater than 50 W have been demonstrated, high throughput, yield, and reproducibility need to be demonstrated. Two schemes have been used to encapsulate the modules. Those with a graphite contact normally require a spacer between the two pieces of glass, while the ones with the metal contacts use EVA/glass encapsulation. The safe handling of Cd has been demonstrated at several facilities that are making CdTe modules within the latest OSHA regulations, and the recycling of broken and spent modules is discussed in the following section.

ES&H and Recycling

A number of environmental, safety, and health issues exist for the manufacture and commercial use of CIS and CdTe. In all cases where CdTe or CdS are present, the presence of Cd complicates the situation. Some of the main concerns are:

- Possible uses of toxic gases
- Manufacturing safety issues, especially those related to Cd or other health threats
- The existence/disposal/recycling of waste modules at various stages during production
- The presence/disposal/recycling of unused materials (CdS, CdTe, Cu, In, Se, CIS, contact metals, etc.) during various process steps
- Cleaners and solvents
- Transportation/regulations concerning various feedstocks and recycling products
- A number of product-level issues, including:
    - Product distribution/transportation
    - Broken/returned product
    - Eventual product disposal/recycling
    - Product safety (e.g., fires, breakage)

Most of these issues closely parallel those for other forms of PV. Moskowitz and co-workers have written extensively on the general subject of PV-related manufacturing issues and on Cd issues.[50-57] Some have written illuminating comparisons between PV and conventional energy sources.[56,57]

An important criterion for waste classification is the U.S. Environmental Protection Agency's (EPA's) EP Toxicity test (or other versions such as the recently adopted TCLP Test). EPA tests are carried out by grinding a module to particle size and then suspending the particles in solution. Various materials (including Se, Ag, Pb, Cd, and others) are then measured against a standard. This test has been conducted on a very limited sample of CIS and CdTe modules. It is important to realize that the results are not necessarily characteristic of future commercial modules. Similarly, it is not clear whether other PV (specifically Si modules with Pb or Ag in them) could pass the test. In the early tests, the CIS modules passed by a wide margin for both Se and Cd (despite the presence of CdS, which, however, is very thin—about 300 Å). The CdS/CdTe modules failed by a small margin for Cd. In fact, the margin was so narrow that future modules (wherein small metallic Cd and CdO inclusions could be excluded) may pass the test. However, because the CdTe manufacturers already assume the need for recycling and for other, similar strategies, passing the TCLP test is not essential.

Moskowitz and co-workers have recently begun collaborations with the National Institute of Environmental Health Sciences (NIEHS) to investigate the oral toxicity of CIS and CdTe. Early results show a surprising tolerance for large oral doses of CIS materials (no toxic effects observed for any of the tested doses and no organ abnormalities) and an expected but very small sensitivity to Cd in CdTe and CdS. The latter caused very few toxic effects because the compounds are much less soluble in biological systems than pure cadmium.[58] These tests are good news for these technologies because they suggest that using cadmium or selenium in PV may turn out to be a way to isolate them from the environment.

Besides clear-cut health issues, the presence of Cd in CdS and CdTe poses some unique and less quantifiable concerns:

- Public reaction to CdTe manufacturing and product use
- International regulatory climate and barriers

These factors have played a visible role in the progress of the polycrystalline thin film technologies, especially the CdTe technology, and may yet play a significant role in their future.

Various reports have been written on Cd-related issues.[50-55,59-61] In most cases, the authors view Cd-related issues as within the normal range expected for any large-scale commercial PV technology. An important reason for this is that the small amount of Cd in thin-film modules (about 5 g per square meter) limits potential impacts. During 1 year of output, a square meter area CdTe module (10% efficient, average U.S. sunlight) would produce about 200 kWh electricity. (Put in terms of output per gram of Cd, the module would produce 40 kWh/g Cd during that same annual period.) The total amount of electricity in the U.S. annually is about 2.5×10^{12} kWh. To generate (using CdTe PV) an amount of electricity equal to the U.S. total would require deploying 12,500 km^2 modules with about 60,000 metric tons (MT) Cd in them. In practice, it would take many years to deploy such a large amount of PV. If one assumed a 30-year period for deployment (probably unrealistically quick), about 2000 to 3000 MT per year Cd would be needed (depending on losses during production). The world's use of Cd today for existing purposes is greater than 15,000 MT.[62] Thus, transforming the U.S. energy infrastructure (a huge undertaking) would not significantly change the global annual use of Cd. For another similar order-of-magnitude example, consider that today the U.S. throws away about 2000 MT per year of Cd from Ni/Cd batteries. Not only would CdTe PV not be thrown away (it would last 30 years and could then be recycled), but its use would have a much greater impact (a transformed world energy infrastructure vs the use of batteries for home electronics). This comparison suggests that the use of Cd in PV is, environmentally, totally different from any other current use of that material. Indeed, cadmium in PV modules can be considered an environmentally positive strategy, in that the cadmium can be safely sequestered within the module for an extended period of time (30 years) and then recycled.

Processes already exist for recapturing and recycling Cd-containing materials.[54] For example, most of the Ni/Cd batteries used in industry are recycled. (Batteries in toys and other consumer items are not often recycled by the public, and those batteries are the source of the Cd entering the waste stream.) As the volume of the industrial Ni/Cd batteries is very large compared to the volume of Cd in CdTe PV, they form a substantial proof-of-concept for Cd recycling.

For those in the CdTe technology, continued efforts toward reduced process wastes, thinner layers, recycling, and bio-monitoring are essential. For those in CIS, another path is possible. To minimize Cd concerns, some CIS groups have worked to replace CdS with other heterojunction partners. This work has been relatively successful (many cells over 10% have been made), indicating that CdS can be replaced and high-efficiency cells can still be fabricated. However, alternatives to CdS are not yet integrated into any CIS commercialization effort.

Thin-Film Module Integration

Table 2 shows a compilation of the most efficient polycrystalline thin film modules of various sizes. In practice, thin films will need commercial efficiencies greater than 8% to have a competitive edge in the existing market. (This assumes that their manufacturing cost, even in low volumes, will give them some incremental advantage.) The efficiencies being seen in the prototypes of Table 2 show that an 8% near-term goal is within reach. Figure 13 shows the largest PTF module, the SCI CdTe module.[63,64] The success of SCI is built on its previous experience in the glass and thin-film businesses (through Glasstech and Glasstech Solar).

The potential cost of thin-film modules depends to a large extent on the processes used to make the thin films. This is because module efficiency, process material utilization, process capital cost, and process yield are determining cost factors for thin films. All are dependent on the processes by which key layers are made. In the case of CdTe, a number of attractive processes are being used: spraying, high-rate evaporation, electrodeposition, and screen printing. In the case of CIS, Siemens Solar has not revealed its processes, although they have patents in the area of sputtering and selenization. Other companies use various combinations of sputtering the metals followed by selenization in hydrogen selenide or selenium (or layer homogenization of Cu, In, Se). CIS commercialization may in part be delayed because the existing processes are not easily adapted to manufacturing. Recent work at NREL is opening up opportunities for new ways to make CIS layers. These are only now being looked at by industry. CdTe manufacturing appears more promising, but experience is so limited that the lack of problems may simply be a reflection of the early stage of development.

Table 2 **Performance of polycrystalline thin-film modules**

Group	Material	Area (cm^2)	Efficiency (%)	(W)
Solar Cells, Inc.	CdTe	6838	7.76*	53.1*
Siemens Solar Ind.	CuInSe$_2$	3883	9.7*	37.8*
Golden Photon	CdTe	3528	7.7*	27.2*
Siemens Solar Ind.	CuInSe$_2$	938	11.1*	10.4*
Matsushita Battery	CdTe	1200	8.7	10.0
BP Solar	CdTe	706	10.1	7.1
Golden Photon	CdTe	832	8.1*	6.8*
ISET	CuInSe$_2$	845	6.9*	5.8*
Energy PVs	CuInSe$_2$	804	5.0	4.1

* NREL measurements; all aperture-area efficiency.

In addition to the key absorber layer, several other semiconductor and metal layers must be deposited in order to make excellent PV modules. Figure 4 in the introductory section shows mainstream CIS and CdTe device geometries and the different layers. Cadmium sulfide is the preferred heterojunction partner for CIS and CdTe, although deposition conditions differ widely in the two cases. No preferred method of CdS deposition exists. The CdTe companies use high-rate evaporation, spraying, electrodeposition, and solution growth to deposit CdS. The CIS companies use solution growth and evaporation. Because the CdS layer is a key part of the device in terms of performance, the issue of finding a low-cost, highly controllable process to make it requires ongoing attention.

All high-efficiency CdTe cells are made in a superstrate configuration (glass/SnO$_2$ first). Layer thicknesses vary, with CdTe thickness being 1 to 6 μ; contacts also vary, with various alloys (most with Ni/Cu) or Cu-doped graphite being usual alternatives.

All high-efficiency CIS-based cells are made with the substrate configuration (glass/ Mo first) and with various layer thicknesses and CIS-based alloys (with S or Ga).

Both CIS and CdTe devices require a TCO top layer to carry photocurrent while being transparent to solar illumination. The problem is solved differently in each case. CIS uses a top layer of conducting ZnO, frequently deposited by sputtering or OMCVD. Because the CdTe technology is almost always made with a glass superstrate geometry, the TCO can be put down first. A major U.S. glass company, Libbey-Owens-Ford (LOF) makes tin oxide-coated float glass for thermal applications. This same low-cost, coated glass can be adapted with little change as the superstrate for CdTe devices. This allows the CdTe technology to take advantage of a ready-made, low-cost supply of glass with the needed TCO coating. LOF uses a form of chemical vapor deposition to deposit their tin oxide, and the deposition is done on a "football field"-sized float-glass

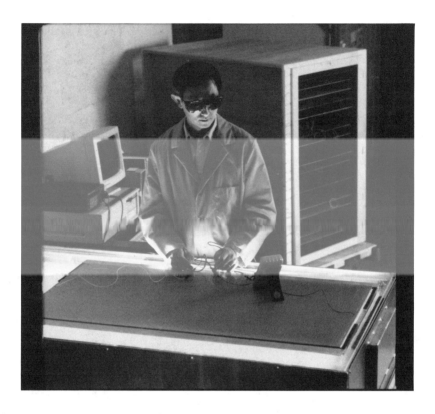

Figure 13 Solar Cells, Inc. is developing a 7200-cm² CdTe module for commercialization. They have reported efficiencies of 8.2% (aperture-area basis) and power output of 56 W (not shown in Table 2).

line, achieving true economies-of-scale. The success of the LOF tin-oxide-glass is an excellent indication of the potential of thin films to meet ambitious low-cost goals.[65]

Various metal layers and metallic interconnects are also needed to make PV modules. In most cases, metal contacts are sputtered from flat magnetrons or from rotating cylindrical magnetrons. The latter can reach higher deposition rates and use target materials more efficiently. Although metals and contacts are relatively straightforward and familiar from previous thin-film processing, they are a nontrivial part of the developmental process. For example, issues with the glass substrate (dirt, imperfections) and with the back contact (Mo berms and adhesion) for CIS have been cited as central to delays in introducing CIS products.[66] These "details" may yet form the focus of major developmental efforts.

Finished modules require some form of interconnection among neighboring cells and then encapsulation. Scribing is the key step in the interconnection process and is relatively slow and demanding. Scribing can introduce defects and areas that are vulnerable to stress (potentially causing degradation). Special concerns exist about scribing Mo prior to CIS deposition in the sense that CIS growth defects can be caused by the Mo debris.[66]

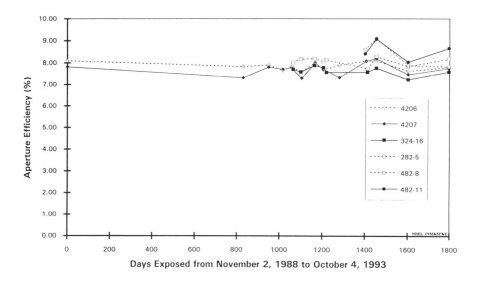

Figure 14 NREL has been testing state-of-the-art ARCO Solar/Siemens Solar CIS modules outdoors for more than 4 years, with no sign of degradation. This is a unique result for all thin films. The ARCO/Siemens Solar CIS modules being tested at NREL include 0.09- and 0.36-m² sizes. The larger sizes were installed more recently, in 1991 and again in 1992.

Encapsulation is an area of much interest, as current schemes seem both over-engineered and too costly. In many cases, two sheets of glass are sealed together to provide the environmental isolation. Frames are usually used, although their need is debated. Future, lower-cost schemes include using an as yet unknown barrier layer as one of the sealing materials. The CdTe superstrate structure could easily be adapted to this approach by using a "spray on" sealant to replace the glass at the back, if an effective sealant could be found. This is a challenge for CdTe, which has been known to be sensitive to performance losses due to the action of water vapor. Although bare CIS cells have been flown successfully in space, it is not at all clear that a transparent top layer to replace glass could be found for terrestrial PV.

We have had encapsulated CIS and CdTe modules outdoors at NREL for several years. In most cases, the CIS modules have been perfectly stable. Data for three sets of Siemens Solar CIS modules (earlier ones, 0.09 m²; later ones, 0.36 m²) are shown in Figure 14. This kind of outdoor stability is still the exception rather than the rule for new PV of any sort, and it is an important strength that sets apart the CIS technology from any other thin film. Some degradation mechanisms have been identified for specific CIS modules that have problems during manufacture. These must be studied and eliminated to assure system-level reliability.

The CdTe technology has a mixed record in terms of reliability. We have much NREL data from stable and unstable modules, and similar data from other sources (e.g., Matsushita Battery and BP Solar[67-69]). Numerous modules appear unchanged after several years outdoors. They are a reasonable indicator that CdTe can be stable if properly manufactured and sealed. We also have many unstable modules. They indicate that serious issues exist. The current hypothesis is that most of the module stability problems are with encapsulation design and because of subsequent poor quality control. Work must be done at both the cell and module level to isolate degradation mechanisms and then to eliminate them or reduce their impact.

THE BOTTOM LINE: LONG-RANGE GOALS AND COSTS

THIN-FILM STATUS

With a minor exception, polycrystalline thin-film technologies based on CIS, CdTe, and Si film are not commercially available. The exception is about 1 MW/year of CdTe cells fabricated by Matsushita Battery for Texas Instruments (TI) calculators and other consumer applications.[70] These cells have been in use for almost a decade. However, in terms of mainstream PV for power applications, no PTF modules can be bought "off the shelf."

Commercialization of PTF power modules is underway along several fronts. Several companies have had agreements with PVUSA to deliver 20-kW systems for the PVUSA "emerging module technologies" program. However, system deliveries for this program have been plagued by significant delays (1 to 3 years). The delays have been caused by a variety of issues, including some related to processes and others having to do with module-level shortcomings. In any case, no PTF modules have been delivered to PVUSA, causing a loss of credibility for the various PTF groups.

A similar set of PTF module delivery agreements have now been made between NREL and various U.S. companies. These call for about 1 kW to be delivered to NREL for testing. In some cases, this testing is part of the PVUSA qualification process. In others, it represents an independent effort to assist industry in overcoming product introduction issues without encountering the kind of public disappointment associated with failing to deliver a system for PVUSA. For example, NREL has had about 150 W of CIS modules from Siemens Solar outdoors at NREL for more than a year, and others (CIS and CdTe) for several years. At this writing, we have received one 1-kW system, a Siemens Solar CIS system (fall 1993).

Current plans exist by two U.S. CdTe companies to manufacture modules. Golden Photon, Inc., has announced a new facility in Golden, CO; and Solar Cells, Inc., has announced a facility in Toledo, OH. No United States CIS company has made an official announcement of a commercial manufacturing facility, and introduction of Si film products awaits further developments at AstroPower. Thus, only CdTe modules are likely to reach the market in the next few years. Indeed, it is unlikely that commercialization of CdTe will go smoothly, as the introduction of a new PV technology has many challenges. Recent experience in CIS (and ongoing difficulties in amorphous silicon, e.g., stability) show that new thin films face numerous barriers.

The economics of PV, and of thin films in particular, do not support an easy transition to successful commercialization. This is because thin films have been developed as a means of reducing the cost of PV to compete for large power markets. However, the initial entry of thin films is at much higher prices that cannot yet compete for these presumed large markets. The problem is that initial production of thin films occurs at 1) lower than expected efficiencies (due to the relative immaturity of the device technology) and 2) at low volumes that fail to take advantage of economies-of-scale. In addition, the start-up costs of many years of R&D (without any offsetting sales revenues) as well as the need to keep a major ongoing R&D staff (to reach potential performance and cost goals) are serious financial burdens. Thus, instead of being inexpensive and immediately penetrating very large power markets (peak shaving in the U.S.; rural electrification in developing countries), thin films must instead compete for traditional PV markets against established PV options.

U.S. government cost-shared programs are playing a substantial role in the development of PTF options.[32,59,71-76] For example, the recent emergence of CdTe as a commercial possibility results from the technological incubation of the last decade, carried out by the private sector with U.S. DOE/NREL support. This support continues through the NREL Thin-film Technology Project (subcontracts with GPI and SCI, and with several U.S. universities) and through potential support from other U.S. DOE-sponsored programs (e.g., funding from PVMaT for CdTe and thin film Si).

Probably, the best hope of the PTF commercialization effort is that it is based on 1) a set of technical options that allow truly low cost to eventually be achieved and 2) a significant track record of technical achievements at the cell, module, and processing levels. These achievements, and the cost question, are addressed in the following sections.

THIN-FILM FUTURE

The long-term cost goals of thin-film modules are in the range of $50/m^2.[75,77] At power outputs of 10 and 15% (100 W_p/m^2 and 150 W_p/m^2, respectively), this would mean that module costs would be about $0.50 and $0.33/W_p$ (peak watt). The $/W costs are simply recalculations of these costs ($/m^2/W_p/m^2$ yields $/W_p$). Combined with systems costs that reflect larger volume production for larger markets, the cost of installed PV could fall to less than $1/W_p$, implying costs of about $0.06/kWh for PV-generated electricity.[75]

Are these cost goals possible? The answer varies for different thin-film options because each has different strengths and weaknesses. Certainly, the PTFs appear to have the necessary performance for long-term cost goals. For example, if the PTF laboratory results reach the market (which is a huge undertaking that will take years), efficiencies of commercial modules should be about 80% of the active area efficiencies of today's best cells. Today's best cells are about 16% efficient, so this implies modules of about 13% efficiency with today's technology. In terms of module production costs, various studies[78-85] of materials' costs, combined with energy inputs, labor, and capital costs, support very ambitious cost projections for all thin films. All of the cited cost studies agree that ultimate manufacturing costs of thin films can be as low as $40 to $50/m^2 (if everything works well). The studies include those on specific PTF technologies provided by U.S. manufacturers to the U.S. DOE/NREL PV Manufacturing Initiative as part of their final reports.[82-85] These provide the most up-to-date information on module cost projections. Since the modules are the major cost-driver for systems (in terms of system capital cost, and energy output per unit cost), progress in module technology is the essential factor behind the cost projections for systems.

However, the road to low-cost PV remains a long one. Numerous challenges remain, not the least of which is the truly high cost of today's PV (Figure 14). Even the success of this generation of new PV technologies, such as the PTF technologies reviewed in this article, does not of itself assure the global impact of PV, because the price of PV will still remain much higher than conventional alternatives until major technological improvements have time to reach commercial products. Technology development work will still be needed to:

1. Make modules in production that are as efficient as today's best cells (up to 15%).
2. Replace many of today's chosen processes with those that are less expensive in terms of material use and capital costs.
3. Invent novel module designs that improve costs through reduced complexity and improved ease of manufacture.
4. Optimize manufacturing lines, given optimal processes and module designs.
5. Fully implement recycling strategies at all stages of the product lifecycle.
6. Accelerate cost reductions in balance-of-system (BOS) components to meet the costs needed to make total system costs equivalently low.

If most of these issues can be addressed, thin-film PV modules would drop below $0.50/W (module efficiencies of 13% and costs of $50/m^2 imply module costs of about

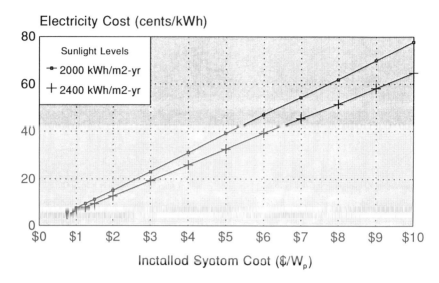

Figure 15 Utility systems installed at various costs (in $/W_p) imply various electricity prices for the consumer (depending also on local sunlight). Note that today's installed systems, at about $8/W_p, imply electricity at $0.5 to 0.6/kWh for utility systems based on analysis from References 78 to 85. Today's high cost of PV is the driving force behind developing new technologies such as those based on thin films. Note that this figure gives no credit for capacity, no environmental credit to PV, and does not take into account any added value for distributed generation that reduces transmission losses or adds value for avoided system upgrades. No value is given for reduced system-level risk (associated with less reliance on fuels, which can fluctuate in price). As such, it is a conservative measure of PV costs/value.

$0.40/W). If systems costs come down in a similar way due to improved designs and economies-of-volume production, installed system costs of about $1 to $1.50/W would be possible. Figure 15 then implies that electricity costs would be about $0.06 to $0.10/kWh, depending on local sunlight.

The goal of truly low-cost PV assumes that there are major markets that will allow PV to evolve toward larger and larger production volumes (to achieve economies of scale) despite PV's relatively high cost. These markets include very large, rural markets in developing countries, peak shaving in sunny/humid developed locations, and general use of PV to reduce reliance on nonrenewable resources (e.g., for intermediate, daytime electricity). The PTF PV technologies should be able to reach the earliest large market (developing countries) once first-generation scale-up has been fully achieved. At that time, module prices should be near $2/W, and BOS may be in that price range as well. Installed system costs could then be about $4/W, and electricity costs would be near $0.2 to $0.30/kWh (see Figure 15).

Entry of PV into the large, developing country market could be politically supported because the use of PV has strong, positive social and economic implications in these countries. By improving farm and non-farm productivity, PV systems for rural use would slow the flight of rural poor to the cities. It would have the kind of social impact that many planners at the United Nations and elsewhere are seeking in terms of reducing world tensions while improving food supply and lifestyles. One can easily imagine a world in which PV fills this role. As such, PV development could become self-sustaining. Given the kind of progress seen in technologies like CIS and CdTe, one could then be confident that PV would achieve its potential and become a revolutionary new element in energy production.

BALANCE-OF-SYSTEMS COSTS
PV modules are typically sized to produce between 50 and 200 W. While the cost and performance of the module are critical determining factors of overall system performance, other components of the system are not insignificant. Some of the first analyses of PV systems pointed out that the system components other than the module, the (BOS), could be lumped into categories that scaled according to either the aperture area of the modules needed for the generating station or to the maximum peak power that the arrays generate. Table 3 reproduces results from one of the more detailed studies with costs presented in 1982 dollars. These systems include a 100-MW central station, a 100-kW intermediate load, and a roof-mounted residential unit. One particularly interesting point that is evident in the numbers shows up in the estimates for land cost. The central station is placed in some remote location. The intermediate load stations, sized to meet some of the high-value utility applications, uses a land cost of nearly $10,000 per acre. The point is that applications that benefit from deferred capital expenditures for rewiring or replacing transmission lines occur only in locations where growth is overtaxing existing facilities. If the PV generating station is to be located in that growth area, the land cost will be far higher than for remote unpopulated land. Despite more than a decade of inflation, many in the PV community expect that these cost estimates are still achievable in current dollars.

Table 3   **Balance-of-systems cost estimate for various PV applications**

Cost Item $ cost related to:	100 MWp Area	100 MWp kW	100 kWp Area	100 kWp kW	Residential (10 kWp) Area	Residential (10 kWp) kW
Array structure	29		32		45	
Foundation	5		6			
Module installation	7		8			
Land	1		24			
Site prep	8		10			
Roads & fences	2		15			
Control & grounding	2		3		1	
Surge protection	4		5			
dc subsystem		24		73		138
Power conditioning		93		103		335
Ac interface		26		15		33
Total	58	143	103	191	46	506

From EPRI report AP-2474, PV Balance-of-System Assessment, Final Report, Bechtel Group, June 1982.

The actual costs from some recent installations are presented in Table 4. Most of these systems use crystalline Si modules. Many of the systems include a large cost for one-time design and installation charges. For example, half of the balance-of-system cost for the 1-MW installation for the Sacramento Municipal Utility District (SMUD) was associated with engineering. In some of the small installations, the transportation of a skilled installer and the components may cost up to $1 per watt. In the example where SMUD is installing 100 systems providing 4 kW from the roofs of residential customers, the relatively low BOS cost can be attributed to installing three very similar systems each day.[86] Without these one-time costs, current experience indicates that both area-related and power-related BOS costs are about 3 to 6 times higher than the target values of $50/m^2 and $150/kWp, respectively.

Reducing the production cost relative to the aperture area of PV modules is an important goal for research and development activities. Two other module performance factors, conversion efficiency and reliability, are also critical. At current BOS costs, increasing the module efficiency by 10% increases the value of the module in $/Wp by about 10%. This same improvement in efficiency increases the allowable production costs in $/m^2 aperture area by about 20%. Increasing the efficiency of the module reduces the area of the field required to meet a given electrical load. This calculation can be pushed down to the cell level. Increasing the density half of the cost of the prevailing module is incurred in encapsulation, the 10% efficiency increase is worth up to a 40% increase in production cost.

Table 4 **Real PV systems**

System	Size (kW)	Year	$/Wp ac	Comments and Details	Ref.
SMUD PV1	1000	1984	9.99	$5.45 modules, $2.13 BOS, $2.41 engineering	86
SMUD	4	1992	>12		86
SMUD roofs	4	1993	7.07	100 systems installed using Siemens M55, $3500 for PCU	86
SMUD building	40	1993	9.11	SEA Corp Concentrators	86
SMUD ground	200	1993	7.70	UPP.	86
PGNE Kermin	500	1992-3	8.90	Seimens	87
PGNE PVUSA	419	1990	4.72	APS amorphous silicon	87
PGNE PVUSA	192	1990	9.34	ARCO Si	87
PGNE PVUSA	195	1990	9.64	Integrated Power/Mobil Solar	87
EPA DSM	4–6	1993	~8	11 systems installed, Siemens, Omnion $4-4.50/Wp dc modules, $0.75-W PCU, $1/W structure and other BOS, $1/W engineering and shipping	88

Thus far, we have discussed PV systems in terms of peak power generated. The value of the PV system will be determined by time. How much energy will it generate in a year? How many years will it remain operational? How much energy does it provide at times of high demand? A system mounted at a fixed tilt in a typical U.S. location will see about 2000 kWh/m^2/year solar energy. The standard insolation used to characterize peak performance is 1000 kW/m^2. Each peak watt of capacity will generate 2 kWh/year for this typical site. If the system costs $8/Wp and is sufficiently reliable to warrant amortizing the investment over 30 years, the electricity it produces will cost between $0.5 and $0.6/kWh. If the cost recovery period must be reduced to 10 years, as it might for a newer technology or one with lower certainty in lifetime projection, the electric energy cost will be about 25% higher.

Systems designed to track the sun capture two additional benefits that can outweigh the added complexity and cost. First, flat-plate tracking systems receive 20 to 30% more total energy from the sunlight than a fixed flat-plate system tilted at the optimum angle. Second, on a daily basis nearly all of this additional energy is collected before 10 a.m. and after 2 p.m. Thus, the system's energy output is usually better matched to the demand of the load. Because the complexity of the system is increased, the area related balance-of-systems costs are higher than for fixed flat-plate systems. Thus, module efficiency becomes even more important.

In international markets in developing countries, or in systems requiring very little energy, the cost structure for the balance of system can be substantially different. System cost may be dominated by charges for transportation. Direct dc electrical output may be desired such that power-related BOS costs can be quite low. In these conditions, the module with the lowest cost per watt may best fit the market niche regardless of the module performance.

SUMMARY

Recent developments in high efficiency and thin-film PV devices give promise for significant advancements in PV generated power. As the technologies are proven manufacturable and balance-of-system costs are reduced, new PV technology should begin to penetrate the market. At this point there is the strong potential that high-efficiency and thin-film PV can eventually compete with conventionally generated power for energy-significant applications.

ACKNOWLEDGMENTS

This work was supported by the U.S. Department of Energy under Contract No. DE-AC02-83CH100093.

REFERENCES

1. **Sopori, B. L.,** Proc. Second Symposium on Defects in Silicon, *Electrochem. Soc.*, 91-9, 1991, 545.
2. **Perichaud, I., Floret, F., and Martinuzzi, S.,** Conference Record of the 23rd IEEE Photovoltaic Specialists Conference, 1993, 243.
3. **Fukui, K., Fukawa, Y., Takahashi, H., Okada, K., Takayama, M., Shirasawa, K., and Watanabe, H.,** Technical Digest, 7th International Photovoltaic Sciences and Engineering Conference, 1993, 87.
4. **Takayama, M., Schichiri, K., Fujii, S., Inomata, Y., Ogasawara, S., Okada, K., Shirasawa, K., and Watanabe, H.,** Technical Digest, 7th International Photovoltaic Sciences and Engineering Conference, 1993, 99.
5. **Narayan, S., Wenham, S. R., and Green, M. A.,** *Appl. Phys. Lett.*, 48, 873, 1986.
6. **Martinuzzi, S., El Ghitani, H., Sarti, D., and Torchio, P.,** Conference Record of the 20th IEEE Photovoltaic Specialists Conference, 1988, 1575.
7. **Rohatgi, A., Chen, Z., Sana, P., Doolittle, A., Crotty, G., and Salami, J.,** Quarterly Technical Report, NREL, July 1992.
8. **Corbett, J. W., Lindstrom, J. L., Pearton, S. J., and Tavendale, A. J.,** *Solar Cells*, 24, 127, 1988.

9. **Sopori, B. L. and Benner, J. P.,** Proc. 1993 Annu. Conf. American Solar Energy Society, 1993, 27.

10. **Sopori, B. L., Deng, X., and Jones, K.,** *Appl. Phys. Lett.,* 61, 2560, 1992.

11. **Estreicher, S. K. and Jones, R.,** Proc. 17th Int. Conf. on Defects in Semiconductors, to be published (1993).

12. **Wang, A., Zhao, J., and Green, M. A.,** *Appl. Phys. Lett.,* 57, 602–604, 1990.

13. **Zhao, J., Wang, A., Taouk, M., Wenahm, S. R., and Green, M. A.,** *IEEE Elec. Dev. Lett.,* 14, 539–541, 1993.

14. **Ciszek, T. F.,** Proc. 20th IEEE Photovoltaic Specialists Conf., IEEE, New York, 1988, 31.

15. **Pang, S. K.,** *J. Electrochem. Soc.,* 137, 1977, 1990.

16. **Higuchi, T.,** *Appl. Phys. Lett.,* 53, 1850, 1988.

17. **Teidje, T., Yablonovitch, E., Cody, G. D., and Brooks, B. G.,** *IEEE Trans. Elec. Dev. E-31,* 5, 711, 1984.

18. **Kim, Y. Kwon, Y. and Nam, H. M., and DiNetta, La,** Proc. 23rd IEEE Photovoltaic Specialists Conf., IEEE, New York, 1993, 294.

19. **Naomoto, H., Hamamoto, S., Takami, A., Arimoto, S., Ishihara, T., Kurnabe, H., Murotani, T., and Mitsui, S.,** Technical Digest, 7th Int. Photovoltaic Sciences and Engineering Conf., 1993, 79.

20. **Bertness, K. A., Friedman, D. J., Kibbler, A.E., Kramer, C., Kurtz, S. R., and Olson, J. M.,** Am. Inst. of Phys. Conf. Proc. of the 12th NREL Photovoltaic Program Review, Denver, to be published, 1994.

21. **Friedman, D. J., Kurtz, S. R., Bertness, K. A., Kramer, C., Emery, K. A., and Olson, J. M.,** Am. Inst. of Phys. Conf. Proc. of the 12th NREL Photovoltaic Program Review, Denver, to be published, 1994.

22. **Wanlass, M. W., Ward, J. S., Emery, K. A., Gessert, T. A., and Osterwald, C. R.,** *Proc. 22nd IEEE Photovoltaic Specialists Conf.,* IEEE, New York, 1991, 38.

23. **Fraas, L. M., Avery, J. E., Sundaram, V. S., Dinh, V. T., Davenport, T. M., Yerkes, J. W., Gee, J. M., and Emery, K. A.,** *Proc. 21st IEEE Photovoltaic Specialists Conf.,* IEEE, New York, 1990, 190.

24. **Kurtz, S. R., Olson, J. M., and Kibbler, A. E.,** *J. Appl. Phys.,* 68, 1890, 1990.

25. **Chu, C., Iles, P., Yoo, H., Reed, B., and Krogen, J.,** *Proc. 22nd IEEE Photovoltaic Specialists Conf.,* IEEE, New York, 1991, 1512.

26. **Vernon, S., Tobin, S. P., Haven, V. E., Geoffrey, L. M., and Sanfacon, M. M.,** *Proc. 22nd IEEE Photovoltaic Specialists Conf.,* IEEE, New York, 1991, 353.

27. **Yang, M., Soga, T., Egawa, T., Jimbon, T., and Umeno, M.,** Tech. Digest 7th Int. Photovoltaic Sci. and Engr. Cong., Nagoya, Japan, 1993, 317.

28. **Sudersena, Nozawa, T. K., and Horaikoshi, Y.,** *Appl. Phys. Lett.,* 62, 153, 1993.

29. **Okada, Y., Ohta, S., Simomura, H., Kawabata, A., and Kawabe, M.,** Tech. Digest 7th Int. Photovoltaic Sci. and Engr. Cong., Nagoya, Japan, 1993, 327.

30. **Venkatasubramanium, R., Timmons, M. L., Sharps, P. R., and Hutchby, J. A.,** *Proc. 23rd IEEE Photovoltaic Specialists Conf.,* IEEE, New York, 1993, 691.

31. **Gale, R. P., McClelland, R. W., Dingle, B. D., Gormley, J. V., Burgess, R. M., Kim, N. P., Mickelson, R. A., and Stanbery, B. J.,** *Proc. 21st IEEE Photovoltaic Specialists Conf.,* IEEE, New York, 1990, 53.

32. **Ullal, H. S., Stone, J. L., Zweibel, K., Surek, T., and Mitchell, R. L.,** 6th Int. Photovoltaic Science and Engineering Conf., New Delhi, India, 1992.

33. **Hedstrom, J., Ohlsen, H., Bodegard, M., Kylner, A., Stolt, L., Hariskos, D., and Schock, H. W.,** *Proc. 23rd IEEE Photovoltaic Specialists Conf.,* Louisville, KY, IEEE, New York, 1993, 364.

34. Devaney, W. E., Chen, W. S., Stewart, J. M., and Mickelson, R. A., *IEEE Trans. Elec. Devices*, 37, 428, 1990.

35. Chen, W. S., Stewart, J. W., Devaney, W. E., Mickelson, R. A., and Stanbery, B. J., *Proc. 23rd IEEE Photovoltaic Specialists Conf.*, Louisville, KY, IEEE, New York, 1993, 422.

36. Albin, D., Carapella, J., Gabor, A., Tennant, A., Tuttle, J., Duda, A., Matson, R., Mason, A., Contreras, M., and Noufi, R., *AIP Conference Proc.*, 11th PV AR&D Meeting, Denver, CO, Vol. 268, 1992, 108.

37. Tuttle, J., Contreras, M., Tennant, A., Matson, R., Duda, A., Carapella, J., Albin, D., and Noufi, R., *AIP Conference Proc.*, 11th PV AR&D Meeting, Denver, CO, Vol. 268, 1992, 186.

38. Contreras, M., Tuttle, J., Du, D., Qu, Y., Swartzlander, A., Tennant, A., and Noufi, R., *Appl. Phys. Lett.*, 1993, in press.

39. Contreras, M., Tuttle, J. and Noufi, R., *Proc. 23rd IEEE Photovoltaic Specialists Conf.*, Louisville, KY, IEEE, New York, 1993, 486.

40. Tuttle, J. R., Contreras, M., Albin, D. S., Tennant, A., Carapella, J., and Noufi, R., *Proc. 23rd IEEE Photovoltaic Specialists Conf.*, Louisville, KY, IEEE, New York, 1993, 415.

41. Chu, T. L. and Chu, S. S., *Progress in PV: Research and Applications*, 1993, in press.

42. Chu, T. L., Chu, S. S., Britt, J., Chen, G., Ferekides, C., Wang, C., Wu, C. Q., and Ullal, H. S., *IEEE Electron Device Lett.*, 13, 303, 1992.

43. Chu, T. L., Chu, S. S., Britt, J., Chen, G., Ferekides, C., Schultz, N., Wang, C., Wu, C. Q., and Ullal, H. S., 11th European PV Solar Energy Conf. Exhibition, Montreux, Switzerland, 1992, 1165.

44. Chu, T. L., Chu, S. S., Wu, C. Q., Britt, J., and Wang, C., *J. Appl. Phys.*, 70, 7608, 1991.

45. Ferekides, C., Britt, J., and Ma, Y., *Proc. 23rd IEEE Photovoltaic Specialists Conf.*, Louisville, KY, IEEE, New York, 1993, 389.

46. Sasasala, R. A. and Sites, J. R., AIP Conf. Proc., 11th PV AR&D Meeting, 1992, 218.

47. Sites, J. R., Personal communication with K. Zweibel dated October 21, 1992 (unpublished).

48. Sites, J. R., *Role of Polycrystallinity in CdTe and CuInSe$_2$ PV Cells*, Annual Report, April 1, 1991 – March 31, 1992, Colorado State University, National Renewable Energy Laboratory, NREL/TP-451-5190 (DE93000045), Nov. 1992.

49. Sasala, R. A. and Sites, J., *Proc. 23rd IEEE Photovoltaic Specialists Conf.*, Louisville, KY, IEEE, New York, 1993, 543.

50. Moskowitz, P. D. and Fthenakis, V. M., *Solar Cells*, 29, 63, 1990.

51. Moskowitz, P. D., Fthenakis, V. M., Hamilton, L. D., and Lee, J. C., *Solar Cells*, 19, 287, 1987.

52. Moskowitz, P. D., Fthenakis, V. M., and Zweibel, K., *Proc. 21st IEEE Photovoltaic Specialists Conf.*, Orlando, FL, IEEE, New York, 1990, 1040.

53. Moskowitz, P. D., Hamilton, L. D., Morris, S. C., Novalm, K. M., and Rowe, M. D., Brookhaven National Laboratory, Upton, NY, BNL-51284, 1990.

54. Moskowitz, P. D. and Zweibel, K., Eds., Golden Colorado, BNL 47787, Brookhaven National Laboratory, Upton, NY, 1992.

55. Moskowitz, P. D., Zweibel, K., and Fthenakis, V. M., SERI/TR-211-3621 (DE90000310), 1990.

56. Meridian Corporation, Environmental Emissions from Energy Technology Systems: The Total Fuel Cycle, prepared for U.S. Department of Energy, Office of Renewable Energy, 1989.

57. **San Martin, R. L.,** Environmental Emissions from Energy Technology Systems: The Total Fuel Cycle, U.S. Department of Energy, 1989.

58. **Moskowitz, P.,** Personal communication March 19, 1993 to A. O. Bulawka, (unpublished).

59. **Zweibel, K. and Mitchell, R.,** *Advances in Solar Energy,* Vol. 5, Boer, K., Ed., Plenum Press, New York (also SERI/TR-211-3571, DE89009503), 1989.

60. **Alsema, E. A. and van Engelenburg, B. C. W.,** *11th European Photovoltaic Solar Energy Conf.,* Montreux, Switzerland, 1992, in press.

61. Federal Register, September 18, 1991, 29 CFR Parts 1910 and 1926, Department of Labor OSHA RIN 1218-AB16, "Occupational Exposure to Cadmium", pp. 47348.

62. **Tolley, W. K. and Palmer, G. R.,** Recovering cadmium and tellurium from CdTe manufacturing scrap, Bureau of Mines, Salt Lake City, Utah, for 1991 AIME Annual Meeting, New Orleans, LA, 1991.

63. **Nolan, J. F.,** *Proc. 23rd IEEE Photovoltaic Specialists Conf.,* Louisville, KY, IEEE, New York, 1993.

64. **Meyers, P. V., Zhou, T., Powell, R., and Reiter, N.,** *Proc. 23rd IEEE Photovoltaic Specialists Conf.,* Louisville, KY, IEEE, New York, 1993.

65. **Gerhardinger, P.,** LOF, private communications, 1992.

66. **Mitchell, K. W. and Eberspacher, C.,** Research on high efficiency, large area CuInSe$_2$-based thin film modules, NREL/TP-413-5332, Annual Subcontract Report 1 May 1992–30 April 1992, Feb. 1993, DE93000077.

67. **Turner, A. K., Woodcock, J. M., Ozsan, E., and Summers, J. G.,** *Proc. 10th EC Photovoltaic Solar Energy Conf.,* Lisbon, Portugal, 1991, 791.

68. **Turner, A. K., Woodcock, J., Ozsan, M., Barker, J., Binns, S., Buchanan, K., Chai, C., Dennison, S., Hart, R., Johnson, D., Marshall, R., Oktik, S., Patterson, M., Perks, R., Roberts, S., Sadeghi, M., Sherborne, J., and Webster, S.,** *Technical Digest of the Int. PVSEC-5,* Kyoto, Japan, 1990, 761.

69. **Woodcock, J. M., Turner, A. K., Ozsan, M. E., and Summers, J. G.,** *Proc. 22nd IEEE Photovoltaic Specialists Conf.,* IEEE, New York, 1991, 842.

70. **Maycock, P. D.,** PV News, 1993.

71. **Johansson, T. B., Kelly, H., Reddy, A. K. N., and Williams, R. L., Eds.,** *Renewable Energy Sources for Fuels and Electricity,* Island Press, Washington, D.C. (especially Chapters 9 and 10 on amorphous and polycrystalline thin films and Chapter 6, Introduction to PV), 1993, 1160.

72. **Zweibel, K.,** Polycrystalline Thin Films FY 1992 Project Report, National Renewable Energy Laboratory, TP-413-5270, DE93000070, 1993.

73. **Zweibel, K.,** *Am. Sci.,* 81, 362, 1993.

74. **Zweibel, K.,** *Harnessing Solar Energy,* Plenum Press, New York, 1990, 319.

75. U.S. Department of Energy, National PV Program, Five Year Research Plan 1987–1991, PV: USA's Energy Opportunity, DOE/CH10093-7, 1987, 33.

76. **Zweibel, K. and Chu, T. L.,** *Cadmium Telluride Photovoltaics in Advances in Solar Energy,* Boer, K. and Prince, M., Eds., American Solar Energy Society, Boulder, CO, 1993, 271.

77. **Taylor, R. W.,** PV Systems Assessment: An Integrated Perspective, EPRI AP-3176-SR, Palo Alto, CA, Electric Power Research Institute, 1985.

78. **Russell, T. W. F., Baron, B. N., and Rocheleau, R. E.,** *J. Vac. Sci. Technol.,* B2(4), 840, 1984.

79. **Jackson, B.,** CdZnS/CuInSe$_2$ Module Design and Cost Assessment, SERI/TP-216-2633, Solar Energy Research Institute, Golden, CO (NTIS DE85016854), September 1985.

80. **Meyers, P. V.,** Polycrystalline Cadmium Telluride n-i-p Solar Cells, SERI Subcontract Report, ZL-7-06031-2, Final Report, Solar Energy Research Institute, March 1990.

81. **Kapur, V. K. and Basol, B.,** *Proc. 21st IEEE Photovoltaic Specialists Conf.*, IEEE, New York, 1990, 467.

82. **Albright, S.,** (Photon Energy, Inc.), Final Report for PV Manufacturing Technology Phase I (Jan.–April 1991), NREL/TP-214-4569 (DE91015032), November 1991.

83. **Stanbery, B. J.,** (Boeing Aerospace & Electronics), Final Report for PV Manufacturing Technology Phase I (Jan.–April 1991), NREL/TP-214-4606 (DE92001176), November 1991.

84. **Jester, T.,** (Siemens Solar Industries), Final Report for PV Manufacturing Technology Phase I (Jan. - April 1991), NREL/TP-214-4481 (DE92001153), November 1991.

85. **Brown, J.,** (Solar Cells, Inc.), Final Report for PV Manufacturing Technology Phase I (Jan.–April 1991), NREL/TP-214-4478 (DE91015027), November 1991.

86. **Collier, D.,** Photovoltaic Performance and Reliability Workshop, NREL Report # CP-410-6033, 1993, 425.

87. **Wenger, H.,** private communication.

88. **Kern, E. C. and Spiegel, R. J.,** *Proc. 1993 Annu. Conf. Am. Solar Energy Soc.*, 1993, 147.

Chapter 10

Solar Processes for the Destruction of Hazardous Chemicals

Daniel M. Blake

CONTENTS

ABSTRACT: Solar technologies are being developed to address a wide range of environmental problems. Sunlight plays a role in the natural destruction of hazardous substances in soil, water, and air. Development of processes that use solar energy to remediate environmental problems or to treat process wastes is underway in laboratories around the world. This chapter reviews progress in understanding the role of solar photochemistry in removing anthropogenic chemicals from the environment and in developing technology that uses solar photochemistry for this purpose in an efficient manner.

INTRODUCTION

Renewable forms of energy have a central role to play in reducing and correcting the negative impact of human activity on the environment. Solar energy is unique among the renewable energy forms in that it can be applied directly to mitigating problems created by past and continuing release of manmade chemicals into the environment. This chapter will briefly review the ways that sunlight degrades hazardous chemicals in the environment and the development of technologies that can use sunlight for that purpose.

The past release of a wide range of substances into the environment as the result of human activity has created conditions at many sites that pose a risk to human health and to the well-being of large and small ecosystems. Existing contamination affects air, water, and soil, and will ultimately cost billions of dollars to correct. Preventing further release of hazardous substances is a high priority for industry as it develops new processes and products.[1]

The involvement of sunlight in the removal of synthetic chemicals from the environment is well documented. A few examples will serve to illustrate how solar energy plays a part in the destruction of a variety of chemicals in the environment under ambient conditions. These naturally occurring processes can be termed "passive" solar photochemistry in the sense that they do not require human intervention. These processes

175

illustrate the kinds of photochemistry that can be performed under solar conditions and the ways in which the chemical reaction path taken depends on the type of substance and the medium.

The classes of photochemical processes that have been observed to occur under solar conditions are briefly described below. For a more detailed discussion of the interaction of light with matter, the reader may consult some of the standard sources on photochemistry.[2] Representative photochemical processes include the following.

1. Direct photochemistry where light is absorbed by a compound, A, and the excited state, A^*, goes on to react with oxygen and/or water to be mineralized to carbon dioxide or to produce small molecules that are, ideally, environmentally benign or can be further broken down.

$$A \xrightarrow{h\nu} A^* \xrightarrow{O_2, H_2O} Products \tag{1}$$

2. Photochemistry of molecules adsorbed on surfaces, A(ads), where the adsorbed compound is excited to a reactive state by absorption of light and is degraded.

$$A + surface \rightarrow A(ads) \xrightarrow{h\nu} A^*(ads) \xrightarrow{O_2, H_2O} Products \tag{2}$$

3. Sensitized formation of singlet oxygen, in the $^1\Sigma_g$ state for example, or hydroxyl radical by energy transfer from an electronically excited donor, D^*, to oxygen in the presence of water.

$$D \xrightarrow{h\nu} D^* \xrightarrow{O_2(^3\Sigma_g)} \rightarrow O_2(^1\Sigma_g) \tag{3}$$

$$\xrightarrow{H_2O, O_2} \rightarrow OH$$

4. Photocatalytic oxidation, where light of wavelength equal to or greater than the band gap of a semiconductor, for example TiO_2, promotes an electron from the valence band to the conduction band. The resulting valence band hole may either directly oxidize a molecule of contaminant or oxidize water to produce a hydroxyl radical.

$$TiO_2(s) \xrightarrow{h\nu} TiO_2(s)^* \xrightarrow{O_2, H_2O} OH + HO_2 \tag{4}$$

5. Photo-Fenton chemistry in which light absorption by iron(+3) hydroxo complexes can result in electron transfer to give a hydroxyl radical and iron(+2).

$$Fe(OH)_{aq}^{+2} \xrightarrow{h\nu} Fe_{aq}^{+2} + OH \tag{5}$$

SOLAR CHEMISTRY FOR DESTROYING HAZARDOUS CHEMICALS

The high-energy cutoff of the solar spectrum at the earth's surface is about 295 nm.[3] This limits the types of organic compounds that will undergo direct photochemical reactions in sunlight to those that have absorption bands extending into the near-UV (300 to 400 nm). In general, these will be substituted aromatic compounds or compounds having extended systems of conjugated double bonds. Adsorption of molecules on surfaces might cause red shifts of the absorption spectra that will make them susceptible to solar photochemistry.[4]

SOLAR CHEMISTRY IN THE ENVIRONMENT

The solar photochemical destruction of chlorinated dibenzo-*p*-dioxins adsorbed on surfaces is believed to be a route for removing such compounds from the environment. This has been shown to occur for compounds adsorbed on soil[5,6] and grass foliage.[7] Similar mechanisms may be important for photochemical oxidation of compounds in air by involvement of minerals in aerosol form.

The solar photochemistry of organic compounds in natural waters is believed to be important in the removal of synthetic compounds. Three kinds of processes have been proposed. The first is photocatalytic oxidation mediated by naturally occurring semiconducting oxides of iron or titanium.[8,9] The second is the sensitized formation of hydroxyl radicals by excited states of humic substances.[10] The third process is Fenton-type chemistry caused by the presence of soluble iron compounds or perhaps iron oxides. This mechanism may also be important in cloud water.[11,12] Hydroxyl radicals are believed to be the active agent responsible for the initial attack on the target compound in each case. The recombination of hydroxyl radicals may be at least partially responsible for the formation of hydrogen peroxide in cloud water.[13]

SOLAR TECHNOLOGY FOR ENVIRONMENTAL REMEDIATION

The role of sunlight in destroying organic compounds in the environment is well established. The development of technology that will use solar energy for this purpose in a controlled manner is a challenge that is being actively pursued in laboratories and at the engineering scale all over the world. Passive solar photochemical destruction of harmful substances in air, water, and soil is free. However, when reactors and solar collection hardware are used, costs are incurred. The goal of research and development being performed worldwide is to develop cost-effective technology that can efficiently utilize solar photochemistry in active processes for destroying hazardous chemicals.

The kinds of processes that have been or are currently being considered include both thermal and photochemical technologies. On the thermal side, pyrolysis, oxidation and catalytic steam reforming of hazardous compounds have been studied, primarily at the laboratory scale.[14] The cutoff of the solar spectrum at about 295 nm precludes direct photochemistry for many classes of organic compounds that are considered environmental hazards. This requires that processes such as 2 through 5, discussed earlier, be considered.

Direct, High-Temperature Photochemistry

The University of Dayton Research Institute working under subcontracts from the National Renewable Energy Laboratory (NREL) has extensively explored the effects of high light flux on the destruction of organic compounds at temperatures in the range 200 to 1000 °C. Compounds such as naphthalene, chlorobenzene, and nitrobenzene that have some absorbance in the near-UV region of the solar spectrum show increased levels of destruction when irradiated compared to the purely thermal reactions at the same temperatures.[15,16]

This effect was tested under solar conditions in a series of experiments conducted by NREL at solar furnaces at White Sands Proving Ground and Sandia National Laboratories. The effects of light on the high-temperature destruction of 1,2,3,4-tetrachloro-*p*-dibenzodioxin are illustrated in Figure 1. This experiment demonstrated increased destruction when the reactor was irradiated with the full solar spectrum compared to the case when a 450-nm cutoff filter was placed in the light path.[17] This work is now being extended to the pilot scale in a joint project funded by the U.S. Departments of Defense and Energy and the Environmental Protection Agency.

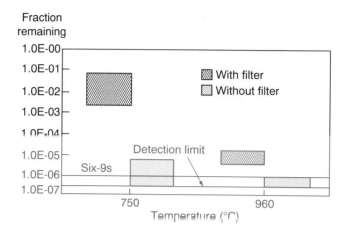

Figure 1   Effect of solar near ultraviolet light on the destruction of 1,2,3,4-tetrachloro-p-dibenzodioxin.

Dye-Sensitized Oxidation

Dye-sensitized formation of singlet oxygen and its application to the destruction of organic compounds and disinfection of effluent from a municipal sewage treatment plant have been investigated extensively. A number of dyes that absorb strongly in the visible region of the solar spectrum sensitize the formation of singlet oxygen. This has been tested on a range of organic compounds and pilot plants operated for both removing organic compounds from contaminated water and for disinfecting sewage-plant effluent to use for crop irrigation.[18-20] Figure 2 shows a pilot system in Israel that was used for the latter purpose.

Photocatalytic Oxidation

Oxidizing organic compounds in water with irradiated semiconductors is the subject of a large body of research. The author has a database containing over 600 references to work on this and related areas. There are a number of reviews that cover aspects of this work[21-23] as well as papers that discuss the mechanisms and kinetics of the processes.[24-26] In general, these processes require a semiconductor, light with higher energy than the band gap of the semiconductor, and a suitable oxidizing agent. Various types of the anatase form of titanium dioxide have been found to be the most effective photocatalyst, and oxygen or hydrogen peroxide are the oxidizing agents of choice.

Solar experiments have been reported by a number of workers.[27-30] Large-scale tests have been conducted by the solar research programs primarily in the U.S. and Europe. The first large-scale outdoor tests used existing solar hardware modified to carry out photochemical reactions.[31] A test on actual groundwater was carried out at a superfund site at Lawrence Livermore National Laboratory (LLNL) by NREL, LLNL, and Sandia National Laboratory. Groundwater contaminated with trichloroethylene (TCE) was treated using a system constructed from parabolic troughs with a borosilicate glass pipe at the focus for a reactor. The system is shown in Figure 3. The contaminated groundwater was mixed with titanium dioxide powder and pumped through the reactor. The TCE level was successfully reduced from about 300–500 ppb (micrograms per liter) to less than 5 ppb.[32]

Experiments have also been done at the European solar test facility at Almeria, Spain. The Spanish have fitted existing, two-axis tracking parabolic troughs with borosilicate glass reactors, shown in Figure 4, and studied the destruction of pentachlorophenol and dichloroacetic acid with this system.[33,34]

Figure 2 Pilot plant for dye-sensitized disinfection of sewage plant effluent at Herzliyya City, Israel.

Figure 3 Pilot plant for photocatalytic removal of chlorinated solvents from ground-water at Lawrence Livermore Laboratory, California.

Figure 4 Pilot plant for studying solar photocatalytic removal of organic compounds from water. Background, two-axis tracking parabolic trough system, foreground, nonconcentrating thin-film reactor. Location: Plataforma Solar, Almeria, Spain.

The mechanism of action of the semiconductor-initiated photochemistry predicts that the rates and efficiencies of the oxidation reactions will increase as the square root of light flux at low organic concentrations.[24] Hence, multiple-sun reactors will be less efficient than reactors using the ambient intensity of sunlight. This was shown to be the case in a test using the LLNL system.[32] For this reason, a second generation of tests has been performed using solar reactors that do not concentrate the sunlight. Typical one-sun (i.e., nonconcentrating) reactors are shown in Figures 4 and 5. The first was tested at the Plataforma Solar in Almeria, Spain in collaboration with a group from the Institute für Solarenergieforschung in Hannover, Germany. The second was developed by American Energy Technologies (AET), Inc. for treating contaminated groundwater at Tyndall Air Force Base in Florida.[35] In addition to avoiding the square-root penalty for increased light flux, the one-sun reactors can use both the direct beam and diffuse components of the near-UV part of the solar spectrum, and they eliminate losses due to reflector surfaces. A number of innovative reactors that operate with low solar flux or at one sun have been evaluated at NREL.[36,37] Examples are shown in Figures 6 and 7. Cost studies have shown

Figure 5 Nonconcentrating reactor designed by AET, Inc. for treating contaminated groundwater at Tyndall Air Force Base, Florida.

that, under most conditions, the cost of solar photons is competitive with the cost of photons from electric lamps.[38-40]

New Developments

The photocatalytic oxidation of low molecular weight hydrocarbons in the gas phase was the subject of considerable attention in the 1970s and 1980s, partially because of the search for solar technologies for the production of chemicals. The goal was to produce oxygenated products such as alcohols, aldehydes, and ketones.[41,42] Interest waned when the price of oil dropped in the latter part of the 1980s. The report that trichloroethylene could be rapidly oxidized in air over irradiated titanium dioxide[43] revived interest in the gas-phase oxidation processes as a potential method for reducing emissions of volatile organic compounds and as a tool for environmental remediation. There is a great deal of research being done, but only bench-scale tests have been reported to date.[44,45]

Other developments include processing water to remove metal ions and a method for cleaning up ocean oil spills. Photocatalytic reduction of metal ions by the conduction-band electrons at irradiated semiconductor surfaces has been demonstrated and considered as a potential solar process for the removal of heavy metals from contaminated water.[46,47] Development of TiO_2-coated hollow glass microspheres for use in decomposition of organic compounds on the surface of water is underway. The coated beads can be applied to oil spills in order to break the oil down into harmless substances when exposed to sunlight.[48]

Cost estimates have been made that compare the cost of solar and conventional photochemical processing for environmental remediation.[38-40] Figure 8 gives a graphical presentation that compares the cost of photons from electric lamps with photons supplied by one-sun solar collectors. As expected, photons from the one-sun systems are less expensive than those from the concentrating systems because of the lower hardware costs.

Figure 6 Reactor with compound parabolic concentrators designed by IST, Inc. and tested at NREL.

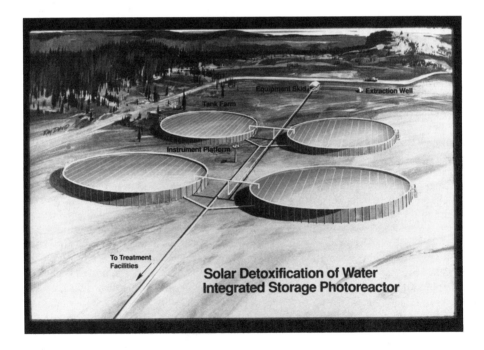

Figure 7 Artist's concept of a solar photocatalytic water detoxification system using ponds or tanks as 1-sun reactors.

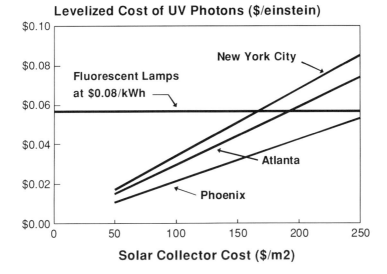

Figure 8 Comparison of the cost of photons from lamps and from the sun as a function of solar collector cost.

The results show that solar photons can be less expensive than photons from lamps in many parts of the U.S. when low, but realistic collector costs are achieved.

CONCLUSION

The destruction of hazardous substances by solar processes has been amply demonstrated. Commercialization of solar technologies requires that they be shown to be cost effective, and that they overcome the barriers normally encountered by new technologies as well as the barriers that are unique to solar processes. These barriers include:

1. The risk associated with adopting a new technology
2. A general lack of familiarity with photochemical processes on the part of industry
3. Using the sun as a light source introduces new process requirements that deal with the intermittent nature of this source of photons

It is likely that cost-competitive solar technologies for environmental clean-up applications will become commercially available in the next few years. Technical developments are being driven by the need for innovative tools to treat the wide range of environmental problems that have been created by past practices.

REFERENCES

1. **Curran, M. A.,** Broad-based environmental life cycle assessment, *Environ. Sci. Technol.*, 27, 430, 1993.
2. **Scandola, F. and Balzani, V.,** *Interaction between light and matter*, in *Photocatalysis Fundamentals and Applications*, Serpone, N. and Pelizzetti, E., Eds., John Wiley & Sons, New York, 1989, chap. 2.
3. **Myers, D. R., Cannon, T., and Webb, J. D.,** Terrestrial solar spectral ultraviolet measurements at the National Renewable Energy Laboratory, in *Proc. SPIE-Int. Soc. Opt. Eng., Soc.*, 1764, Photo-Opt. Inst. Eng., Bellingham, WA, 1992, 350.

4. **Kalyanasundraram, K.,** *Photochemistry in Microheterogeneous Systems*, Academic Press, New York, 1987, chap. 10.
5. **Dougherty, E. J., McPeters, A. L., Overcash, M. R., and Carbonell, R. G.,** Theoretical analysis of a method for *in situ* decontamination of soil containing 2,3,7,8-tetrachlorodibenzo-p-dioxin, *Environ. Sci. Technol.*, 27, 505, 1993.
6. **Kleatiwong, S., Nguyen, L. V., Hebert, V. R., Hackett, M., Miller, G. C., Mille, M. J., and Mitzel, R.,** Photolysis of chlorinated dioxins in organic solvents and on soils, *Environ. Sci. Technol.*, 24, 1575, 1990.
7. **McCrady, J. K. and Maggard, S. P.,** Uptake and photodegradation of 2,3,7,8-tetrachlorodibenzo-*p*-dioxin sorbed on grass foliage, *Environ. Sci. Technol.*, 27, 343, 1993.
8. **Cunningham, K. M., Goldberg, M. C., and Weiner, E. R.,** The aqueous photolysis of ethylene glycol adsorbed on goethite, *Photochem. Photobiol.*, 41, 409, 1985.
9. **Korman, C., Bahnemann, D. W., and Hoffman, M. R.,** Environmental photochemistry: is iron oxide (hematite) an active photocatalyst? A comparative study: α-Fe₂O₃, ZnO, TiO₂, *J. Photochem. Photobiol., A: Chem.*, 48, 161, 1989.
10. **Haag, W. R. and Hoigne, J.,** Photo-sensitized oxidation in natural water via OH radicals, *Chemosphere*, 14, 1659, 1985.
11. **Zuo, Y. and Hoigne, J.,** Evidence for photochemical formation of H_2O_2 and oxidation of SO_2 in authentic fog water, *Science*, 260, 71, 1993.
12. **Faust, B. C., Hoffmann, M. R., and Bahnemann, D. W.,** Photocatalytic oxidation of sulfur dioxide in aqueous suspensions of α-Fe_2O_3, *J. Phys. Chem.*, 93, 6371, 1989.
13. **Kormann, C., Bahnemann, D. W., and Hoffmann, M. R.,** Photocatalytic production of H_2O_2 and organic peroxides in aqueous suspensions of TiO_2, ZnO, and desert sand, *Environ. Sci. Technol.*, 22, 798, 1988.
14. **Nimlos, M. R. and Milne, T. A.,** Direct mass spectroscopic studies of the destruction of hazardous wastes. 1. Catalytic steam re-forming of chlorinated hydrocarbons, *Environ. Sci. Tecnhnol.*, 26, 545, 1992.
15. **Graham, J. L. and Dellinger, B.,** Solar thermal/photolytic destruction of hazardous organic wastes, *Energy*, 12, 303, 1987.
16. **Graham, J. L., Dellinger, B., Klosterman, D., Glatzmaier, G., and Nix, G.,** Disposal of toxic wastes by using concentrated solar radiation, *Emerging Technologies in Hazardous Waste Management II. ACS Symposium Series No. 468*, Tedder, D. W. and Pohland, F. G., Eds., ACS, Washington, D.C., 1991, chap. 6.
17. **Glatzmaier, G. C., Nix, R. G., and Mehos, M. S.,** Solar destruction of hazardous chemicals, *J. Environ. Sci. Health*, A25, 571, 1990.
18. **Acher, A. and Saltzman, S.,** Photochemical inactivation of organic pollutants from water, *Ecological Studies — Toxic Organic Chemicals in Porous Media*, Gerstl, Z., Chen, Y., Mingelgrin, U., and Yaron, B., Eds. 73, Springer-Verlag, New York, 1989, chap. 15.
19. **Saltzman, S., Acher, A. J., Brates, N., Horowitz, M, and Geveleberg, A.,** Removal of phytotoxicity of uracil herbicides in water by photodecomposition, *Pestic. Sci.*, 13, 211, 1982.
20. **Acher, A., Fischer, E., Zellingher, R., and Manor, Y.,** Photochemical disinfection of effluents — pilot plant studies, *Wat. Res.*, 24, 837, 1990.
21. **Ollis, D. F., Pelizzetti, E., and Serpone, N.,** Photocatalyzed destruction of water contaminants, *Environ. Sci. Technol.*, 25, 1522, 1992.
22. **Fox, M. A. and Dulay, M. T.,** Heterogeneous photocatalysis, *Chem. Rev.*, 93, 341, 1993.
23. **Matthews, R. W.,** Photocatalytic oxidation of organic contaminants in water: an aid to environmental preservation, *Pure Appl. Chem.*, 64, 1285, 1992.

24. **Webb, J. D., Blake, D. M., Turchi, C., and Magrini K.,** Kinetic and mechanistic overview of TiO$_2$-photocatalyzed oxidation reactions in aqueous solution. *Solar Energy Mat.*, 24, 584, 1991.

25. **Gerischer, H. and Heller, A.,** The role of oxygen in the photooxidation of organic molecules on semiconductor particles, *J. Phys. Chem.*, 95, 5261, 1991.

26. **Turchi, C. and Ollis, D.,** Photocatalytic degradation of organic water contaminants: Mechanisms involving hydroxyl radical attack, *J. Catal.*, 122, 178, 1990.

27. **Reeves, P., Ohlhausen, R., Sloan, D., Pamplin, K., Scoggins, T., Clark, C., Hutchinson, B., and Green, D.,** Photocatalytic destruction of organic dyes in aqueous TiO$_2$ suspensions using concentrated simulated and natural solar energy, *Solar Energy*, 48, 413, 1992.

28. **Pelizzetti, E., Pramauro, E., Minero, C., and Serpone, N.,** Sunlight photocatalytic degradation of organic pollutants in aquatic systems, *Waste Management*, 10, 65, 1990.

29. **Matthews, R. W. and McEvoy, S. R.,** Destruction of phenol in water with sun, sand and photocatalysis, *Solar Energy*, 49, 507, 1993.

30. **Muszkat, L., Halmann, M., Raucher, D., and Bir, L.,** Solar photodegradation of xenobiotic contaminants in polluted well water, *J. Photochem. Photobiol., A, Chem.*, 65, 409, 1992.

31. **Pacheco, J., Prairie, M., and Yellowhorse, L.,** Photocatalytic destruction of chlorinated solvents with solar energy, *ASME-JSME-JSES International Solar Energy Conference*, Reno, NV, March 17, 1991.

32. **Mehos, M., Turchi, C., Pacheco, J., Boegel, A. J., Merrill, T., and Stanley, R.,** Pilot-scale study of the solar detoxification of VOC-contaminated groundwater, *American Institute of Chemical Engineers 1992 Summer Annual Meeting*, Minneapolis, MN, NREL/TP-432-4981, August 9, 1992.

33. **Blanco, J. and Malato, S.,** Influence of solar irradiation over pentachlorophenol solar photocatalytic decomposition, in *Proc. 1st Int. Conf. TiO$_2$ Photocatalytic Purification and Treatment of Water and Air*, London, Ontario, Canada, November 8–13, 1992.

34. **Bockelmann, D., Goslich, R., Nogueira, R. F. P., Weichgrebe, D., Jardim, W. F., and Bahnemann, D.,** Solar detoxification of polluted water: Comparing the efficiencies of parabolic trough reactor and a novel thin-film fixed-bed reactor, in *Proc. 1st Int. Conf. TiO$_2$ Photocatalytic Purification and Treatment of Water and Air*, London, Ontario, Canada, November 8–13, 1992, 187.

35. **Turchi, C. S., Klausner, J. F., Goswami, D. Y., and Marchand, E.,** Field test results for the solar photocatalytic detoxification of fuel-contaminated groundwater, NREL/TP-471-5345, presented at *Chemical Oxidation: Technologies for the Nineties, Third Int. Symp.*, Nashville, TN, February 17–19, 1993.

36. **Turchi, C. S. and Mehos, M. S.,** Solar photocatalytic detoxification of groundwater: Development of reactor designs, *Chemical Oxidation: Technologies for the Nineties, Second Int. Symp.*, Nashville, TN, February 19–21, 1992.

37. **Pacheco, K., Watt, A. S., and Turchi, C. S.,** Solar Detoxification of water: Outdoor testing of prototype photoreactors, *ASME/ASES Joint Solar Energy Conference*, Washington, D.C., April 4–8, 1993.

38. **Turchi, C. S. and Link, H. F.,** Relative cost of photons from solar or electric sources for photocatalytic water detoxification, *Int. Solar Energy Soc. 1991 Solar World Congress*, Denver, CO, 17 August 1991.

39. **Schertz, P., Kelly, D., and Lammert, L.,** Analysis of the cost of generating or capturing ultraviolet light for photocatalytic water detoxification systems, Final Report, Subcontract No. AF-2-11252-1, National Renewable Energy Laboratory, Golden, CO, 1992.

40. **Turchi, C. S., Mehos, M. S., and Link, H. F.,** Design and cost of solar photocatalytic systems for groundwater remediation, NREL Technical Paper, TP-432-4865, May, 1992.

41. **Pichat, P., Herrmann, J.-M., Disdier, J., and Mozzanega, M.-N.,** Photocatalytic oxidation of propene over various oxides at 320K. Selectivity, *J. Phys. Chem.*, 83, 3122, 1979.

42. **Sakata, T. and Kawai, T.,** Photosynthesis and photocatalysis with semiconductor powders, *Energy Resources through Photochemistry and Catalysis*, Gräetzel, M., Ed., Academic Press, New York, 1983, 332.

43. **Raupp, G. B. and Dibble, L. A.,** Gas-solid photocatalytic oxidation of environmental pollutants, United States Patent 5,045,288, 1991.

44. **Peral, J. and Ollis, D. F.,** Heterogeneous photocatalytic oxidation of gas-phase organics for air purification: Acetone, 1-butanol, butyraldehyde, formaldehyde, and *m*-xylene, *J. Catal.*, 136, 554, 1992.

45. Nimlos, M. R., Milne, T. A., Riley, T. R. et al. Mass spectrometric studies of the destruction of hazardous wastes. 2. Gas-phase photocatalytic oxidation of trichloroethylene over TiO$_2$: Products and mechanisms, *Environ. Sci. Technol.*, 27, 732, 1993.

46. **Foster, N. S., Noble, R. D., and Koval, C. A.,** Reversible photoreductive deposition and oxidative dissolution of copper ions in titanium dioxide suspensions, *Environ. Sci. Technol.*, 27, 350, 1993.

47. **Prairie, M. R., Evans, L. R., Stange, B. M., and Martinez, S. L.,** An investigation of TiO$_2$ photocatalysis for the treatment of water contaminated with metals and organic chemicals, *Environ. Sci. Technol.*, 27, 1776, 1993.

48. **Heller, A. and Brock, J. R.,** Materials and methods for photocatalyzing oxidation of organic compounds on water, United States Patent 4,997,576, March 5, 1991.

Chapter 11

The Detoxification of Waste Water Streams Using Solar and Artificial UV Light Sources

James R. Bolton, Ali Safarzadeh-Amiri, and Stephen R. Cater

CONTENTS

INTRODUCTION

Traditionally, the removal of pollutants from wastewater streams has utilized either air stripping or granulated activated carbon filters. These methods are generally effective in removing pollutants from a given waste stream; however, they only transfer the pollutants from one phase to another and do not destroy them. In addition, relatively slow-acting oxidants, such as hydrogen peroxide and ozone, have been used in the dark for slow oxidative breakdown of some classes of pollutants.

Over the past decade, a series of new methods, called Advanced Oxidation Processes (AOP), have been developed that are able to destroy (i.e., mineralize) organic pollutants *in situ*, so that the ultimate products are CO_2, H_2O, and a small amount of inorganic salts. For example, 2,4-dichlorophenol ($C_6H_4OCl_2$) reacts according to:

$$C_6H_4OCl_2 + 5H_2O + 6O_2 \rightarrow 8H^+ + 6HCO_3^- + 2Cl^- \tag{1}$$

Almost all reactions of this type are energy releasing; however, they are usually kinetically slow in the absence of initiators. The AOP technologies almost all rely on the generation of very reactive free radicals, such as the hydroxyl radical ($\cdot OH$), which then reacts with most organic compounds either by abstraction of a hydrogen atom, for example,

$$\cdot OH + CH_3OH \rightarrow \cdot CH_2OH + H_2O \tag{2}$$

or by addition, for example,

$$\cdot OH + CCl_2 = CH_2 \rightarrow \cdot CCl_2 - CH_2OH \tag{3}$$

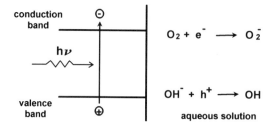

Figure 1 Energy-level diagram of the valence and conduction bands in TiO$_2$ showing the light generation of excess holes and electrons.

These radicals then react with oxygen to initiate a series of degradative oxidation reactions that ultimately lead to the products of mineralization. UV-driven AOP systems can be divided into two major classes: homogeneous and heterogeneous. *Homogeneous processes* include UV photolysis of H$_2$O$_2$, O$_3$, or proprietary absorbers to generate ·OH radicals. In some cases, direct photolysis of the pollutant (e.g., N-nitrosodimethylamine) is effective without any additives or sensitizers. *Heterogeneous processes* are almost all based on UV irradiation of TiO$_2$ (anatase) particles. TiO$_2$ is a semiconductor with a band gap of 3.2 eV and thus only absorbs below ca. 380 nm. Photon absorption causes the creation of excess holes in the valence band and excess electrons in the conduction band. As shown in Figure 1, the holes can react with surface-bound hydroxyl groups to generate ·OH radicals, which then attack organic pollutants in the same manner as in homogeneous systems.

The UV portion of the terrestrial solar spectrum extends only down into the near-UV (≥300 nm), since atmospheric ozone absorbs almost all solar radiation below 300 nm. Thus, any solar detoxification system must rely on an absorber that is active above 300 nm. This rules out the use of hydrogen peroxide or ozone, which are popular absorbers with detoxification systems using artificial UV light sources. Fortunately, TiO$_2$ absorbs in the desired wavelength range and has proven effective in solar detoxification systems. A considerable body of research has been conducted to date that shows that TiO$_2$ (in its anatase crystalline form) is effective as a photocatalyst for photodegradation reactions of a wide variety of pollutants; however, the overall efficiency is rather small.

In this chapter, we describe and compare solar detoxification AOP systems with those using artificial UV light sources in terms of efficiency and capital costs. We also introduce a "new" solar detoxification process called **Solaqua®**.

UV LIGHT SOURCES

We begin our analysis with a comparison of artificial and solar UV light sources.

ARTIFICIAL UV LIGHT SOURCES

Artificial (i.e., commercial) UV light sources can be divided into three major classes.

Low-pressure Mercury Lamps

These are largely used as germicidal lamps to sterilize water and air. They are also used in commercial AOP systems that rely on the photolysis of O$_3$ to generate ·OH radicals. These commercially available lamps are characterized by:

- Low energy density (ca. 1 W cm^{-1})
- High conversion efficiency of electricity-to-photon energy (ca. 30%)
- Approximately 80% of the emission is in a narrow range around 254 nm
- Long life (>5000 h)

A modification of this lamp, with a phosphor emitting in a moderate range around 360 nm, is used for AOP systems utilizing TiO$_2$ as a photocatalyst.

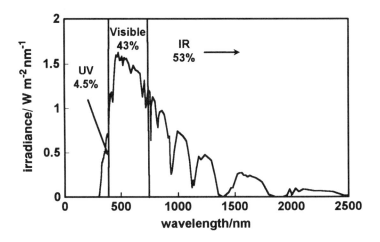

Figure 2 Solar spectral irradiance distribution calculated for the standard AM 1.5G global solar spectrum. (From Hulstrom, R., Bird, R., and Riordan, C., *Solar Cells*, 15, 365, 1985. With permission.)

Medium-pressure Mercury Lamps
These are used in UV curing of coatings and in some commercial AOP water treatment systems. These commercially available lamps are characterized by:

- Moderate energy density (ca. 125 W cm^{-1})
- Moderate electrical-to-photon energy efficiency (ca. 15% for the range 200–300 nm)
- Broad spectral output, but not much below 250 nm
- Moderate life (>2000 h)

Advanced Proprietary Medium-Pressure Mercury Lamps
Some companies (e.g., Solarchem) have available advanced medium-pressure mercury lamps in their AOP systems. These lamps have increased emission in the desired UV range (200–300 nm) and thus considerably improve the performance with absorbers such as O_3 or H_2O_2 and for direct photolysis. They are characterized by:

- High energy density (250 W cm^{-1})
- High electrical-to-photon energy efficiency (ca. 30% for the range 200–300 nm)
- Strong output below 250 nm
- Long life (>3000 h)

THE SOLAR UV LIGHT SOURCE
Figure 2 shows the solar spectral irradiance distribution for the standard AM 1.5 solar spectrum, which is characteristic of solar radiation as received at the earth's surface. Only 4.5% of the solar irradiance is in the UV (<400 nm) and none is below 300 nm. Since H_2O_2 and O_3 do not absorb significantly above 300 nm, the solar light source cannot be used with these absorbers.

Figure 3 shows the integrated solar irradiance up to a given wavelength, where the total irradiance is 1000 W/m^{-2}. TiO$_2$ absorbs only up to 380 nm and thus utilizes only ~3% of the available solar irradiance. Also shown in Figure 2 is a "new" absorber that absorbs all solar radiation up to ~500 nm and thus is able to utilize ~18% of the available solar irradiance.

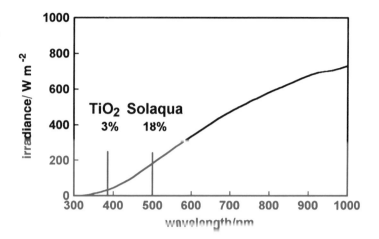

Figure 3 Integrated solar irradiance up to a given wavelength. The solar spectrum used is the same as that used for Figure 2.

ECONOMICS OF ARTIFICIAL AND SOLAR AOP SYSTEMS

Since both artificial and solar UV light sources generate ·OH radicals, we have chosen to base our economic analysis on the capital cost required to create the capacity to generate 1 mole of ·OH radicals per hour.

ARTIFICIAL AOP UV SYSTEM
For the artificial AOP UV system we make the following assumptions:

1. 30% of the input electrical energy to the lamp is available as useful UV light (200–300 nm)
2. The average UV wavelength is 250 nm
3. Photolysis of H_2O_2 is the primary photochemical reaction with a quantum yield of 1.0 per ·OH radical generated[1,2]
4. The capital cost is $1500 per kW of UV lamp capacity for a large installation (>200 kW)*

The moles of UV photons generated per hour per kW of electrical input is computed** from

$$\text{moles of photons per hour} = \frac{3.6 \times 10^6 \, \text{J} \times 0.30}{hcL/\lambda} = 2.26 \qquad (4)$$

where h is Planck's constant, c is the speed of light, and L is Avogadro's number. Since the quantum yield for generation of ·OH radicals is 1.0, the number of moles of ·OH generated per hour is also 2.26. Thus, the capital cost per kilowatt of power installation per mole of ·OH per hour is 1500/2.26 = $665.

TIO₂ SOLAR AOP SYSTEM
We make the following assumptions for a TiO_2 solar AOP system:

---

* This is the approximate capital cost of an installed Solarchem system.
** The value obtained by integrating every nanometer over the spectral output of the lamp is 2.02 moles photons per hour per kilowatt electrical energy input.

1. Only solar energy with $\lambda \leq 380$ nm is absorbed, representing 3% of the total solar irradiance.
2. The average wavelength is 360 nm.
3. The quantum yield for generation of ·OH radicals on the surface of the TiO_2 particles is 0.10.[*]
4. A flat-plate solar collector is used where the average (24-h) solar irradiance is 250 W/m².
5. The capital cost of the solar collector is $200 per square meter.[**]

The moles of UV photons generated per square meter per hour is computed[***] from:

$$\text{moles of photons per hour per m}^2 = \frac{250 \text{ W} \times 3600 \text{ s/hr} \times 0.03}{hcL/\lambda} = 0.0813 \qquad (5)$$

Since the quantum yield for generation of ·OH radicals is assumed to be 0.10, the number of moles of ·OH generated per hour is 0.00813. Thus, the capital cost per mole of ·OH per hour is 200/0.00813 = $25,000.

"NEW" SOLAR AOP SYSTEM
To illustrate the potential to improve the solar option, we intoduce a "new" solar AOP system that Solarchem has recently developed, called **Solaqua®**. This process has the following characteristics:

1. A proprietary absorber is used such that solar energy with $\lambda \leq 500$ nm is absorbed, representing 18% of the total solar irradiance.
2. The average wavelength is 400 nm.
3. The quantum yield for generation of ·OH radicals is very high (>0.8).
4. A flat-plate solar collector is used where the average (24 h) solar irradiance is 250 watts per square meter.
5. The capital cost of the solar collector is $200 per square meter.

As before, the moles of UV photons generated per square meter per hour is computed[****] from:

$$\text{moles of photons per hour per m}^2 = \frac{250 \text{ W} \times 3600 \text{ s/hr} \times 0.18}{hcL/\lambda} = 0.54 \qquad (6)$$

---

[*] The quantum yield for ·OH generation on TiO_2 has not been determined. The value assumed is based on the determination that the quantum yield for the bleaching of methylene blue on TiO_2 is about 0.05 and the assumption that it requires at least two ·OH radicals to bleach a methylene blue molecule. See Valladares, J. E. and Bolton, J. R., A method for the determination of quantum yields in heterogeneous systems: The TiO_2 photocatalyzed bleaching of methylene blue, in Proc. First Int. Conf. on TiO_2 Photocatalytic Purification and Treatment of Water and Air, in press.

[**] The capital cost of a flat-plate solar collector for domestic hot water use is ~$150/m². $50/m² is the estimated cost of the pumps, lines, and tanks that would be necessary for a solar detoxification system. All cost figures in this chapter are given in U.S. dollars.

[***] The value obtained by integrating every nanometer up to 380 nm over the solar spectral irradiance distribution for the standard global air mass 1.5 solar spectrum is 0.0907 moles photons per hour for a 250-W average total spectral irradiance.

[****] The value obtained by integrating every nanometer up to 500 nm over the solar spectral irradiance distribution for the standard global air mass 1.5 solar spectrum is 0.504 moles photons per hour for a 250-W average total spectral irradiance.

Table 1 **Comparison of artificial light and solar light AOP systems**

	Artificial Light System	TiO$_2$ Solar System	"New" Solar System
Capital Cost	$665 per mole ·OH per hour	$25,000 per mole ·OH per hour	$465 per mole ·OH per hour
Footprint	0.05 m^2/kW	4 m^2/kW	4 m^2/kW

Since the quantum yield for generation of ·OH radicals is assumed to be 0.8, the number of moles of ·OH generated per hour is 0.43. Thus, the capital cost per mole of ·OH per hour is 200/0.43 = $465. Table 1 summarizes and compares the economic characteristics of the three AOP systems considered.

POTENTIAL APPLICATIONS

Solar detoxification systems could have many potential applications if the economics are favorable. Some possible applications are:

1. Industrial waste ponds
2. Leachate runoffs from toxic dumps
3. Agricultural runoffs
4. Off-shore drilling rigs
5. Groundwater remediaton

CONCLUSIONS

- Solar detoxification systems based on TiO$_2$ have too high a capital cost compared to artificial systems.
- The capital cost of the "new" **Solaqua®** process is comparable to that of an artificial AOP system.
- More research and development are necessary to prove the feasibility of solar detoxification, but the potential looks promising.

REFERENCES

1. **Hunt, J. P. and Taube, H.,** The photochemical decomposition of hydrogen peroxide. Quantum yields, tracer and fractionation effects, *J. Am. Chem. Soc.*, 74, 5999, 1952.
2. **Weeks, J. L. and Matheson, M. S.,** The primary quantum yield of hydrogen peroxide decomposition, *J. Am. Chem. Soc.*, 78, 1273, 1956.

Chapter 12

Photoelectrochemical Production of Hydrogen

A. J. Nozik

CONTENTS

INTRODUCTION

A very attractive, but long-term approach to the economical production of hydrogen from renewable energy is the photoelectrolysis of water. Such systems are based on photoactive semiconductor materials that operate efficiently to split water photolytically into its elements using solar radiation as the energy source. The semiconductor materials can be in the form of planar electrodes in photoelectrochemical cells or monolithic structures termed photochemical diodes or photocatalytic particles. In this chapter, we shall first summarize the fundamental principles of the photoelectrochemical generation of hydrogen, then summarize the progress and prognosis for photoelectrolysis technology. Finally, we will present some economic analyses of the technology that will compare solar photoelectrolytic hydrogen production costs with hydrogen generated from fossil and nuclear fuels.

PRINCIPLES OF PHOTOELECTROCHEMISTRY

All phenomena associated with photoelectrochemical systems are based on the formation of a semiconductor-electrolyte junction when an appropriate semiconductor is immersed in an appropriate electrolyte. The junction is characterized by the presence of a space-charge layer in the semiconductor adjacent to the interface with the electrolyte. A space-charge layer generally develops in a semiconductor on contact and equilibration with a second phase whenever the initial chemical potential of electrons is different for the two phases. For semiconductors, the chemical potential of electrons is given by the Fermi level in the semiconductor. For liquid electrolytes, it is determined by the redox potential of the redox couples present in the electrolyte; these redox potentials are also identified with the Fermi level of the electrolyte.[1-3]

If the initial Fermi level in an n-type semiconductor is above the initial Fermi level in the electrolyte, then equilibration of the two Fermi levels occurs by transfer of electrons from the semiconductor to the electrolyte. This produces a positive space-charge layer in the semiconductor (also called a depletion layer because the region is depleted of majority charge carriers). As a result, the conduction and valence band edges are bent so that a potential barrier is established against further electron transfer into the electrolyte (see Figure 1).

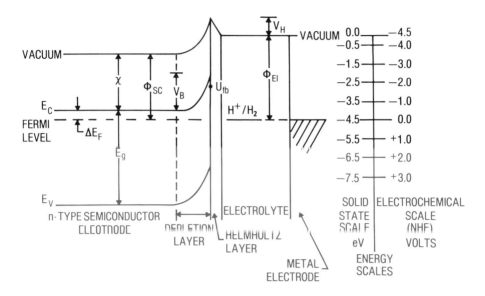

Figure 1 Energy level diagram for semiconductor-electrolyte junctions.

The inverse but analogous situation occurs with p-type semiconductors having an initial Fermi level below that of the electrolyte.

The width of the depletion layer w is given by

$$w = \left(2\varepsilon\varepsilon_o V_B/qN\right)^{1/2} \tag{1}$$

where V_B is the amount of band bending in the depletion layer, N is the charge carrier density in the semiconductor, q is the electronic charge, ε is the dielectric constant of the semiconductor, and ε_o is the permittivity of free space. In semiconductors, w can be quite large depending on the conductivity and the band bending; in typical cases, it ranges from 100 Å to several micrometers. In metal electrodes, the space-charge layer is infinitesimally small.

As discussed earlier, a charged layer also exists in the electrolyte adjacent to the interface with the solid electrode — the well-known Helmholtz layer. The potential drop across the Helmholtz layer depends on the specific ionic equilibrium at the surface.

A very important consequence of the presence of the Helmholtz layer for semiconductor electrodes is that it markedly affects the band bending that develops in the semiconductor when it equilibrates with the electrolyte. Without the Helmholtz layer, the band bending would be expected to equal the difference in initial Fermi levels between the two phases (i.e., the difference between their respective work functions). However, the potential drop across the Helmholtz layer modifies the net band bending as shown in Figure 1.

In Figure 1, the energy scales commonly used in solid-state physics and in electrochemistry are shown for comparison. In the former, the zero reference point is vacuum; while in the latter, it is the standard redox potential of the hydrogen ion-hydrogen (H⁺/H₂) redox couple. It has been shown[3,4] that these scales are related in that the effective work function or Fermi level for the standard H⁺/H₂ redox couple at equilibrium is –4.5 eV with respect to vacuum. Hence, using this scheme, the energy levels corresponding to any given redox couple can be related to the energy levels of the valence and conduction bands of the semiconductor electrode.

To make the connection between the energy levels of the electrolyte and the semiconductor, it is necessary to introduce the flat-band potential U_{fb} as a critical parameter characterizing the semiconductor electrode. The flat-band potential is the electrode potential at which the semiconductor bands are flat (zero space charge in the semiconductor); it is measured with respect to a reference electrode, usually either the normal H^+/H_2 redox potential (NHE) or the standard calomel electrode (SCE). Hence, the band bending is given by

$$V_B = U - U_{fb}, \tag{2}$$

where U is the electrode potential (Fermi level) of the semiconductor. At equilibrium in the dark, U is identical to the potential of the redox couple in the electrolyte.

The effect of the Helmholtz layer on the semiconductor band bending is contained within the flat-band potential. This important parameter is a property both of the bulk semiconductor and the electrolyte, as seen from the following relationship:

$$U_{fb}(NHE) = \left(\chi + \Delta E_F + V_H\right) - 4.5 = \left(\phi_{SC} + V_H\right) - 4.5 \tag{3}$$

where χ is the electron affinity of the semiconductor, ΔE_F is the difference between the Fermi level and majority carrier band edge of the semiconductor, V_H is the potential drop across the Helmholtz layer, ϕ_{SC} is the work function of the semiconductor, and 4.5 is the scale factor relating the H^+/H_2 redox level to vacuum.

In the absence of a large density of semiconductor surface states, the charge density and capacitance of the Helmholtz layer will be much larger than that of the semiconductor depletion layer. Inasmuch as the capacitances of the Helmholtz and depletion layers are in series, any externally applied voltage will appear primarily across the depletion layer; the potential drop across the Helmholtz layer, V_H, will be constant with applied voltage. Consequently, the band edges of the semiconductor electrode will be fixed with respect to the redox potentials of the electrolyte; this means that the band edges and the flat-band potential will be independent of electrode potential.

On the other hand, the situation is quite different if the semiconductor has a large density of surface states. If these surface states are charged, then the applied voltage can drop across the Helmholtz layer rather than the depletion layer. This produces a change in V_H with electrode potential, and a consequential movement of the band edges with electrode potential; the semiconductor electrode behaves like a metal in this situation.

In photoelectrochemistry, the above condition is commonly termed band-edge unpinning or band-edge movement; the flat-band potential then becomes dependent on the electrode potential. In solid-state physics, the analogous situation for semiconductors with metal, air, or vacuum interfaces is termed *Fermi level pinning*. This is because the Fermi level of the semiconductor equilibrates with the surface states and becomes independent of the nature of the second phase at the interface.

In addition to being produced by high surface-state densities, high surface-charge densities and consequent band-edge movement can also be caused by charge-carrier "inversion" of the surface in the presence of large band bending under reverse bias, or through majority carrier "accumulation" under forward bias of the semiconductor electrode.

The flat-band potential can also change if the electrolyte composition is changed such that the equilibrium distribution of adsorbed surface charge changes. This occurs for many semiconductors when the pH of the electrolyte is changed. In aqueous solutions, the dominant equilibrium controlling the surface charge is

$$S - OH^- \rightleftharpoons H^+ + S - O^{2-} \tag{4}$$

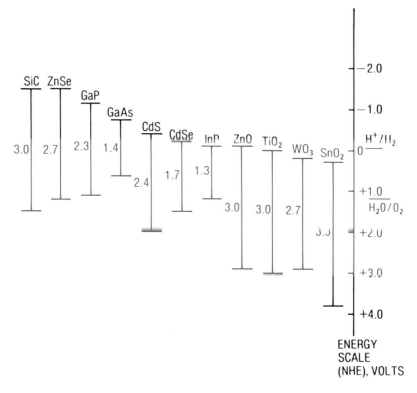

Figure 2 Position of valence and conduction band edges for several semiconductors in contact with aqueous electrolyte at pH = 1.0. For water-splitting with a single band gap semiconductor, the conduction and valence band edges must straddle the redox potentials for the hydrogen evolution and oxygen evolution redox reactions. The redox potentials for these reactions are indicated in the diagram.

where S represents the semiconductor surface. This condition generally leads to a 59-meV positive increase in flat-band potential per unit decrease in pH.

The classic method for experimentally determining flat-band potentials is based on the Schottky-Mott equation, and involves measuring the differential capacitance (C) of the semiconductor electrode as a function of applied voltage. For an ideal system, a plot of $1/C^2$ vs electrode potential (called a Schottky-Mott plot) yields a straight line with an intercept at $1/C^2 = 0$ equal to U_{fb}, and a slope related to the carrier density. In the ideal case, the measured capacitance is equal to the capacitance of the space charge layer, and it is frequency independent. This holds for a simple equivalent circuit in which the Helmholtz layer capacitance (C_H) is in series with two parallel capacitances due to the space-charge layer (C_{SC}) and to semiconductor surface states (C_{SS}), and where C_H is much greater than C_{SC} and C_{SS} is negligible compared to C_{SC}.

Capacitance-voltage data have been obtained for many semiconductors by many workers, and the Schottky-Mott plots often exhibit a frequency dependence. This effect has been analyzed in detail.[5-7] When the Schottky-Mott plots have different slopes for different frequencies but converge to a common intercept, the Schottky-Mott plot yields a valid value for the flat-band potential. A compilation of flat-band potentials for several semiconductors is presented in Figure 2 at pH = 1.0.

When the semiconductor-electrolyte junction is illuminated, photons having energies greater than the semiconductor band gap (E_g) are absorbed and create electron-hole pairs.

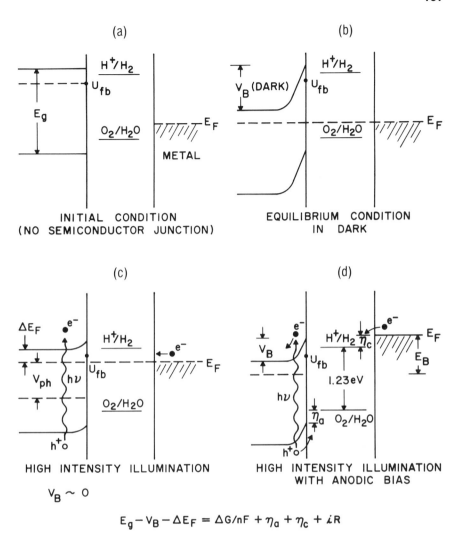

Figure 3 Sequence of energy level diagrams for Schottky-type photoelectrolysis cells from the initial condition to the final condition at equilibrium under conditions of photoelectrolysis of water with an external bias. Energy balance is written for the case of zero bias; if bias is present, E_B is added to the left-hand side of the equation.

Photons absorbed in the depletion layer produce electron-hole pairs that separate under the influence of the electric field present in the space-charge region. Electron-hole pairs produced by absorption of photons beyond the depletion layer will separate if the minority carriers can diffuse to the depletion layer before recombination with the majority carriers occurs.

The photoproduction and subsequent separation of electron-hole pairs in the depletion layer cause the Fermi level in the semiconductor to return toward its flat-band potential (see Figure 3). Under open-circuit conditions between an illuminated semiconductor electrode and a metal counter-electrode, the photovoltage produced between the electrodes is equal to the difference between the Fermi level in the semiconductor and the redox potential of the electrolyte. Under closed-circuit conditions, the Fermi level in the

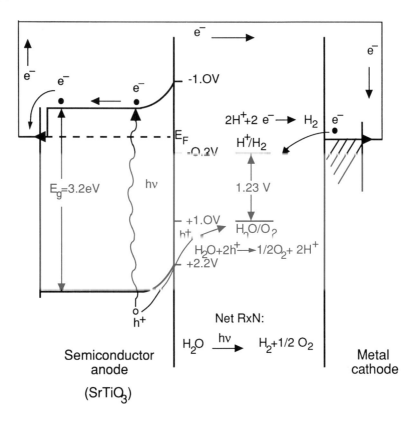

Figure 4 Energy level diagram for the photoelectrolysis of water using the single, wide, band gap semiconductor strontium titanate (SrTiO₃). The photoelectrolysis reaction can proceed without an external bias because the conduction and valence band edges straddle the redox potentials for the hydrogen- and oxygen-evolution reactions.

system is equalized and no photovoltage exists between the two electrodes. However, a net charge flow does exist.

Photogenerated minority carriers in the semiconductor are swept to the surface where they are subsequently injected into the electrolyte to drive a redox reaction. For n-type semiconductors, minority holes are injected to produce an anodic oxidation reaction; while for p-type semiconductors, minority electrons are injected to produce a cathodic reduction reaction. The photogenerated majority carriers in both cases are swept toward the semiconductor bulk, where they subsequently leave the semiconductor via an ohmic contact, traverse an external circuit to the counter-electrode, and are then injected at the counter-electrode to drive a redox reaction inverse to that occurring at the semiconductor electrode. This situation is shown in Figure 4 for the case of the photoelectrolysis of water into H_2 and O_2 using SrTiO₃ photoanodes with an external bias voltage. This figure is discussed further in the section entitled "Photoelectrolysis of Water: Energetics".

Two very different types of photoelectrochemical cells can be created depending on the nature of the electrolyte. If the electrolyte contains only one effective redox couple, then the oxidation reaction at the anode is simply reversed at the cathode and no net chemical change occurs in the electrolyte. The electrode reactions merely serve to shuttle charge across the electrolyte and maintain current continuity. In this case, the cell behaves as an electrochemical photovoltaic cell, and the incident radiant power is converted into the electrical power in the form of the external photocurrent and photovoltage.

If the electrolyte contains two effective redox couples, then the respective oxidation and reduction reactions at the anode and cathode are different. This leads to a net chemical change in the electrolyte, and the cell behaves as a photoelectrosynthetic cell. Here, the incident optical energy is converted into the chemical free-energy characteristic of the net oxidation-reduction reaction generated in the electrolyte (see Figure 4). The most prominent example of photoelectrosynthesis is the photoelectrolysis of water into H_2 and O_2.

The photogenerated holes and electrons in semiconductor electrodes are generally characterized by strong oxidizing and reducing potentials, respectively. Instead of being injected into the electrolyte to drive redox reactions, these holes and electrons can oxidize or reduce the semiconductor itself and cause decomposition. This possibility is a serious problem for practical photoelectrochemical devices because photodecomposition of the electrode leads to inoperability or to short electrode lifetimes.

In a simple model of electrode stability, the redox potential of the oxidative and reductive decomposition reactions are put on an energy-level diagram such as that in Figure 4. The relative positions of the decomposition reactions are compared with those of the desired redox reactions in the electrolyte and with those of the semiconductor-valence and conduction-band edges. Absolute thermodynamic stability of the electrode is assured if the redox potential of the oxidative decomposition reaction of the semiconductor lies below (has a more positive value on the SCE scale) the valence band edge, and if the redox potential of the reductive decomposition reaction lies above (has a more negative value on the SCE scale) the conduction band edge. Typically, however, one or both of the redox potentials of the semiconductor oxidative and reductive decomposition reactions lie within the band gap and, hence, become thermodynamically possible. Electrode stability then depends on the competition between thermodynamically possible semiconductor decomposition reactions and thermodynamically possible redox reactions in the electrolyte. This composition is governed by the relative kinetics of the two possible types of reactions.

In photoelectrochemistry, it appears that the more thermodynamically favored redox reactions also become kinetically favored, so these reactions predominate. The origin of this effect has been generally attributed to the existence of surface states within the semiconductor band gap. These surface states allow photogenerated minority holes in n-type electrodes to rise to the highest redox level, where efficient and isoenergetic hole transfer can then occur; an inverse process occurs with p-type electrodes. This effect has been used to stabilize semiconductor electrodes by establishing a redox couple in the electrolyte with a redox potential more negative than the oxidative decomposition potential (or more positive than the reductive decomposition potential), such that this electrolyte-redox reaction occurs preferentially to the decomposition reaction and scavenges the photogenerated minority carriers. However, this stabilization technique can only be used for electrochemical photovoltaic cells.

PHOTOELECTROLYSIS OF WATER

ENERGETICS

Two types of photoelectrolysis cells can be distinguished. In the first type, one electrode is a semiconductor and the second is a metal. In the second type, one electrode is an n-type semiconductor and the second is a p-type semiconductor. The energetics of the former is represented in Figures 3 and 4 for the photoelectrolysis of water into H_2 and O_2 by using n-type electrodes (analogous analyses apply to p-type electrodes). Inasmuch as there are two redox couples in the electrolyte, the initial Fermi level in the electrolyte can be anywhere between them, depending on the initial relative concentrations of H_2 and O_2 in the cell. In Figure 3, the initial Fermi level in the electrolyte is arbitrarily drawn just above the O_2-H_2O redox level. After equilibration in the dark, the Fermi level in the semiconductor equilibrates with the electrolyte Fermi level, producing a band bending, V_B, in accordance with Equation 1.

Under illumination (Figure 3), the Fermi level in the semiconductor rises toward U_{fb}, producing a photovoltage, V_{ph}. This voltage can be measured between the two electrodes with an external potentiometer. However, the value of V_{ph} depends on the initial metal-electrode potential, which depends on the initial relative concentrations of H_2 and O_2; V_{ph} can vary from zero to some finite value. Except under very special circumstances (initial metal electrode potential and the valence-band edge of the semiconductor electrode both at the O_2/H_2O redox potential), V_{ph} is not the potential energy available for photoelectrolysis.

With the two electrodes shorted together, the maximum Fermi level possible in the cell is the flat-band potential. In Figure 3, U_{fb} is below the H^+/H_2 redox potential. Hence, even with illumination intensity sufficient to flatten completely the semiconductor bands, H_2 could not be evolved at the counter-electrode because the Fermi level is below the H^+/H_2 potential (Figure 3). To raise the Fermi level in the metal counter-electrode above the H^+/H_2 potential, an external anodic bias, E_b, must be applied, as shown in Figure 3. This bias also provides the overvoltage at the metal cathode, η_e, required to sustain the current flow, and it increases the band bending in the semiconductor to maintain the required charge-separation rate.

The situation depicted in Figure 3 is the one that describes most of the n-type semiconductors studied to date. For these semiconductors, an external bias is required to generate H_2 and O_2 in a Schottky-type photoelectrolysis cell; the further U_{fb} lies below H^+/H_2, the greater the bias. The bias can be applied either by an external voltage source or by immersing the anode in base and the cathode in acid (the two compartments being separated by a membrane).

Several semiconductors such as $SrTiO_3$, $KTaO_3$, Nb_2O_5, and ZrO_2 have U_{fb} above the H^+/H_2 potential; therefore, no external bias is required to generate H_2 and O_2 in a Schottky-type cell (see Figure 4). This has been confirmed for $SrTiO_3$ and $KTaO_3$. Unfortunately, these oxides all have large band gaps (3.4 to 3.5 eV), which result in very low solar absorptivity; hence, they are ineffective in systems for solar energy conversion.

For purposes of discussion of Figure 3, the difference between the O_2/H_2O redox potential and the valence-band edge at the interface is defined as the intrinsic overpotential of the semiconductor anode, η_a. This overpotential is not the usual overpotential or overvoltage of conventional electrochemistry because it is current independent and is determined only by the band gap, the flat-band potential, and the redox potential of the electrolyte donor state.

A simple identity can be derived from the energy diagram of Figure 3, and it is given at the bottom of the figure. This identity can be considered to represent an energy balance describing the distribution of the energy produced by the absorption of a photon by the semiconductor. In the absence of a bias, the input energy of an absorbed photon is equal to the semiconductor band gap. Energy loss terms in the semiconductor resulting from the movement of electrons and holes from their point of creation to the respective electrolyte interfaces are the band bending (V_b), and the difference between the conduction-band edge in the bulk and the Fermi level (ΔE_F). The net electron-hole pair potential available at the electrode interfaces is thus E_g—V_b—ΔE_F. In the electrolyte, a portion of this potential energy is recovered as the free energy ($\Delta G/nF$) of the net endoergic reaction in the electrolyte. The rest of the potential energy is lost through the irreversible, entropy-producing terms η_a, η_e, and iR (ohmic heating).

For the case where both electrodes are n- and p-type semiconductors, the most interesting situation occurs when the two electrodes are made of different semiconduc-

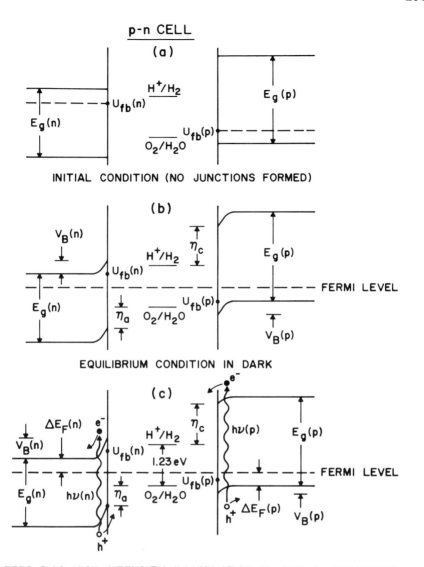

Figure 5 Sequence of energy level diagrams for p/n photoelectrolysis cells.

tors. If the electron affinity of the n-type electrode is greater than that of the p-type electrode, then the available electron-hole potential for driving chemical reactions in the electrolyte is enhanced when both electrodes are illuminated. The energetics of this system are shown in Figure 5.

As seen from the energy balance presented at the bottom of Figure 5, the net electron-hole pair potential available at the electrode surfaces is the sum of the semiconductor band gaps minus the difference in their flat-band potentials and minus the Fermi level terms, ΔE_F, for each semiconductor. The sum of the band bending values produced at each semiconductor is equal to the difference between their flat-band potentials. The amount of potential enhancement is maximized if the difference in flat-band potentials is minimized. The minimum difference in flat-band potentials is determined by the minimum band bending required in each semiconductor electrode to produce efficient charge-carrier separation. The total band bending present in the two electrodes is independent of light intensity. However, the distribution of the total band bending between the electrodes is dependent on light intensity and the carrier densities of each electrode.

In the cell, two photons must be absorbed (one in each electrode) to produce one net electron-hole pair for the cell reaction. This electron-hole pair consists of the minority hole and minority electron from the n type and p type electrodes, respectively, and it has a potential energy greater than that available from the absorption of one photon. The majority electrons and holes recombine at the ohmic contacts.

An important advantage of the double-electrode cell is that, for a given cell reaction, it may allow the use of smaller band gap semiconductors. Because the maximum photocurrent available from sunlight increases rapidly with decreasing band gap, higher conversion efficiencies could be produced. In principle, the maximum theoretical efficiency has been estimated to range from 30 to 40% (see section on "Photoelectrolysis of Water: Efficiency of Photoelectrolysis").

PHOTOCHEMICAL DIODES/PARTICLES

The enhanced electron-hole potential available from double-semiconductor electrode cells can eliminate the bias required for single-semiconductor electrode cells when V_{fb} is below the H^+/H_2 redox potential. This was shown for the first time[8,9] with an n-TiO$_2$/p-GaP cell; photoelectrolysis of water into H_2 and O_2 was achieved at zero external bias with simulated sunlight.[10]

Further studies[11-13] have been made with n-TiO$_2$/p-CdTe, n-SrTiO$_3$/p-CdTe, n-SrTiO$_3$/p-GaP, n-Fe$_2$O$_3$/p-Fe$_2$O$_3$, n-TiO$_2$/p-LuRhO$_3$, n-GaAs/p-InP, and n-GaAs/p-Si. The last two systems involved n-GaAs anodes that were reported[12,13] to be stabilized by an Mn$_2$O$_3$ coating; also, the conversion efficiencies were reported[12,13] to be quite high (8 to 11%). However, efforts[14] to repeat this work were unsuccessful, and it was found that the Mn$_2$O$_3$-coated n-GaAs electrode suffered severe photocorrosion. In addition, the band alignments for n-GaAs and p-InP are completely incorrect for the two-photon "upconversion" effect shown in Figure 5.

The elimination of bias requirements for photoelectrolysis cells through the use of double-electrode systems leads to a very interesting configuration variation. This configuration has been labeled "photochemical diodes",[10] and comprises photoelectrolysis cells that are collapsed into monolithic particles containing no external wires. In one simple form, a photochemical diode consists of a sandwich of either a semiconductor and a metal, or of an n-type and a p-type semiconductor, connected through ohmic contacts.

To generate the relevant redox reactions, photochemical diodes are simply immersed in an appropriate electrolyte, and the semiconductor faces are illuminated. An energy-level diagram for photochemical diodes consisting of heterotype p- and n-type electrodes is shown in Figure 6. A comparison of this energy-level scheme with that for biological photosynthesis reveals very interesting analogies (see Figure 7). Both systems require the absorption of two photons to produce one useful electron-hole pair. The potential energy of this electron-hole pair is enhanced so that chemical reactions requiring energies greater than that available from one photon can be driven. Furthermore, the n-type semiconductor is analogous to Photosystem II, the p-type semiconductor is analogous to Photosystem I, and the recombination of majority carriers at the ohmic contacts is analogous to the recombination of the excited electron from Excited Pigment II with the photogenerated hole in Photosystem I.

The size of a photochemical diode is arbitrary, provided that there is sufficient optical absorption, and a space-charge layer exists that is consistent with efficient charge-carrier

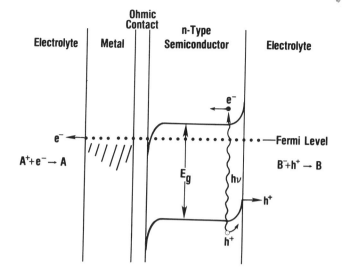

p-n TYPE PHOTOCHEMICAL DIODE

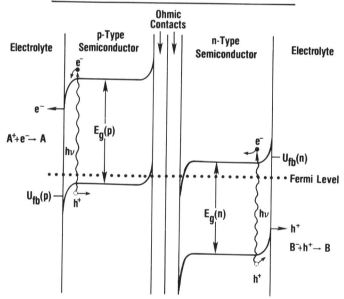

Figure 6 Energy level diagrams for (A) Schottky-type photochemical diode and (B) p/n-type photochemical diode.

separation. Within these constraints, the particle size may approach colloidal, or perhaps macromolecular dimensions, and the diodes can possibly operate as a colloidal dispersion or as a solute in solution, thereby greatly simplifying the design and operation of photoelectrochemical reactors using sunlight.

Kraeutler and Bard[15] conducted the first studies of powdered semiconductors that were platinized and used as photocatalysts to decompose carboxylic acids to alkanes, hydrogen, and carbon dioxide. These platinized semiconductor powders were also used to drive

204

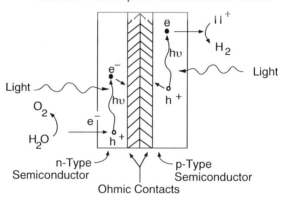

Electron Flow in Photosynthesis

Electron Flow in p-n Photochemical Diodes

Figure 7 Comparison of electron flow in biological photosynthesis (top) and electron flow in p/n photochemical diodes (bottom). In both cases, "up-conversion" of the photon energy is produced so that two photons add their photopotentials together to produce a net single electron-hole pair with a photopotential that is greater than that created by a single photon. This allows high-energy reactions, such as water splitting or CO_2 reduction, to be driven by low-energy, red photons.

other photoreactions, such as the oxidation of CN^- to OCN^-.[15] The platinization of semiconductor powders is a method for producing semiconductor-metal-type photochemical diodes having the energy-level scheme shown in Figure 6.

PHOTOELECTROLYSIS VS PHOTOVOLTAICS AND ELECTROLYSIS

Cascade-type or tandem multijunction semiconductor systems are also being actively pursued in photovoltaic (PV) research to produce high-efficiency solar-to-electrical power conversion.[16,17] Recent work at NREL[17] reports a 27.3%-efficient device based on a two-junction system comprising GaAs and $GaInP_2$. An important question is how photoelectrolysis compares with such high-efficiency PV cells that are coupled to dark electrolysis. Because the efficiency of dark electrolysis can easily be in the range of 80%, a coupled PV-electrolysis system could show efficiencies of 10 to 22%.

The coupled PV-electrolysis system could be either two separate devices electrically connected or an integrated monolithic device. We refer to the latter system as a PV cell operated in the photoelectrochemical (PEC) mode. The band diagram for such a mono-lithic PV/PEC cell is shown in Figure 8A, and is compared to that for a PEC photochemi-cal diode (PCD) in Figure 8B.

There are several important differences between these two systems: 1) the PV/PEC cell has twice as many semiconductor layers as the PEC/PCD; 2) for the PV/PEC cell the photoactive junctions are p-n junctions between two semiconductors, while in the PEC/PCD, the photoactive junctions are between semiconductors and aqueous solutions; 3) in the PV/PEC cell, the n-type region of the device covered with a metal layer becomes a cathode while the p-type region covered with a metal layer becomes an anode (i.e., it behaves like a majority-carrier device); in the PEC/PCD, the opposite is true — it is a minority-carrier device with the n-type region acting as an anode and the p-type region acting as a cathode; and 4) the PV/PEC system is a two-step process with a somewhat lower theoretical efficiency than the one-step PEC/PCD device. The PV/PEC cell must be covered on the illuminated side with a transparent conductor that forms an ohmic contact and is catalytic for the relevant gas evolution reaction; while this is an additional complication, the metallic coatings may also serve to stabilize the photoelectrodes against corrosion.

The GaInP$_2$/GaAs PV/PEC system that is currently being investigated at NREL[17] is shown in Figure 9. This cell can deliver 2.29 V, and it is expected that efficiencies for solar-powered water splitting greater than 20% will be achieved in the next few years.[17b]

EFFICIENCY OF PHOTOELECTROLYSIS

There is controversy about the methodology for reporting the efficiency of photoelectrolysis; this is especially true when an external bias is required. This problem has been addressed in detail by Parkinson.[18]

The most direct method is to subtract the required bias power from the output hydrogen power and divide by the input solar power. This leads to the following expression:

$$\eta = \frac{\left[(\Delta G) - V_{Bias}\right]i_{ph}}{P_{hv}A} \times 100 \tag{5}$$

where η is the conversion efficiency, ΔG is the free energy change of reaction (1.23 V for H$_2$O splitting), V_{Bias} is the bias voltage, P_{hv} is the solar irradiance, i_{ph} is the photocur-rent, and A is the irradiated area. This expression is consistent in that all energies are expressed as thermodynamic free energy, and only net output in the form of hydrogen free energy is considered. Other methods include calculating the power saved compared to a conventional dark electrolysis cell.[18]

The maximum efficiencies permitted by thermodynamics for photoelectrolysis of water have been analyzed by many authors.[19-21] For a single band gap absorber system, the upper limit is 31%; while for a two band gap system, it can be 42%. As the number of band gaps in a tandem, multiphoton device increases, the efficiency increases; the ultimate limit is 67% with an infinite number of tandem layers. The dependence of the theoretical efficiency on the number of absorber layers is shown in Figure 10. The curve quickly approaches the limiting efficiency after four or five layers. Thus, for practical reasons, only systems with two or perhaps three layers are being investigated; the three-layer system would have a theoretical efficiency of 52%.

Another way to reach the high efficiency of tandem, multi-band gap systems is through hot-carrier injection.[21,22] Here, the photogenerated electrons and holes are in-jected into the solution from a single band gap absorber before they equilibrate with the lattice and give up their excess kinetic energy as heat. This requires fast electron transfer

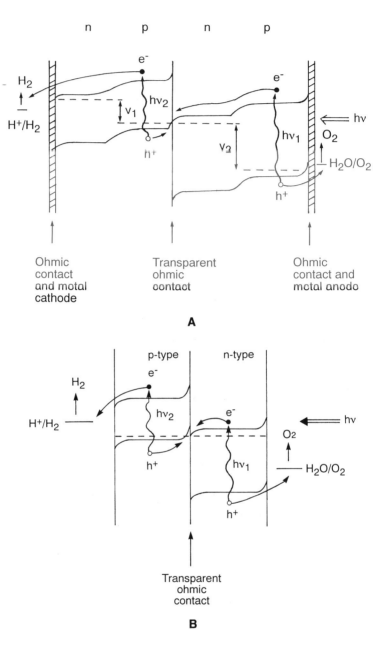

Figure 8 (A): Energy level diagram for a monolithic tandem multiphoton, multijunction, photovoltaic device for the photolytic splitting of water. (B): Energy level diagram for a monolithic, multilayered photochemical diode for the photoelectrolysis of water. Note that in the PV device, twice as many semiconductor layers are needed as for the photochemical diode structure.

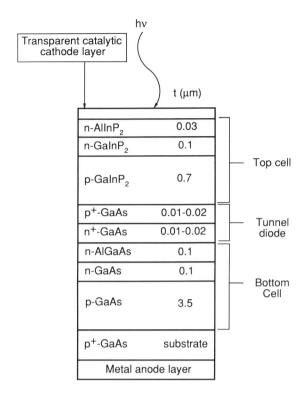

Figure 9 Configuration of the high-efficiency GaInP$_2$/GaAs tandem PV structure for the photolytic splitting of water.

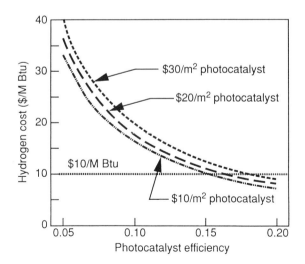

Figure 10 Hydrogen cost in dollars per million BTU as a function of photocatalyst efficiency and photocatalyst cost.

208

across the semiconductor-liquid interface and a system that slows down the rate of hot-carrier relaxation. The use of quantum-well electrodes has been reported[23,24] to accomplish such a decrease in hot-electron cooling rates. The ultimate theoretical conversion efficiency for a perfect hot-carrier photoconverter is the same (67%) as that for an infinite number of semiconductor layers.[21]

ECONOMICS OF PHOTOELECTROLYSIS

There have been several studies of the estimated costs of hydrogen produced by the photoelectrolysis of water.[25,26] The two essential factors entering into these calculations are the conversion efficiency of the photoconverter and the cost per unit area of the photoelectrolysis cell. The former determines the required area per unit of hydrogen produced, and the latter determines the capital investment required per unit of hydrogen produced. A target cost of $10 per million BTU (MBTU) (constant 1988 dollars) has been set by the U.S. DOE/NREL hydrogen program for the year 2020. Achievement of this cost goal would make hydrogen competitive with the current primary method of production, which is the steam reforming of natural gas.

Cost calculations of photoelectrolysis systems show that to attain the $10/MBTU goal, it will be necessary to achieve a photoconverter efficiency of 15% at a photocatalyst cost of $10/m² (see Figure 10); the total system cost in this calculation is limited to $125/m². The hydrogen cost is very sensitive to conversion efficiency, but not very sensitive to photocatalyst cost. Thus, at 15% conversion efficiency, the hydrogen cost increases from $10/MBTU to only $12.50/MBTU as the photocatalyst cost triples from $10/m² to $30/m². On the other hand, decreasing the photocatalyst efficiency from 15 to 10% increases the hydrogen cost from $10/MBTU to $15/MBTU — essentially, the hydrogen cost is inversely proportional to conversion efficiency.

Currently, photoconverter efficiencies of about 10% are possible, with 20% efficiency predicted to be achieved in the near term. However, the cost per unit area is orders of magnitude above the $10-$30/m² level. Additional research is required to achieve the required high efficiency and low photocatalyst cost, together with long photocatalyst lifetime (10 to 15 years).

REFERENCES

1. **Gerischer, H.,** in *Advances in Electrochemistry and Electrochemical Engineering*, Delahay, P., Ed., Interscience, New York, 1, 139–232, 1961.
2. **Gerischer, H.,** in *Physical Chemistry: An Advanced Treatise*, Eyring, H., Henderson, D., and Jost, W., Eds., Academic, New York, 9A, 463–542, 1970.
3. **Gerischer, H.,** *J. Electroanal. Chem. Interfacial Electrochem.*, 58, 263, 1975.
4. **Lohmann, F.,** *Z. Naturforsch.*, 22, 843, 1967.
5. **Dutoit, E. C., Van Meirhaeghe, R. L., Cardon, F., and Gomes, W. P.,** *Ber. Bunsenges. Phys. Chem.*, 79, 1206, 1975.
6. **DeGryse, R., Gomes, W. P., Cardon, F., and Vennik, J.,** *J. Electrochem. Soc.*, 122, 711, 1975.
7. **Madov, M. J., Cardon, F., and Gomes, W. P.,** *J. Electrochem. Soc.*, 124, 1623, 1977.
8. **Nozik, A. J.,** *Appl. Phys. Lett.*, 29, 150, 1976.
9. **Yoneyama, H., Sakamoto, H., and Tamura, H.,** *Electrochim. Acta*, 20, 341, 1975.
10. **Nozik, A. J.,** *Appl. Phys. Lett.*, 30, 567, 1977.
11. **Ohashi, K., McDann, J., and Bockris, J.,** *Nature (London)*, 266, 210, 1977.
12. **Kainthla, R. C., Zelenay, B., and Bockris, J. O'M.,** *J. Electrochem. Soc.*, 134, 841, 1987.

13. **Bockris, J. O'M., Martin, A., Sankar, P. S., and Lin, G. H.,** *in Proc. DOE/NREL Hydrogen Program Reviews*, Washington, D.C., 1991.
14. **Peterson, M. W. and Nozik, A. J.,** *in Proc. DOE/NREL Hydrogen Program Review*, Golden, CO, 1990.
15. **Kraeutler, B. and Bard, A. J.,** *J. Am. Chem. Soc.*, 100, 2239, 1978.
16. **Fan, J. C. C., Tsaur, B. Y., and Palm, B. J.,** in *Proc. 16th IEEE Photovoltaics Specialists Conf.*, IEEE, New York, 1982, p. 692.
17a. **Olson, J. M. and Kurtz, S. R.,** in *FY89 Annual Report, Solid State Photovoltaic Research Branch, NREL*, Sept. 1990, pp. 204–240.
17b. **Montgomery, D., Kocha, S., Turner, J., and Nozik, A. J.,** in *Proc. DOE/NREL Hydrogen Program Review*, Washington, D.C., 1991.
18. **Parkinson, B.,** *Accts. Chem. Res.*, 17, 431, 1984.
19. **Archer, M. D. and Bolton, J. R.,** *J. Phys. Chem.*, 94, 8028, 1990.
20. **Bolton, J. R., Strickler, S. J., and Connolly, J. S.,** *Nature (London)*, 316, 495, 1985.
21. **Ross, R. T. and Nozik, A. J.,** *J. Appl. Phys.*, 53, 3813, 1982.
22. **Cooper, G., Turner, J. A., Parkinson, B. A., and Nozik, A. J.,** *J. Appl. Phys.*, 54, 6463, 1983.
23. **Eddstein, D., Tang, C. L., and Nozik, A. J.,** *Appl. Phys. Lett.*, 51, 48, 1987.
24. **Parsons, C. A., Thacker, B. R., Szmyd, D. M., Peterson, M. W., and Nozik, A. J.,** *J. Chem. Phys.*, 94, 3381, 1990.
25. Solar Photochemical Production of Hydrogen, Report No. TDA-90-SERI002, TDA Research Inc., Wheat Ridge, CO 80033, 1990.
26. **Hoagland, W.,** Renewable Hydrogen Program, NREL Report for U.S. DOE, 1990.

Chapter 13

Renewable Energy, Today and Tomorrow— An Overview

Thomas D. Bath

Renewable energy technology (RET) development has continued apace during the past decade despite federal government funding constraints. Impressive gains have been made in the conversion efficiencies of all forms of wind and solar energy, and significant cost reductions have occurred on all fronts. In this chapter, we review the progress of the past decade and summarize the renewable energy penetration expected to occur over the next four decades. These projections are the result of studies done by the U.S. Department of Energy (DOE) national laboratories as part of the development of a National Energy Strategy (NES).

In September 1991, President Bush elevated the Solar Energy Research Institute (SERI) to a DOE national laboratory and designated it the National Renewable Energy Laboratory (NREL) in recognition of the importance of RET to the nation and the outstanding contributions of its staff to the advancement of RETs over the past 14 years. In addition, the creation of a national laboratory devoted specifically to renewable energy signifies the critical role of renewables in the nation's future energy supply mix.

The contribution of renewable energy to U.S. energy needs over the next four decades will be driven by the economic and societal attractiveness of technologies to convert renewable resources to useful energy forms and by our success in reducing institutional barriers to the adoption of attractive but novel technology systems. This chapter addresses the key factors appropriate to determining the relative attractiveness of these systems— resource characteristics and intensity, conversion technology performance and reliability, environmental societal values, and compatibility with demand—and discusses their significance regarding future energy contributions from renewable resources. The institutional factors that bear on broad acceptance of these systems are reviewed and incorporated with system and market factors to provide a basis for assessing future contributions by these technologies to U.S. energy needs. The results of this assessment suggest that as much as 28% of U.S. energy needs can be provided by these resources in 2030. A key factor in attaining this result is progress in technology performance that is expected to occur in the coming decade.

Renewable energy will make increasing contributions to U.S. energy supply. However, it will be necessary to continue progress on the research, development, and deployment (R,D&D) front to assure that these technologies achieve their potential as soon as possible. Greater reliance on RETs can provide a secure domestic supply of energy that can be used with minimal environmental impact. Continued reductions in cost can ensure that the rapid introduction of RETs is accomplished without economic penalty.

The contribution of renewable energy to the U.S. energy supply has been growing since 1970 and will continue to grow. In 1970, essentially all renewable energy came from hydroelectric power (used to generate electricity) and wood (used for space heating, industrial processes, and power generation), and represented approximately 6% of the energy supply. In 1990, renewable energy supplied 6.8 quadrillion (10^{15}) Btus (quads) or approximately 8% of the nation's energy needs (see Figure 1). In contrast to the situation in 1970, in 1990 there were substantial contributions from geothermal power plants, municipal waste energy plants, wind plants, PV systems, solar thermal devices, and alcohol fuels.

National Energy Supply (1988)
Renewables Provide 7 of 82 Quads

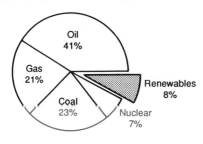

U.S. Renewable Energy Supply (1988)
(7 Quads)

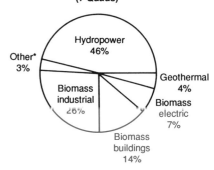

*Other: wind, alcohol fuels, solar thermal, and PV.

Figure 1 National energy supplies and the renewable contribution.

Part of the move to use renewable energy is due to the fact that these technologies are generally considered environmentally benign. Most of the technologies use sunlight, rainwater, or wind as the energy resources and, therefore, do not produce emissions or waste materials (including CO_2 or greenhouse gases) that would have a negative impact on the environment. Many of the most promising RETs, however, are still in the early stages of development. In order to reduce the environmental impact of the nation's energy use, additional technological advances and cost reductions must be achieved. Great progress in both cost and reliability of a number of the emerging RETs was made during the past decade, despite constraints in the federal renewable energy R,D&D budgets. In addition, development efforts during the 1980s provided a much better understanding of the remaining technical hurdles which must, and we believe can, be surmounted to bring the renewable energy systems to market. A number of innovative means of efficient conversion of resources into usable energy have been identified and are being explored, including thin film PV cells, high-performance wind turbines, fast pyrolysis of biomass, anaerobic digestion of biomass, simultaneous saccharification and fermentation of cellulosic materials, and genetically engineered biofuels production.

For each demand sector—electric power, process and building energy, and transportation fuels—there are several RETs being developed. Thus, if any one RET fails to meet the technological and economic goals needed for its demand sector, at least one other RET will be available for that sector. Some of the technologies are already well established; others require more R,D&D effort before they will become proven commercial commodities.

Wind energy has expanded its role in electricity generation in the past 2 decades, particularly in California. Windmills have long been used for pumping water, grinding grain, and other tasks, and more recently for generating power for such tasks as powering radios at remote locations. In the early 1980s, wind equipment capital costs were $2200/kW, compared to less than $1000/kW today. O&M costs have dropped from $0.04/kWh to $0.01/kWh over the same period. Wind turbine availability is now greater than 95% for the roughly 1500 MW installed in the U.S. vs 50 to 60% availability factors in 1980. The cost of wind-generated electricity is $0.07 to $0.09 today compared with $0.30/kWh in 1980. The federal government investment of more than $240 million has been matched with more than a $3 billion investment by private industry. The intensive R,D&D effort of recent years has brought wind-generated power to the commercial stage, particularly for larger remote stand-alone applications and for electric utility supply in locations where wind resources are particularly good. Researchers have identified improvements in blade airfoil design which will improve efficiency, allowing greater output of power from better wind sites and also making lower-intensity wind resources economical to develop. In addition, improvements in gear and bearing mechanisms, blade materials, and support structures are expected to reduce costs and increase reliability. As a consequence, the cost of wind-generated electricity is expected to drop from today's level to about $0.03/kWh in 2030, assuming an acceleration of R,D&D efforts. At such a cost level, wind power would cost sufficiently less than oil- or gas-generated electricity so that utility systems might be designed to use substantial wind-generating capacity and to shut in fossil fuel capacity before wind when loads dictated. On a smaller scale, wind generator systems are currently economic and are being sold, both in the U.S. and overseas, for use in stand-alone applications, usually with either battery or diesel generator back-up capacity. As the technology improves, in part the result of R,D&D on utility-scale systems, the costs are likely to drop and the systems to become more competitive.

PV systems are becoming a more significant source of power. Smaller-scale systems are being widely used for powering remote facilities, such as communications and lighting, and for providing power in developing countries. The costs have dropped to the point that utilities (or their customers) can better afford to install PV systems with battery back-up where the alternative would be a long extension (1 mile or more) from an existing distribution line. Meanwhile, costs of utility-scale systems have also dropped from $1/kWh in 1980 to $0.25–$0.30/kWh today and are expected to reach the $0.04–$0.05 kWh range by 2030, particularly if R,D&D is accelerated. Research is being focused on improving the efficiency of the PV cells, reducing the costs of the cells and extending their life, and in reducing the cost of the supporting systems. Multijunction cells provide a particularly attractive option to increase module efficiency to the 13 to 17% range. A number of alternative technologies are being investigated; success in meeting program cost objectives is not dependent on success in any one pathway.

PV systems, in a number of potential forms, might be used to provide power generation at the outskirts of utility systems, thereby reducing the transmission and distribution investments otherwise required. PV modules integrated in roof tiles, for example, might provide the bulk of a household's power, with the utility's distribution system as back-up.

Solar thermal energy is used for both power generation and provision of industrial process heat, often in a co-generation mode. California is the site of several large installations of parabolic trough solar collector systems that are used to produce electric power. High-intensity systems are also being devised that would focus reflected sunlight on a collecting fluid, such as liquid salt, thereby providing storage of heat as well as an efficient transfer system to a central generating unit. A by-product of this work is a solar incinerator that can effectively destroy some toxic materials, such as dioxin precursors. On a smaller scale, Cummins and others are developing a 4- to 20-kW stand-alone generating system that uses a parabolic-dish mirror (made of a thin-film plastic) to focus sunlight on tubes filled with a working fluid, which drives a Stirling engine. The utility-scale systems are expected to reach a cost of $0.04/kWh by 2030 (down from $0.10/kWh today), including provisional energy storage to permit base-load operation. The smaller stand-alone units may be competitive for remote applications by the turn of the century.

Solar water heating, one of the first applications of solar technologies to be commercialized, depended heavily upon tax benefits. Unfortunately, many of the firms initially involved failed to design and/or install the systems properly. The resulting problems that consumers encountered left a negative legacy. At current energy prices, the market for systems to heat household water or swimming pools is small. However, if energy prices rise as forecast, roof-top water heaters will become increasingly common after 2000.

Solar building energy applications involve a wide range of technologies. In some cases, such as the use of electrically activated window coatings, which selectively admit sunlight, the technologies could be considered conservation investments. The efficient use of sunlight for heating, interior lighting, insulation materials and systems, and desiccant cooling systems are some of the technologies being pursued.

Biomass and biofuels represent perhaps the largest potential commercial energy source among the RETs. Biomass is used in four major ways: direct combustion, electric power generation, conversion to gas for use as a fuel or chemical feedstock, and conversion to liquid fuels. Biomass resources include waste and by-product materials, such as municipal solid waste (MSW) and forest product industry wastes, and materials produced purposely for energy use, such as trees.

Direct combustion of biomass, such as wood, was the first application of renewable energy, many millennia ago. The more convenient form of fossil fuels led to their displacing wood as a fuel in most circumstances. Today, new biomass stoves and heaters have improved efficiency. Advances are being made in materials and waste-product handling to improve costs. About 3 quads of energy are being provided in the U.S. today by wood and combustible materials in the combustion of wood heat power. Most of this is from the use of waste materials in the lumber and pulp and paper industries. MSW combustion provides a small amount of process heat also. Expanded use of wood for household purposes is expected, particularly if fossil fuel prices increase significantly as projected. At present, availability of low-priced wood is a key constraint.

If costs of producing biomass for energy use can be reduced to $2/million Btu by 2010 as targeted, dedicated power plants using wood may become more commonplace. Although coal-fired power plants may be more economic in most circumstances (absent major constraints on use of fossil fuels) in the case of smaller plants, wood may become an attractive choice. High transportation costs of smaller quantities of coal and the high cost of smaller-scale SO_2 removal provide advantages for use of wood. (Other uses of wood will compete for the material, however.) MSW-fired power plants are also becoming more common, although the necessity of waste disposal (and unavailability of acceptable alternatives) is the key driving force, not the economics *per se*.

Gasification of biomass may provide access to significant markets. The targeted cost of $3.50/million Btu for production of methane from biomass would make the process a competitive source of methane for the nation's natural gas systems by 2000. Once the cost of dedicated production of biomass is brought down to $2/million Btu, the potential for gasification of biomass will be very large. (Conversion of dedicated biomass production to liquid fuels, however, may be a higher national priority.) Similarly, the conversion of a large portion of MSW and sewage sludge to methane via anaerobic digestion may provide an attractive alternative means of disposing of such wastes.

Liquefaction of biomass for use as transportation fuel may be the largest single potential use of renewable energy. Significant progress is being realized in research in several areas. Today, the production of liquid fuels from biomass is primarily via fermentation of corn constituents to produce ethanol, entailing the use of high-cost raw materials. The advanced technologies will utilize lower-cost materials—wood or other cellulosic materials, oil seeds, or microalgae—as feedstocks. Methanol is projected to be producible at a cost of $0.55/gal from biomass by 2000 ($10/million Btu), assuming the feedstock cost goals are achieved or waste materials are available. Ethanol is projected to be available at a cost of $0.60/gal (or $7/million Btu) shortly after 2000 (the timing largely dependent on whether R,D&D is accelerated), using an enzymatic hydrolysis process and an improved fermentation system. Increased yields, reduced residence times, and higher viable concentrations of ethanol in the fermentation reactors are among the key R,D&D targets. Fast pyrolysis of biomass to produce gasoline via a syncrude is expected to be commercialized by 2020, producing gasoline at a cost of $0.90/gal ($7.00/million Btu). Accelerated R,D&D would move the date up to between 2005 and 2010. R,D&D on producing diesel fuels from plant seeds or from microalgae is also progressing, offering alternative sources of renewable liquid fuels; these technologies have similar cost objectives, but it may take a decade longer to reach them.

Biomass feedstock may be the most important area where acceleration of R,D&D is needed. The achievement of the cost goals for liquid fuels is highly dependent on the availability of feedstock at a cost of $2/million Btu. If a substantial portion of the nation's liquid fuels is to be replaced with biofuels, the feedstock has to be available over wide areas at the low costs indicated. More than in the case of the other RETs, widespread, large-scale demonstrations must be made to farming and forestry interests to convince them that land should be withdrawn from other crops and animal production and dedicated instead to energy crops.

Further in the future lie the prospects for hydrogen production and ocean thermal energy conversion (OTEC) technologies. Production of hydrogen via electrolysis of water is well understood. Cheap electricity is the key. Production of hydrogen using single-step photoconversion devices is being investigated. So is production of hydrogen through genetically engineered plant species. More remains to be done in developing the technologies for handling and storing hydrogen. Fuel cells may prove to be the best mechanism for using hydrogen as a transport fuel (such cells could also use alcohols or other biofuels, possibly more efficiently overall than the internal combustion engine). Hybrid fuel cell/electric battery vehicles may evolve by the middle of the next century. If the efficiency of OTEC can be improved adequately, it may provide a number of opportunities, such as power and water production for island economies, hydrogen or methanol production for shipment to consumers, etc.

The potential contribution of renewable energy systems is very large, as noted earlier. However, the dispersed nature of the basic solar resources results in competition with other activities for the use of land where the resource is available. Hence, the pressures for efficient conversion of solar energy into a usable form are intensified. Land suitable for growing biofuel crops is usually also valuable for food or forest product production. The scenic areas where wind, hydroelectric, and geothermal resources are abundant may call for difficult choices. The capture of sunlight for power or heat production takes space; many such installations have been mounted on rooftops, allowing multiple use of land. Integrated roofing and power recovery systems are likely to be developed. Efficiency of conversion is therefore an environmental issue as well an economic one.

The development of domestic renewable energy industries would also provide other contributions in the areas of energy security, economic development, foreign trade, technological spin-offs, etc. The largest impact, and perhaps more important economically than the total energy supplied by renewable energy systems, may be the restraining impact of renewables on the prices of oil and other energy.

A recent study, *The Potential of Renewable Energy—An Interlaboratory White Paper* (SERI, 1990), provides an evaluation of the present and projected performance status of RETs and their potential contributions to the nation's energy requirements over the next 4 decades. It was prepared as background for the NES by the U.S. DOE national laboratories in response to a request from the DOE Office of Policy, Planning and Analysis. Based on a consensus-forming approach to the assessment of many difficult questions, the study attempts to convey a sense of what *is* and *is not* known about the range of technical and analytical issues surrounding future RET deployment. Although it is apparent that uncertainties are large, it was felt that an expert consensus with visible uncertainty was preferable to an *apparently* precise analysis based on nonvalidated modeling constructs. The study identifies broadly the R,D&D thrusts that, if undertaken, would remove the key technological constraints on the utilization of these energy resources and thereby enhance the contribution renewable energy can make. The study also identifies a number of other constraints, largely institutional, that can be reduced or removed with appropriate action. It does not address the question of what is the appro-

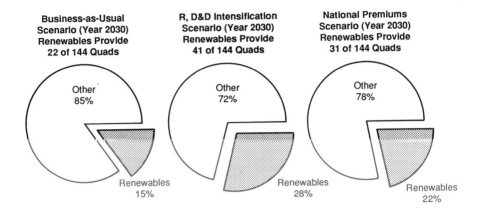

Figure 2 The contribution of renewable energy in 2030, based on the Business-As-Usual Scenario, the R,D&D Intensification Scenario, and the National Premiums Scenario.

priate national energy policy, nor does it suggest what policy actions might be undertaken to address the needs discussed here.

In collaborating to produce the study, the participants saw themselves as representatives of a larger community of scientists and engineers who have worked to improve RETs for more than a decade. Thus, although the study focuses on the impacts of federal policy, it does so in the context of a sustained federal/industry partnership. Ultimate success in this area depends on use of RETs in the private sector. The best way to ensure that the developed technology is used is to involve users in its development.

The market penetration potential of RETs was assessed by reviewing expected system performance against demand and market competition constructs, which are described in the study. This effort, which considered regionalized energy demand and price projections, and took into account competition among the renewable options, was designed to produce mid-range penetration estimates. The 40-year time horizon for these estimates introduces very large uncertainties, both with regard to the evolution of the performance and costs associated with technologies that provide energy from renewable resources and with regard to the evolution of the markets in which these technologies compete. However, all energy technologies—and the NES process itself—must cope with substantial uncertainty that increases over time. For simplicity, RET performance and penetration projections were approached from a mid-range point of view over three scenarios of RET deployment described below. The projections based on those three scenarios are presented in Figure 2. It would have been preferable to apply upper and lower bounds and use a range of variables to derive a band of outcomes, but it was not feasible to conduct the study at this level of sophistication because of time and resource limitations.

The study projects that the contribution of renewable energy to total energy requirements in competition with other energy supply systems should grow over the next four decades, assuming that research and development expenditures continue at present levels, which reflect major reductions from those at the peak of the "oil crisis". In the Business-as-Usual (BAU) Scenario, it is projected that by the year 2030, renewable energy will provide a U.S. domestic contribution of 22 quads, or 15% of projected total energy needs. This assumes that federal funding continues at current levels and energy/environmental policy is unchanged. In this scenario, industry eventually provides much of the development impetus as the technologies become more clearly economical. If federal R,D&D

Table 1 **Business as usual**

U.S. total (quads)	1988	2000	2010	2020	2030
Hydropower	3.1	3.4	3.4	3.5	3.5
Geothermal electric	0.2	0.3	0.5	0.7	0.9
Solar thermal electric	<0.1	<0.1	0.3	0.9	2.0
Photovoltaics	<0.1	<0.1	0.2	0.7	1.8
Wind power	<0.1	0.2	1.0	1.9	2.9
OTEC	0.0	0.0	<0.1	<0.1	<0.1
Biomass-electric	0.5	1.0	1.6	1.9	2.2
Biomass-buildings	1.0	1.0	1.5	1.8	2.2
Biomass-industrial	1.8	2.2	2.9	3.3	3.8
Biomass-liquid fuels	<0.1	0.2	0.4	0.8	2.2
Biomass (total)	**3.3**	**4.4**	**6.3**	**7.8**	**10.3**
Solar heat-buildings	<0.1	0.2	0.3	0.4	0.5
Solar heat-industrial	<0.1	<0.1	<0.1	<0.1	0.2
Geothermal-heat	<0.1	<0.1	0.1	0.2	0.2
Total RET	**6.7**	**8.6**	**12.1**	**16.1**	**22.3**
% of total energy demand	**8**	**9**	**11**	**13**	**15**

expenditures are expanded by a factor of 2 to 3, or some $3 billion over the next two decades, the projected contribution of renewable energy is nearly double that in the BAU Scenario: 41 quads by 2030. Such levels of expenditure have been shown by past efforts to be sustainable and effective and would pursue well-defined objectives along identified R&D pathways. Much of the gain projected in the R,D&D Intensification Scenario is due to the acceleration of the technological improvements that private industry would otherwise pursue independently much later in time as economics become more favorable and risks smaller. This estimate of a substantially greater contribution suggests that large economic and security benefits will derive from this investment because of the existence of a greater range of economical energy alternatives. Additional benefits would accrue because of the generally environmentally attractive character of the RETs.

In addition to the "technology push" implied by the R,D&D Intensification Scenario, the study explored the implications of a "market pull" approach, called the National Premiums Scenario. This scenario assumed that a substantial market incentive would be applied in order to capture environmental and other national values relevant to renewables. The scenario also assumed that this market incentive would begin to be applied in 1990 to all new RET deployment decisions and continue at that level through 2030. Thus, the market incentive largely affects the deployment of technologies that are at or near market competitiveness. The results of this analysis suggested a 9-quad increase (over BAU) in the U.S. renewable energy contribution by 2030. The relatively small size of this increase is largely due to the inability of the premium to substitute for the relatively large increases in RET performance necessary for several of the technologies to enter the competitive range in projected markets. The detailed results of the three scenarios are provided in Tables 1 to 3.

Both the R,D&D Intensification Scenario and the National Premiums Scenario are artificial constructs, structured for analytic simplicity rather than to detail the realities of the market. Any effort designed to explore "technology push" or "market pull" mechanisms to increase the deployment of renewable supplies would consider cost effectiveness

Table 2   **R,D&D intensification**

U.S. total (quads)	1988	2000	2010	2020	2030
Hydropower	3.1	3.4	4.0	4.7	5.1
Geothermal electric	0.2	0.4	0.8	2.0	3.7
Solar thermal electric	<0.1	0.2	1.0	2.0	2.6
Photovoltaics	<0.1	<0.1	0.7	1.7	2.5
Wind power	<0.1	0.4	2.2	4.2	5.3
OTEC	0.0	0.0	<0.1	<0.1	<0.1
Biomass-electric	0.5	1.1	1.9	2.4	2.9
Biomass-buildings	1.0	1.2	2.0	2.6	3.2
Biomass-industrial	1.8	2.3	3.3	3.9	4.6
Biomass-liquid fuels	<0.1	0.3	2.4	4.4	8.4
Biomass (total)	3.3	4.9	9.6	13.3	19.1
Solar heat-buildings	<0.1	0.2	0.5	0.7	0.9
Solar heat-industrial	<0.1	<0.1	0.1	0.2	0.3
Geothermal-heat	<0.1	<0.1	0.3	0.7	1.6
Total RET	**6.7**	**9.7**	**19.2**	**29.6**	**41.1**
% of total energy demand	**8**	**10**	**18**	**24**	**28**

Table 3   **National premiums**

U.S. total (quads)	1988	2000	2010	2020	2030
Hydropower	3.1	3.5	4.2	4.9	5.5
Geothermal electric	0.2	0.4	0.6	0.8	1.2
Solar thermal electric	<0.1	0.2	0.9	1.7	2.9
Photovoltaics	<0.1	<0.1	0.4	1.4	2.7
Wind power	<0.1	0.4	1.7	2.8	4.1
OTEC	0.0	0.0	<0.1	<0.1	<0.1
Biomass-electric	0.5	1.5	2.1	2.4	2.7
Biomass-buildings	1.0	1.0	1.5	1.8	2.2
Biomass-industrial	1.8	2.7	3.4	3.8	4.3
Biomass-liquid fuels	<0.1	0.3	1.0	1.5	2.8
Biomass (total)	**3.3**	**5.5**	**7.9**	**9.5**	**11.9**
Solar heat-buildings	<0.1	0.3	0.9	1.2	1.5
Solar heat-industrial	<0.1	<0.1	<0.1	0.1	0.2
Geothermal-heat	<0.1	0.2	0.4	0.9	1.5
Total RET	**6.7**	**10.5**	**16.9**	**23.4**	**31.6**
% of total energy demand	**8**	**11**	**15**	**19**	**22**

in choosing policies, probably applying a mix of policy mechanisms over time to achieve desired goals. Therefore, it is likely that neither scenario represents an upper limit on the potential contribution of renewables.

For purposes of comparison, the results of the study under discussion (the *White Paper*) can be contrasted with the results of two recent reports [*The National Energy Strategy, Powerful Ideas for America* (DOE, 1991), and *America's Energy Choices, Investing in a Strong Economy and a Clean Environment* (Union of Concerned Scientists,

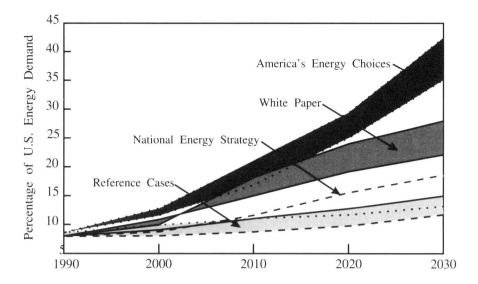

Figure 3 A comparison of projections of the future U.S. renewable energy contribution.

1991); henceforward identified as *NES* and *Choices*, respectively]. Owing to different assumptions about energy demand and economic growth, this comparison is best drawn by plotting the U.S. renewable contribution (in percent) over the 40-year planning horizon common to all three studies. The results of this comparison are shown in Figure 3. Estimates from the two *White Paper* policy scenarios are banded by those of the *NES* and *Choices* studies. The estimates of renewables in the reference cases from all three studies are within the limits of uncertainty for such estimates.

The contributions from renewable energy presented in the *White Paper* reflect a mid-range assessment of economic accomplishments, given the inherent uncertainties of the R,D&D process, as reflected in the comparison just shown. The study was unable to address many other uncertainties, such as those surrounding the demand and price projections used in the study. Neither did it assess how conservation and other energy supply systems would respond to the impact of the expanded deployment of renewable energy projected in the R,D&D Intensification and National Premiums cases. The study also assumed that no other significant environmental or energy policy changes are implemented that will have impacts on renewable or other energy systems. Most importantly, the study was unable to assess the impact of expanded renewable energy deployment on the projections of market prices of other energy forms. If the markets available for competitive energy are smaller than those reflected in our base-case assumptions, market forces are likely to result in lower prices for oil, gas, and coal.

The projected potential contribution of renewable energy rests on a sound base. As shown in Figure 4, the estimated available U.S. domestic renewable resources are huge compared with the remaining available fossil fuel resources projected by the U.S. Geological Survey (USGS). Although today's proven reserves of renewable energy are small relative to proven fossil energy reserves, the relative size of the accessible resource shows the large potential benefit of successful RET development. This estimate of the accessible resource considers potential limitations of access to renewable resources and competitive uses for land and other resources. In many cases, more than one RET is expected to be competitive in the electric power generation and liquid fuels markets. Hence, the overall projection of the contribution of RETs is neither a maximum potential, nor does it require

220

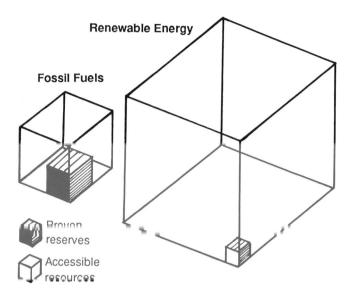

Renewable Energy

Fossil Fuels

Proven reserves

Accessible resources

Figure 4 Available U.S. energy resources—fossil fuels and renewable energy.

the success of all technological pathways. Finally, it is anticipated that renewable energy will capture additional energy market share *after* 2030 as technological progress continues and institutional constraints are overcome.

The R,D&D efforts needed to accomplish the projected contribution of RETs are relatively well known. The efforts of the past two decades have permitted key technological constraints and most promising opportunities to be identified, thereby allowing future work to be focused efficiently in the areas that will be most productive in making RETs competitive with conventional supplies. Expanded efforts would also be targeted to accelerate the deployment of renewable energy, with the government effectively underwriting the risks that the private sector would be unable to justify at this stage of the industry's development. Finally, increased domestic R,D&D should also allow the U.S./Canadian manufacturing sector to capture a larger share of renewable energy equipment markets in North America and overseas.

The study concludes that there are advantages, other than supply diversity, to expanding use of renewable energy. These include environmental, security, and indirect economic benefits.

In recent years, many industrialized countries have been increasing their R,D&D contributions in support of RETs. They are emphasizing the technologies that appear to have the greatest national potential. For example, Denmark has promoted wind energy, Germany and Japan are investing in PV systems, and China is pursuing low-head hydropower. If the U.S. does not take similar positive actions, foreign manufacturers may gain the advantage in international and even local markets when the technologies reach commercial maturity.

As noted earlier, technological constraints are not the only ones that have impacts on the deployment of RETs. A number of institutional constraints also act to slow the expansion. Some of these are a matter of timing; some of education or technology transfer; some are products of the technological and economic structure of energy-using systems (e.g., autos, buildings); some emanate from financial and tax systems; and others are products of regulatory structure. Many of these institutional constraints can be eliminated, or their impacts reduced, thereby further accelerating the implementation of RETs. The following problem areas, then, present opportunities for appropriate remedies:

- *Integration into existing energy systems.* Renewable technologies face a number of problems associated with integration into electric power and fuel supply systems. Remote electric power sources such as geothermal systems, for example, may not have access to transmission facilities because of location, competitive demands on available facilities, ownership issues, etc., even if there is a market demand for the power. Power may not be provided with the desired degree of dispatchability and storage, or other load-leveling capacities may not be available. Alcohol fuels may require special fleets of vehicles or special distribution facilities; the demand will not exist until the user facilities are in place. The potential fuel supplier may be unable to remove such obstacles without assistance.

- *Access to resources.* The dispersed character of renewable energy resources introduces a number of institutional impacts. Real or perceived competition for land use with agricultural, recreational, scenic, or other uses frequently evokes strong reactions from government agencies or other groups. Often, complex trade-offs are involved. In the absence of a comprehensive, detailed assessment of the environmental impacts of renewable energy systems, we have developed our projections based on the generally accepted position that renewable energy is less environmentally intrusive than other options.

- *Perceived risks.* Since most of these new technologies are being developed with heavy government involvement, potential users may understand less about their nature and background than they do about the conventional energy technologies and their recent modifications. Demonstration projects and other evidence of reliability and operability acceptable to industry may be needed to a greater extent for renewables than for evolving conventional technologies.

- *Treatment of risk factors in comparing technologies.* Low initial capital costs can make almost any energy system *appear* attractive in comparison to high-capital-cost technologies. However, over time, price fluctuations in fuel markets, escalations in plant decommissioning costs, or other uncertain future events can put a greater burden on future costs than the one originally estimated. For example, in the electric utility industry, several U.S. utilities installed base-load, oil-fired facilities before the oil supply difficulties of the 1970s. Thus, the decision to use oil for fuel exposed utilities and customers to higher prices and availability problems that were not quantified initially. RETs, having low or no fuel costs, offer protection from such fuel price fluctuations, but may be given little economic credit in decision making for this important risk-reducing quality. There are pressures on the regulated utility market, which involves many state utility commissions, to keep near-term rates as low as possible, thereby implicitly deferring the risk of uncertain future events. The difficulty of evaluating and weighing such risk factors makes technology comparisons complex and places RETs at a disadvantage.

- *The difficulty of quantifying environmental values.* Many of the environmental costs of fossil fuels are not reflected in today's market prices. Consequently, discussions of the market penetration of renewable energy based on prices alone understate the total potential value of renewables, when nonmarket environmental damages are included. The incentives ($.02) used in the National Premiums Scenario very likely underestimate these nonmarket values.

As indicated earlier, the study not only presents results, but also identifies areas where there are gaps in data and analysis. These are summarized as follows:

- A key area of uncertainty has to do with the evolution of energy markets (how they respond to renewable-energy-related policies and other energy policy changes) and, most significantly, the evolution of RETs. The study authors were experts in these technologies and made their best predictions about expected advances in RETs, but the success of R,D&D depends on creativity and occasional good luck as well as good management and technical know-how.

- The R,D&D on renewables conducted in other developed nations has come to exceed current U.S. efforts. The impact that this will have on technology status and competition for markets was not addressed in the study, but it represents a significant area of uncertainty.
- Current data on RET deployment and energy market trends relevant to future deployment (both in bulk and dispersed applications) are inadequate for any assessment of market behavior with respect to these new technologies.
- Nonmarket externalities, such as environmental benefits, are very important vis-à-vis the RETs. A major need is an acceptable means of assessing these externalities to support the formulation of an energy policy.
- More accurate means of evaluating resource availability are needed. Examples are questions about sustainable levels of biomass productivity, land-use alternatives to provide biomass resources, and site-specific characteristics of wind resources.

The key to realizing the potential contributions of renewable energy in the near term is to intensify R,D&D activities. If market factors—higher prices or special markets—appear earlier than projected, the industry will respond in force only after the prerequisite R,D&D has been carried out. The impacts of U.S. R,D&D must be considered from a global perspective. Where substantial foreign investment is taking place, a lack of U.S. efforts would not preclude technological advancements. However, a lack of U.S. R,D&D may result in U.S. companies importing RETs for deployment in North America. If it becomes a goal of U.S. energy policy to see large quantities (>20%) of renewable energy deployed in U.S. markets before 2030, intensified R,D&D is essential. This is not to say that institutional constraints are not important now and will not be more important in the future; however, R,D&D is the key to moving toward that goal. It provides a cost-effective way of providing substantial additional economic energy supplies from domestic sources by 2010 and thereafter, given the base-case projection of conventional energy prices.

The annual use of energy in the U.S. is expected to grow from its current value of roughly 80 quads to 144 quads by 2030. Meeting that energy demand without significant reliance on imported oil and without environmental penalties will be problematic. However, renewable energy represents a large, domestic energy resource that has been virtually undeveloped and carries relatively minor environmental penalties. Intensified R,D&D could result in the availability of as much as 41 quads of renewable energy per year by 2030. As demonstrated by the research efforts reviewed in this volume, providing the advances needed to achieve that level of renewable energy use is a challenge that researchers at NREL and throughout the country are prepared to meet.

BIBLIOGRAPHY

Department of Energy, National Energy Strategy Report, 1st ed., Office of the Secretary, 1991/1992.

Solar Energy Research Institute, The Potential of Renewable Energy—An Interlaboratory White Paper, Report for U.S. Department of Energy, Office of Policy, Planning and Analysis. Contributions from Idaho National Engineering Laboratory, Los Alamos National Laboratory, Oak Ridge National Laboratory, Sandia National Laboratories and Solar Energy Research Institute; SERI Report SERI/TP-260-3674, 1990.

Union of Concerned Scientists, America's Energy Choices, Investing in a Strong Economy and a Clean Environment, contributions from Alliance to Save Energy, American Council for an Energy-Efficient Economy, Natural Resources Defense Council, and Union of Concerned Scientists, 1991.

Chapter 14

Mitigation of Environmental Impacts at Hydroelectric Power Plants in the United States

Glenn F. Čada and James E. Francfort

CONTENTS

INTRODUCTION

Hydroelectric power is the most mature and widely implemented of the renewable energy technologies. The energy of flowing water has been used to perform work directly since ancient times, and the use of hydropower turbines to generate electricity traces back to the 19th century. Two commonly used turbine types, the Francis and Kaplan turbines, are essentially refinements of the simple reaction turbine of Hero of Alexandria, dating from about 100 B.C.[1]

Hydroelectric power production (conventional, excluding pumped storage) provides over 10% of the net electrical generation in the U.S., more than petroleum or natural gas and far more than the other renewable energy technologies combined (Figure 1). On a regional basis, hydroelectric power represents 14% of the net electrical power generation in the Rocky Mountain states and nearly 63% along the Pacific Coast (Figure 2). Those states that have the largest percentages of their electricity generated by hydropower (e.g., Idaho, Oregon, Montana, and Washington) also tend to have the lowest average cost of electricity per kilowatt-hour (Figure 3).

ENVIRONMENTAL ISSUES

Hydroelectric power generation is largely free of several major classes of environmental impacts associated with other types of energy production. For example, with the exception of dust and equipment emissions during project construction, there are no air emissions. Thus, hydroelectric power does not produce sulfur and nitrogen oxides that can cause acidic precipitation, nor does it produce carbon dioxide, methane, or other "greenhouse gases" that have been implicated in global climate change. Unlike most other forms of energy production, hydropower generates little solid waste. Small amounts of land may be required to dispose of dredge spoils (mainly during construction), but neither large amounts of land (e.g., for continuing coal ash and slag disposal) nor long-

224

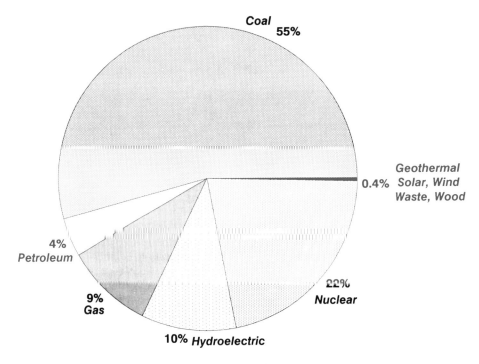

Figure 1 Renewable and nonrenewable sources of net electrical generation in the U.S., 1991. *(From Table 3 in Electric Power Monthly, Energy Information Administration, December 1992, with permission.)*

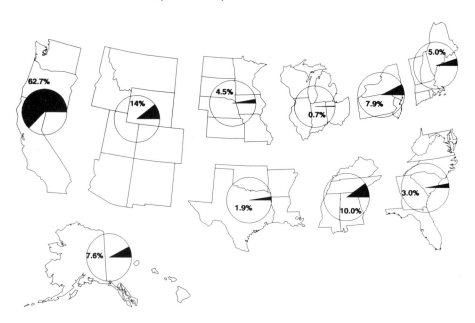

Figure 2 Contribution of hydropower to net electrical generation by census divisions, 1991. *(From Table 14 in Electric Power Monthly, Energy Information Administration, December 1992, with permission.)*

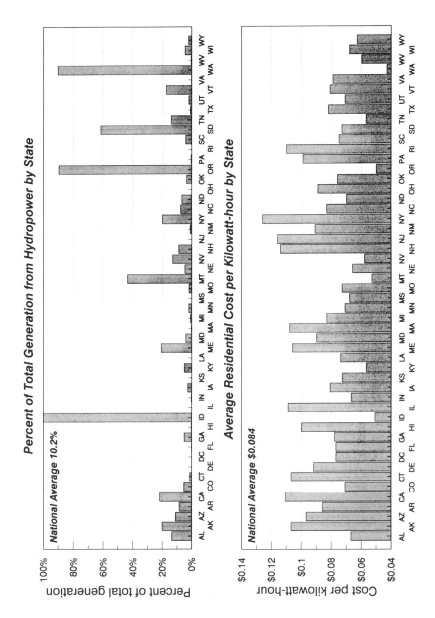

Figure 3 Comparison of the percent total electricity generation from hydropower and the average residential cost per kilowatt-hour, by state. *(From Tables 14 and 60, in Electric Power Monthly, Energy Information Administration, December 1992, with permission.)*

Table 1 **Major potential ecological impacts of construction and operation of hydroelectric power plants**

Class of Impacts	Construction	Operation
Alteration of terrestrial habitats	Inundation of wetlands and terrestrial vegetation Displacement of wildlife	Desiccation of riparian (streamside) vegetation below the diversion
Alteration of aquatic habitats	Conversion of a free-flowing river to a reservoir	Alteration of flow patterns below the diversion Retention of sediments and nutrients in the reservoir Development of aquatic weeds and eutrophication Alteration of water quality
Direct effects on aquatic organisms	Replacement of riverine aquatic communities with reservoir communities	Entrainment in the turbine Impingement on intake screens Interference with fish movements

term storage facilities (e.g., for hazardous or nuclear wastes) are needed. Finally, hydroelectric power production does not bear many of the external costs that characterize other energy technologies, i.e., potential environmental and health impacts associated with other aspects of the overall fuel cycle.[2] These externalities include impacts of resource extraction (e.g., coal mining), preparation/beneficiation, transportation (e.g., oil spills), and waste disposal.

However, hydropower does have a number of potential ecological impacts, many related to the localized alteration of aquatic and terrestrial habitats (Table 1). Impoundment of a river to create a water storage reservoir will inundate terrestrial habitat and replace the free-flowing river environment with a reservoir environment. These habitat changes result in the loss of terrestrial vegetation (often including wetlands), the displacement of wildlife, and conversion of riverine fish communities to lentic (reservoir) communities.[3]

Operation of the hydropower project can have additional impacts on terrestrial and aquatic biota, especially as mediated through habitat alterations (Table 1). Diversion of water out of the stream channel and into a penstock (or withholding reservoir releases in order to store water for future electrical generation) can desiccate riparian (streamside) vegetation.[4] Insufficient instream flow releases can also degrade habitat for fish and other aquatic organisms in the affected river reach below the dam. The relatively stagnant water in the reservoir will trap water-borne sediments and nutrients and encourage the undesirable proliferation of aquatic weeds and algae (eutrophication). Stratification of water impounded by the dam can lead to alteration of temperature and dissolved gas concentrations both in the reservoir and in water discharged from the turbines.[5]

Finally, hydropower operations can have direct effects on aquatic organisms.[6] The dam can block upstream movements of fish, which can have severe consequences for anadromous fish (e.g., salmon, steelhead, and American shad) that must migrate upstream to fresh water to reproduce. Downstream-moving fish may be drawn into the turbine intake (entrainment) and suffer injury or mortality when passing through the turbine. Screens used to exclude debris and other materials from the turbines can also trap fish (impingement).

MITIGATION OF ENVIRONMENTAL ISSUES

Many of these environmental impacts can be minimized or mitigated by appropriate construction or operational measures. For example, the size of the storage impoundment can be reduced in order to minimize the inundation of terrestrial habitat and to reduce the nutrient- and sediment-trapping potential. Water can be released in ways that will reduce extreme flow fluctuations and minimize reservoir stratification and its attendant water quality changes. Fish ladders or lifts can be incorporated into the design of the project to allow migrating fish to ascend the dam and gain access to upstream spawning areas. Appropriately designed intake screens or other exclusion measures can prevent fish from being drawn into the turbine intake. These types of mitigative measures are often recommended by water quality regulatory agencies and/or fish and wildlife resource agencies, and are incorporated as requirements into the operating licenses of hydroelectric power projects.

Environmental mitigation requirements have an important and growing effect on hydroelectric power production in the U.S.[7] Non-federal hydropower development is regulated by the Federal Energy Regulatory Commission, which determines environmental and other requirements under the authority of the Federal Power Act. Section 401 of the Clean Water Act requires hydropower licensees to obtain a water quality certification from the state or U.S. Environmental Protection Agency to ensure that water quality is not degraded by construction or operation. Section 404 of the Clean Water Act authorizes the U.S. Army Corps of Engineers to issue permits for the discharge of dredged or fill material at hydro sites. Most recently, the Electric Consumers Protection Act of 1986 significantly strengthened the role of fish and wildlife agencies in determining environmental mitigation requirements for hydropower development. The purpose of most environmental mitigation at hydroelectric projects is to minimize adverse effects of construction and operation on water quality and aquatic organisms. Concomitant with these benefits, however, mitigation practices usually entail costs arising from required studies, capital structures, reporting, and operational changes. The impact of environmental mitigation on hydropower developers is reduced profits; the impact to society is reduced production of renewable energy, which may be replaced by energy produced by fossil fuels with their inherent environmental and economic costs.

The U.S. Department of Energy's Hydropower Program has been engaged in a multiyear study of environmental mitigation practices at hydroelectric projects. This study is intended to characterize the status of mitigative measures and to develop an understanding of both the benefits and costs of these measures. Volume I of the Environmental Mitigation Study[8] examined current, nationwide practices for mitigating adverse impacts of hydropower production on water quality [specifically, dissolved oxygen (DO)], instream flows, and upstream and downstream fish passage. Input from the hydropower industry and regulatory and resource agencies indicated that these issues were considered the most significant and most in need of mitigation. Information was collected about the types and frequencies of mitigation methods in use, their environmental benefits and effectiveness, and their costs. This information is discussed below.

Information on mitigation practices at non-federal hydroelectric power projects was obtained directly from three sources: (1) existing records of the Federal Energy Regulatory Commission, the agency responsible for licensing non-federal hydropower projects; (2) new information provided by the non-federal hydropower community; and (3) new information obtained from the natural resource agencies involved in hydropower regulation. Information on mitigation practices was obtained from 280 projects; these represented more than 40% of all the projects licensed during the 1980s that could be identified *a priori* as having relevant mitigation requirements. Instream flow requirements are the most common of the environmental mitigation requirements among projects licensed in the 1980s, followed by downstream fish passage, DO mitigation, and upstream fish passage facilities. The proportion of newly licensed projects with mitigation requirements has increased significantly during the past decade.

Costs were examined for mitigation methods associated with each of the four issues. Overall costs were divided into capital, study, operations and maintenance, and annual reporting costs. Costs were expressed in 1991 dollars in the following ways: average cost per project, average cost per kilowatt of capacity (for capital and study costs), and average cost per kilowatt-hour (for operations and maintenance and annual reporting costs). Because of the huge cost ranges for similar mitigation measures at different sites, costs were also presented by capacity categories. This allows the developer to compare environmental mitigation costs of a current or potential project to those incurred by similarly sized projects.

Occasionally, hydropower projects are required to make some contribution to environmental projects or enhancement of fishery resources not associated directly with the hydroelectric project to compensate for unavoidable environmental damage. Off-site compensation was reported at an average level of $307,000 per site (four projects; 469 MW average capacity).

INSTREAM FLOWS

Minimum instream flows are released from water diversion projects to maintain various instream water benefits, including maintenance of fish and other aquatic organisms, water quality, recreation, and streamside vegetation. This study considered only instream flows designed for the protection of aquatic organisms, i.e., minimum flow releases for recreation were not directly examined. Providing flows below a diversion is simple; water can simply be spilled from the dam instead of diverted to a penstock or stored in the reservoir. Inasmuch as water released to support downstream aquatic and riparian communities is water that is frequently unavailable for generation of electricity, hydropower developers are interested in providing sufficient, but not excessive releases. More than one method to determine adequate instream flow releases was reported to have been used at many projects (Figure 4). Of the established and documented methods used to determine instream flows, the most frequently applied formal methodology was the complex and expensive Instream Flow Incremental Methodology (IFIM). Half of the project operators reported that the professional judgment of resource agency staff was the key element in at least one of the methods used to set instream flows. Professional judgment was often cited in conjunction with the IFIM. Thus, many licenses for hydroelectric power projects have been issued in which the requirements for minimum instream flow releases have been determined without the use of formal or documented methods.

Monitoring sufficient to evaluate the consequent effects of instream flow releases on fish populations is not common. Among operating projects with instream flow requirements in their licenses, only about 50% measure the flow releases at all, whether continuously, daily, or less frequently; about 20% have collected fish data that might be used to verify the success of the instream flow mitigation.

The costs of providing instream flows vary widely among projects. At diversion projects (where flows for power generation are diverted through a pipeline and around a stream reach), instream flow requirements in the bypassed reach must be subtracted from that available for generation. Storage projects that generate without a bypassed stream reach can release instream flows through their turbines. Operators of such projects frequently reported no costs associated with the instream flow releases. However, even the requirements for instream flows below storage projects can have significant costs to the operators because of forced sales of electricity at base rates instead of at peak rates. The capital costs of providing devices to release and measure instream flows averaged

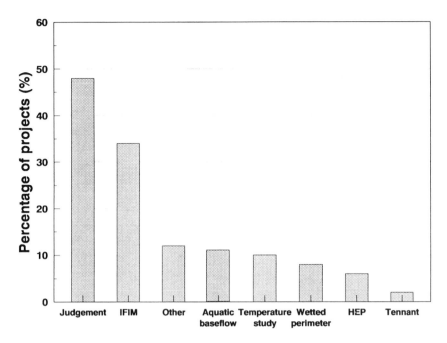

Figure 4 Methods used to determine instream flow requirements, based on informa-tion provided by developers of 185 non-federal hydropower projects. *(From Sale, M. J., et al., Environmental Mitigation Requirements at Hydroelectric Projects, Vol. 1, DOE/ID-10360, U.S. Department of Energy, Idaho Field Office, Idaho Falls, ID, 1991, with permission.)*

$99,000 per project (58 projects; 19 MW average capacity). Instream flow studies averaged $100,000 per project (31 projects; 9 MW average capacity).

DISSOLVED OXYGEN

Water released from reservoirs, including those at hydroelectric projects, can have low DO concentrations, especially during the summer and at large projects with deep reser-voirs, low flushing rates, or warm climates. In response to the need to maintain adequate DO, which is necessary for respiration of aquatic organisms, numerous methods have been developed to improve the quality of hydropower releases.[8] These include techniques to aerate water discharged into the tailrace below the dam (weirs, surface aerators, and diffusers), aeration within the powerhouse (turbine venting and draft tube aeration), and operational techniques (spill flows and turbine operating schedule adjustments).

Spill flows are the most commonly used of the DO mitigation technologies (Figure 5). Over 60% of all responding projects use spill flows, 8% use intake level controls, and nearly 30% use turbine aeration. Twenty-five percent of the projects use more than one mitigation method. Spill flows are used twice as often as other methods at projects with design heads of less than 15 m. This may be because fewer mitigation options are available to small projects, which are constrained by cost and design considerations. Hydropower operators reported that projects with design heads greater than 15 m were as likely to use spill flows as any other method.

About 65% of the projects that reported DO mitigation indicated that water quality is monitored (most commonly, water temperature and DO concentration), but biological monitoring is rarely conducted. Consequently, the actual biological benefits of DO mitigation are usually unknown.

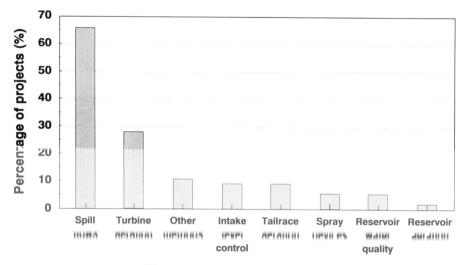

Dissolved oxygen mitigation methods

Figure 5 Methods used to increase dissolved oxygen concentrations in releases from non-federal hydroelectric power projects, based on information provided by developers of 56 non-federal hydropower projects. *(From Sale, M. J., et al., Environmental Mitigation Requirements at Hydroelectric Projects, Vol. 1, DOE/ID-10360, U.S. Department of Energy, Idaho Field Office, Idaho Falls, ID, 1991, with permission.)*

Capital costs for DO mitigation equipment averaged $162,000 per project (15 projects; 56 MW average capacity). The energy generation lost because of environmental requirements based on water quality averaged 107,000 kWh per project annually (11 projects; 63 MW average capacity; 244,994 MWh average energy production).

UPSTREAM FISH PASSAGE

Blockage of upstream fish movements by a hydroelectric dam can have serious impacts on species whose life history requires spawning migrations. Anadromous fish (e.g., salmon, steelhead, American shad, blueback herring, and striped bass), eels, and some resident fish (e.g., trout, white bass, and sauger) have spawning migrations that may be constrained by hydroelectric dams. Maintaining or enhancing populations of such fish may require facilities for upstream fish passage.[9]

Fish ladders are the most commonly reported means of passing fish upstream at non-federal hydroelectric dams (Figure 6). Fish elevators are a less common and relatively recent mitigative measure. Trapping and hauling (via trucks or barges) of fish to upstream spawning locations is used at some older dams.

Preconstruction studies, performance monitoring, and detailed performance criteria for upstream passage facilities are frequently lacking;[8] 40% of the projects with upstream fish passage mitigation had no performance monitoring requirements. Those projects that do monitor the success of upstream passage generally quantify fish passage rates (e.g., fishway counts) or, less commonly, fish populations.

The costs of upstream fish passage mitigation are relatively easy to determine. In addition to the capital costs of constructing the fishway, there are operation and maintenance costs (e.g., for clearing debris from the fish ladder or elevator and for electrical power to operate a fish elevator), lost power generation resulting from flow releases needed to operate a fish ladder or elevator (including attraction flows), and any monitoring and reporting costs. Capital costs of fish ladders averaged $7,586,000 per project

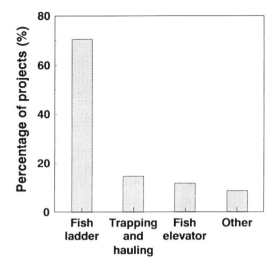

Upstream passage measures

Figure 6 Relative frequency of upstream fish passage measures at non-federal hydroelectric power projects, based on information provided by developers of 34 non-federal hydropower projects. *(From Sale, M. J., et al., Environmental Mitigation Requirements at Hydroelectric Projects, Vol. 1, DOE/ID-10360, U.S. Department of Energy, Idaho Field Office, Idaho Falls, ID, 1991, with permission.)*

(8 projects; 236 MW average capacity). Other costs of upstream fish passage averaged $51,000 for environmental studies (6 projects; 6 MW average capacity), $26,000 for annual reporting (11 projects; 220 MW average capacity), and $80,000 per year for additional operation and maintenance (10 projects; 80 MW average capacity).

DOWNSTREAM FISH PASSAGE

A variety of downstream fish-screening devices are employed to prevent fish from being drawn into turbine intakes. The simplest passage measure, the spill flows used to increase DO concentrations or to provide instream flows, can also transport fish below the dam. Techniques that utilize more sophisticated technology are under development, but are not widely used.[10] For example, physical screening and light- or sound-based guidance measures are being studied as ways to pass migrating fish downstream with a minimal loss of flow for power generation. The most frequently reported downstream fish passage device is the angled bar rack, in which the trash rack is set at an angle to the intake flow and the bars may be closely spaced (Figure 7). This device is commonly used in the Northeast. Other frequently used fish screens range from variations of conventional trash racks (e.g., use of closely spaced bars) to more novel designs employing cylindrical, wedge-wire intake screens. Intake screens usually have a maximum approach velocity requirement and a sluiceway or some other type of bypass as well.

As with upstream fish passage measures, performance monitoring and detailed performance criteria for downstream passage facilities are relatively rare. There are no performance monitoring requirements for 82% of the projects. Operational monitoring studies of passage rates or mortality rates have been conducted at a few of the projects.

In addition to the capital costs of constructing the downstream fish passage facility, costs typically include those for cleaning closely spaced screens or maintaining traveling screens, lost power generation resulting from flow releases needed to operate sluiceways

232

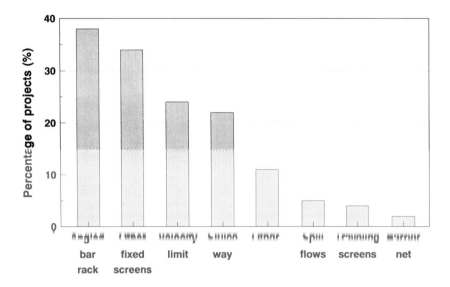

Figure 7 Relative frequency of downstream fish passage measures at non-federal hydroelectric power projects, based on information provided by developers of 85 non-federal hydropower projects. *(From Sale, M. J., et al., Environmental Mitigation Requirements at Hydroelectric Projects, Vol. 1, DOE/ID-10360, U.S. Department of Energy, Idaho Field Office, Idaho Falls, ID, 1991, with permission.)*

or other bypasses, and monitoring and reporting. Downstream fish protection/passage mitigation averaged $959,000 per project for capital costs (37 projects; 55 MW average capacity) and $14,000 per year for operation and maintenance (32 projects; 6 MW average capacity). Studies and annual reporting costs averaged $639,000 (21 projects; 93 MW average capacity) and $2,000 (23 projects; 5 MW average capacity), respectively.

ONGOING RESEARCH

Volume I of the Environmental Mitigation Study[8] identified both technical and economic problems associated with the most common mitigative measures used to protect important aquatic resources. All of the measures can have high costs and, with few exceptions, there is little information available on their effectiveness. The findings of the report pointed to a critical need for additional study and monitoring to address the efficacy of each type of mitigation and the valuation of benefits.

Volume II of the Environmental Mitigation Study[11] contains a more detailed examination of the costs and benefits of fish passage and protection measures. Capitalizing on the nationwide assessment of fish passage mitigation that was accomplished for Volume I, analyses in Volume II are focused on 16 particular case studies that span a wide range of measures, geographic settings, and fish species. For each case study, the following costs are quantified: initial capital and study costs, annual reporting costs, operation and maintenance costs, and rebuilding or modification costs. Evaluation of the environmental benefits associated with each case study quantifies, to the extent possible, the performance of the mitigative measure, as well as the subsequent effects on fish populations and recreational and/or commercial fisheries.

SUMMARY

Hydroelectric power production is free of several classes of environmental impacts that severely constrain nonrenewable (and some renewable) energy sources, i.e., air emissions, solid wastes, and significant fuel cycle externalities. Unlike many other sources of renewable energy, hydropower is a well-developed technology that is already contributing substantially to U.S. electricity needs. In order to ensure that hydropower continues to play an important role in the U.S. electricity mix, the often unique environmental issues must be resolved.

Our assessment of mitigative measures at non-federal hydropower projects in the U.S. found that license requirements associated with the most common environmental issues increased during the 1980s, most notably in the area of downstream fish passage/protection. Numerous innovative concepts and designs are being considered to mitigate adverse impacts of hydroelectric generation, but adequate performance monitoring has been rare. The ecological impacts of hydroelectric generation can be serious, but they are not insurmountable. Mitigative measures are available to deal with these issues, and the challenge is to develop an understanding of the true costs and benefits of the most effective measures.

ACKNOWLEDGMENTS

We thank Sigurd W. Christensen and Michael J. Sale for their reviews of this manuscript. This work was performed for the Hydropower Program, Office of Energy Efficiency and Renewable Energy, U.S. Department of Energy. The Oak Ridge National Laboratory (ORNL) is managed by Martin Marietta Energy Systems, Inc. under Contract No. DE-AC05-84OR21400 with the U.S. Department of Energy. Idaho National Engineering Laboratory is managed by EG&G Idaho, Inc. under Contract No. DE-AC07-761D01570 with the U.S. Department of Energy. This is publication no. 4217 of the Environmental Sciences Division, ORNL.

REFERENCES

1. National Academy of Sciences (NAS), *Energy for Rural Development: Renewable Resources and Alternative Technologies for Developing Countries*, National Academy of Sciences — National Research Council, Washington, D.C., 1976, 306.
2. Oak Ridge National Laboratory and Resources for the Future (ORNL/RFF), U.S.-EC Fuel Cycle Study: Background Document to the Approach and Issues, ORNL/M-2500, U.S. Department of Energy, Washington, D.C., 1992.
3. **Čada, G. F. and Zadroga, F.,** Small-scale hydroelectric power for developing countries: Methodology of site-selection based on environmental issues, *Environ. Conservation*, 9, 329, 1982.
4. **Petts, G. E.,** *Impounded Rivers: Perspectives for Ecological Management*, John Wiley & Sons, New York, 1984, 326.
5. **Čada, G. F., Kumar, K. D., Solomon, J. A., and Hildebrand, S. G.,** An analysis of dissolved oxygen concentrations in tail waters of hydroelectric dams and the implications for small-scale hydropower development, *Water Resources Res.*, 19, 1043, 1983.
6. **Olson, F. W., White, R. G., and Hamre, R. H.,** *Proc. Symp. on Small Hydropower and Fisheries*, American Fisheries Society, Bethesda, MD, 1985, 497.
7. **Mattice, J. S.,** Ecological effects of hydropower facilities, *Hydropower Engineering Handbook*, Gulliver, J. S. and Arndt, R. E., Eds., McGraw-Hill, New York, 1991, chap. 8.

8. **Sale, M. J., Čada, G. F., Chang, L. H., Christensen, S. W., Railsback, S. F., Francfort, J. E., Rinehart, B. N., and Sommers, G. L.,** Environmental Mitigation Requirements at Hydroelectric Projects. Volume 1. Current Practices for Instream Flow Needs, Dissolved Oxygen, and Fish Passage, DOE/ID-10360, U.S. Department of Energy, Idaho Field Office, Idaho Falls, ID, 1991.

9. **Orsborn, J. F.,** Fishways — Historical assessment of design practices, *American Fisheries Soc. Symp., 1*, Dadswell, M. J., Klauda, R. J., Moffitt, C. M., Saunders, R. L., Rulifson, R. A., and Cooper, J. E., American Fisheries Society, Bethesda, MD, 1987, 122.

10. **Taft, E. P.,** *Assessment of Downstream Migrant Fish Protection Technologies for Hydroelectric Application,* AP-4711, Electric Power Research Institute, Palo Alto, CA, 1986.

11. **Francfort, J. E., Čada, G. F, Dauble, D. D., Hunt, R. T., Jones, D. W., Rinehart, B. N., Sommers, G. L., and Costello, R. J.,** Environmental Mitigation at Hydroelectric Projects. Volume II. Benefits and Costs of Fish Passage and Protection, DOE/ID 10360(V2), U.S. Department of Energy, Idaho Operations Office, Idaho Falls, ID, 1994.

Chapter 15

An Overview of Wind Energy in the United States

R. Gerald Nix

CONTENTS

ABSTRACT: Wind energy has the potential to significantly impact electricity generation in the U.S. during the next several decades, especially during the next 10 to 15 years. The wind is a proven energy source; it is not resource-limited in the U.S., and it has no insolvable technical constraints. Orderly development paths exist, leading to a point where the cost of wind-generated electricity is competitive with that of fossil fuel-generated electricity. This chapter describes the potential for wind energy, its current status, challenges, and efforts underway to make wind-generated electricity a cost-competitive option for bulk electricity.

THE WIND RESOURCE

Researchers at the Pacific Northwest Laboratories (PNL) estimate that there is enough wind potential in the U.S. to displace at least 45 quads of primary energy used to generate electricity.[1] This estimate is based on "class 4" winds or greater and the judicious use of land. For comparison, the U.S. currently uses about 28 quads of primary energy to generate electricity. (One quad is equivalent to 10^{15} Btu or about 167,000,000 barrels of oil.)

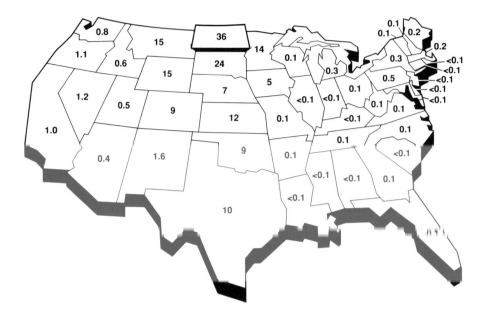

Figure 1 Available wind resource in the United States. Percent of total U.S. electricity demand each state could produce from wind energy.

Winds are grouped into seven classes according to power density: class 1 is the lowest and class 7 is the greatest. Class 4 winds have an average power density in the range of 320 to 400 W/m², which corresponds to a moderate speed of about 5.8 m/s (13 mph) measured at a height of 10 m. Figure 1 shows a summary of the wind resource in the U.S. based on PNL assessments.

While almost all of the U.S. installed wind electric generation capacity is in California, the resource is virtually untapped in the Great Plains region. Approximately 90% of the wind energy resource in the contiguous U.S. is contained in 11 states in the Great Plains. The chapter by Schwartz in this book provides much more detail on this wind resource.[2]

CONVERSION TECHNIQUES

A wind turbine harnesses the kinetic energy in wind to turn a rotor, producing mechanical energy that drives a generator and produces electricity. With the conventional windmill, the mechanical energy is used directly for other purposes, such as pumping water. For utility applications, wind turbines are routinely grouped together in clusters called wind parks, wind plants, or wind farms.

Various types of wind turbines differ greatly in their conversion efficiency. Modern turbines are primarily horizonal- or vertical-axis machines that make full use of lift-generating airfoils (ancient windmills relied primarily on drag forces). Modern wind turbines have efficiencies in the range of 35 to 40%, with availabilities greater than 97%. An excellent description of various types of wind turbines is found in Reference 3.

Figure 2 shows typical horizontal wind turbines. These have two or three blades that can be mounted on a rigid or teetered hub. The main parts of the turbine are the rotor assembly (blades and hub), gearbox, generator, nacelle or housing enclosing the gearbox and generator, and tower. The turbines also incorporate mechanical control systems (mechanical brakes, overspeed controls), electric control systems, aerodynamic controls, and power conditioning equipment. Most existing machines run at constant rotational

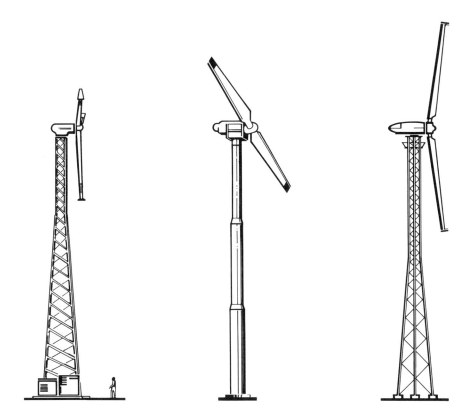

Figure 2 Typical horizontal-axis wind turbines.

speed, although newer machines may operate at variable speed, which can lead to greater efficiency.

HISTORY

Since the middle of the 19th century, more than six million windmills and wind turbines have been installed in the U.S.; most are windmills with a rating of less than 1 hp. The most common application has been for pumping water, particularly on remote farms and ranches. Many windmills are still in operation in remote areas that are not serviced by the electric grid. These windmills are extremely durable and rugged, some having run for many years with little or no maintenance.

Small wind turbines, usually rated at 1 kW or less, originally were used to supply electricity to remote sites. A typical example is the Jacobs machine, of which tens of thousands were produced for farm applications from 1930 to 1960. By 1960, the production of wind turbines in the U.S. had essentially come to a standstill as many of the rural areas had been electrified via a grid of wires carrying electricity from more cost-effective central fossil fuel-fired generating stations.

The first large wind turbine was the Smith-Putman unit. Erected in southern Vermont during World War II, it was rated at 1.25 MW of alternating-current electricity and used a two-bladed metal rotor 53.3 m (175 ft) in diameter.

The oil embargoes and fuel price escalations of the 1970s brought a flurry of activity to develop cost-effective wind turbines. The DOE and National Aeronautic and Space

Figure 3 *SeaWest* San Gorgonio Wind Park near Palm Springs, California.

Administration (NASA) led this activity by developing very large machines rated up to 4.5 MW. While they proved the feasibility of many technical innovations, these large machines had mechanical and structural problems, and only two are currently operating in U.S. utility systems. The WTS-4, rated at 4 MW, is in operation at Medicine Bow, Wyoming, and the Mod-5, with a rotor diameter of 320 ft, is generating power for the island of Oahu in Hawaii.

Numerous other machines rated at 30 to 300 kW were developed and installed to produce electricity that was fed into the electricity grid; smaller machines (1 to 10 kW) were used for remote applications. Most of the utility-size machines (100 to 300 kW) were installed in California under favorable power purchase agreements and investment tax credits. The three primary locations are Altamont Pass near San Francisco and Tehachapi and San Gorgonio near Palm Springs. The machines were of widely differing quality, as were the developers and operators of the wind parks. After a sorting-out period, well-managed and well-operated wind parks remained. Figure 3 shows the SeaWest San Gorgonio wind park near Palm Springs.

CURRENT STATUS
The wind energy industry in the U.S. has shown the technology to be technically viable and economically competitive with high-value electricity. Today, more than 16,000 wind turbines are installed in California, with a total generating capacity exceeding 1600 MW. Approximately 40,000 MW of wind-generated electricity is required to displace 1 quad of primary energy consumption for fossil fuel power generation. About 1% of the electricity used in California is generated from wind energy.

In 1992, wind plants in California produced approximately 3 billion kWh electricity, compared to about 750 million kWh in 1985, a fourfold increase in output. It is interesting to note that more than half of the wind energy capacity in California was installed after the expiration of the tax credits. This shows that wind energy was cost competitive under the economic conditions of the time. Further, it shows that major operating problems with existing machines have been systematically fixed and that the average annual energy production per wind turbine has increased.

There are obstacles to the widespread application of wind energy. The life cycle cost of energy from wind for current technology is between $0.06-0.09/kWh, depending on

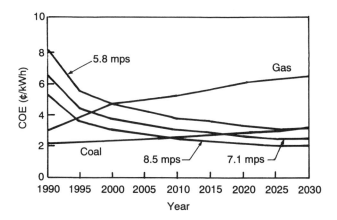

Figure 4 Projected costs for wind-generated electricity.

the machine and the available wind resource. Contracts signed by the utilities in the 1980s agreed to pay wind park operators $0.10-0.15/kWh in the 1990s. As these contracts expire, they are being replaced with new contracts that promise to pay only the avoided cost for wind-produced electricity. This price can be as low as $0.02-0.04/kWh. Simply put, the nation's energy policy will have to change or technical advances will have to further reduce the cost of wind energy before a major expansion in the use of wind energy can become reality. On the brighter side, the required technical advances appear achievable.

COSTS AND GOALS
Wind turbines in California currently produce electricity at a cost of $0.06-0.09/kWh, depending on the location, design, and operating policy. This life cycle cost of energy is for constant 1992 dollars, assuming utility financing. When the near-term advanced wind turbines currently being developed begin operation in the mid-1990s, the electricity-production cost is expected to drop to about $0.05/kWh at class 4 wind sites. Around the year 2000, when the innovative next-generation wind turbines begin operating, the cost of wind-generated electricity is expected to drop to about $0.04 or less per kWh.[4] By the year 2030, DOE estimates that wind-generated electricity will displace about 3 to 4 quads of primary energy through the installation of 120,000 to 160,000 MW of wind turbine capacity. Figure 4 is a summary chart of projected costs for wind-generated electricity under various scenarios.

TECHNICAL CHALLENGES
Some of the technical challenges that must be addressed are shown schematically in Figure 5. The objective is to make wind turbines more robust so they will require less maintenance and will have longer lifetimes while performing better and having lower initial costs. These challenges may be summarized as follows:

- Better characterization of the resource
- More efficient airfoils
- Better understanding of the aerodynamics of wind turbines
- Development of theoretical models and computer codes
- Better understanding of fatigue and structures
- Better control techniques

Research and development to meet these challenges will provide better design tools and will result in optimized drivetrains, optimized tower designs, use of innovative mate-

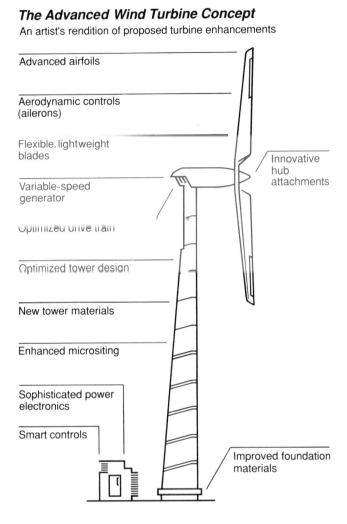

The Advanced Wind Turbine Concept
An artist's rendition of proposed turbine enhancements

Advanced airfoils

Aerodynamic controls
(ailerons)

Flexible, lightweight
blades

Innovative
hub
attachments

Variable-speed
generator

Optimized drive train

Optimized tower design

New tower materials

Enhanced micrositing

Sophisticated power
electronics

Smart controls

Improved foundation
materials

Figure 5   Some opportunities for turbine improvements.

rials, enhanced micrositing, use of sophisticated power electronics, and use of smart
controls.

RESEARCH AND DEVELOPMENT ACTIVITIES
Advanced wind turbines must be more efficient, more robust, and less expensive. The
DOE has the responsibility of developing new technology to enable the U.S. wind
industry to be more viable and to enhance the use of wind energy. DOE, its national
laboratories, and the wind industry are working together to accomplish these goals
through various R&D programs.

The DOE-funded programs include research programs at NREL, Sandia National
Laboratories (SNL), and Pacific Northwest Laboratories (PNL). Each program is aimed
at specific goals ranging from improving the current generation of turbines and components to designing and testing the innovative turbines of the next century. Specific
activities supported by DOE wind energy R&D programs are described below.

Better characterization of the resource. This involves taking better measurements of wind characteristics, especially within wind farms, and developing better siting methods. Turbulence within wind farms is greater (by up to 50%) than that in open terrain, resulting in structural and fatigue loads that limit lifetime or dictate maintenance of turbines and components like gearboxes. NREL and PNL staff lead these activities, with significant interaction with wind-park operators, local weather agencies, and state energy agencies. More information can be found in the articles by Kelley and Wright,[5] Veers,[6] and Elliott et al.[1]

More efficient airfoils. NREL has developed airfoils tailored to meet the specific demands of wind turbines. This has resulted in greater efficiency of energy capture than was possible with the existing airfoils, which were based on designs for helicopters. Major problems with the conventional airfoils include a decrease in efficiency when the airfoil's leading edge becomes fouled, and generator burn-out because of excessive energy capture from wind gusts. The NREL airfoils are the first of a new generation of airfoils that will significantly improve performance and make wind energy more competitive. Efficiency gains of up to 30% have been accomplished with the NREL airfoils. Details may be found in the article by Tangler et al.[7]

Better understanding of aerodynamics. Aerodynamic phenomena in a time-variant, three-dimensional wind-flow field are significantly more complex than those observed in steady, two-dimensional wind tunnel tests. NREL researchers are working to generate better field data to provide an enhanced understanding of the basic phenomena. NREL maintains significant interactions with university, industry, and foreign researchers in the area of fundamental aerodynamics. The result will be better design methods. The article by Butterfield et al.[8] contains significant details on aerodynamics R&D projects.

Development of theoretical models and computer codes. Substantial effort is being devoted to allowing computer characterization of every component in the integrated wind turbine. As a result, better designs of components and systems will arise to gain longer lifetimes and allow cost reductions where it is possible to reduce the structural requirements of components. A good reference is the report by Wright and Thresher.[9]

Better understanding of fatigue and structures. Work is under way to better predict design impacts on the fatigue-life of components. The goal is to produce more robust and innovative designs. Numerous advanced experimental and analytical techniques are being used, as typified in the article by Musial et al.[10]

Better control techniques. This involves using power electronics to increase efficiency and generate higher quality electricity. An example is the use of variable-speed turbines, rather than fixed-speed turbines, because they perform better and experience less stress. Details can be found in the article by Vachon.[11]

Most of the activities listed below are collaborative efforts between DOE and members of the wind energy industry. Most of the subcontracted activities are significantly cost-shared by industry, with a typical goal of DOE providing half of the funding and the industrial partner providing the other half. These programs serve to transfer technology developed in the research programs to industry. They also allow the wind industry the opportunity to leverage their resources to attack critical problems. While all of the DOE-sponsored R&D activities are in direct response to wind industry needs, about two thirds of the DOE wind energy budget is dedicated to these collaborative activities that are cost-shared by industry.

242

Cooperative Research Program. The objective is to help the wind energy industry solve technical problems with existing wind turbines to make them more efficient, to reduce maintenance, to reduce component replacement costs, and to increase lifetimes. The goal is to improve the competitiveness of existing wind technology.

Value-Engineered Turbine Program. The objective of this program is to provide a domestic source of very robust turbines based on existing technology, with improvements as necessary to enhance cost competitiveness. The goal is to improve domestic turbine manufacturing capability and to enhance the competitiveness of American turbine manufacturers. The costs of the Value-Engineered Turbine Program are substantially shared by the industrial participants, with participation limited to existing manufacturers or windfarm operators.

Advanced Wind Turbine Program. This program has two parts: (1) development of near-term (mid-1990s) advanced turbines and (2) development of next-generation (post-2000) turbines. The goal is to develop and deploy wind turbines that produce electricity at less than $0.05/kWh for the near term and less than $0.04/kWh for the next generation turbines for class 1 wind sites.

Distributed Wind Energy Generation for Utility Applications Program. This is an activity led by utilities to define and validate the benefits of placing wind turbines at selected locations along an overloaded transmission line to enhance the power quality and to perhaps mitigate the need for replacement or renovation of the line.

ECONOMIC CHALLENGES
The competitiveness of wind energy in the generation of electricity is very dependent on economics and U.S. energy policy. At this time in the U.S., the principal measure of competitiveness is the cost of energy. This simple use of electricity cost per kilowatt-hour as a comparative parameter does not fully take into account the effects of fuel subsidies or depletion allowances, cost of pollution abatement, or the long-term effects of the loss of nonrenewable fossil fuels.

As discussed above under the current economic scenario, the cost of wind-generated electricity must be decreased for long-term cost competitiveness. Wind-generated electricity is currently cost competitive with electricity generated from gas and residual oil at many locations. A cost reduction to about $0.035/kWh would make wind-generated electricity cost competitive with electricity generated from coal. The most recent bids to install wind turbines in good wind areas would produce electricity at about $0.05/kWh.

The cost competitiveness of wind energy could be very favorably impacted by some of the energy policy changes being considered by the current administration. These include energy taxes or Btu taxes on fossil fuels, proper accounting of external costs such as pollution, or renewable energy production incentives.

INTEGRATION ISSUES
The California wind farms successfully produce electricity that is fed into the utility grids, proving that there are no insurmountable issues to integrating wind-generated electricity into the electric supply grid. Power conditioning for wind-produced electricity is sufficient, so that there is little concern for power quality. Wind farms are typically designed with local substations and control systems that automatically remove the wind park from the grid if the quality of electricity produced does not meet utility specifications. However, this type of shutdown is rare.

Wind energy is not currently considered dispatchable because of the variable nature of the resource. The addition of cost-effective energy storage to wind parks could change this perspective. Storage would allow wind-produced electricity to be used to meet electricity demands even at times of low wind speed, so that it can be considered a firm generating capacity that is dispatchable. Batteries, pumped hydro, compressed air, and superconducting magnets are candidate storage techniques. The most viable near-term storage technique is pumped hydro where it is feasible. When storage is integrated with wind farms, the value of wind-generated electricity will be much higher and may provide significantly improved economics.

Transmission line access is important, especially when very substantial wind resources are located in states like Montana that are sparsely populated. Wind park access to transmission lines may be enhanced by building fossil-fueled plants nearby. This allows joint use of the transmission lines and provides the justification for constructing new lines when needed. With new transmission lines typically costing $600,000 to 1,000,000 per mile, it is important to optimize line use.

ENVIRONMENTAL ISSUES

Wind energy is environmentally benign; however, some concerns must be addressed. The death of birds flying into operating turbines is a concern, especially when the birds are raptors such as golden eagles. The problem of avian collisions with wind turbines has been observed at Altamont Pass. There have been no directly observed instances of birds flying into rotors or towers. Data are deduced indirectly from necropsies on carcasses found in the vicinity, leading to significant uncertainty.

There are several potential ways to mitigate avian collisions, including painting towers and rotors appropriate colors, using acoustic warning devices, eliminating avian prey around the towers, controlling vegetation around towers to minimize avian hunting in the vicinity, or more careful siting to avoid migratory flyways. Although the extent of the problem is not well defined, wind park operators are addressing it as a significant issue.

Another concern is aesthetics. Wind farm siting and layout are important as what is beautiful to an engineer may simply be ugly to others. Public education is important to enable citizens to understand the benefits of wind parks and to mitigate concerns about aesthetics. In some of the areas of the California wind parks, low-power AM radio transmission is used to inform passing citizens about the wind turbines and the benefits of wind energy.

CONCLUSIONS

Wind energy will be one of the most important and applied of the renewable energy forms during the next decade or two. A viable wind energy industry exists and is capable of substantial expansion to displace 3 to 4 quads of primary fossil-fuel energy use by the year 2030. There are substantial challenges to be met, but all appear solvable. Orderly development paths have been established to meet the challenges.

REFERENCES

1. **Elliott, D. L., Wendell, L. L., and Gower, G. L.,** U.S. areal wind resource estimates considering environmental and land-use exclusions, presented at Windpower '90 Conf., Washington, D.C., September 1990.
2. **Schwartz, M. N. and Elliott, D. L.,** Areal wind resource assessment of the United States, in *Alternative Fuels and the Environment*, Lewis Publishers, Boca Raton, FL, 1994, chap. 17.
3. **Eldridge, F. R.,** *Wind Machines*, van Nostrand Reinhold, New York, 1980.
4. **Hock, S. M., Thresher, R. W., and Williams, T. W.,** The future of utility-scale wind power, *Adv. in Solar Energy*, 7, 1992.

5. **Kelley, N. D. and Wright, A. D.,** A comparison of predicted and observed turbulent wind fields present in natural and internal wind park environments, presented at Windpower '91 Conf., Palm Springs, CA, September 1991.

6. **Veers, P. S.,** Three-dimensional wind simulation, presented at 8th ASME Wind Energy Symposium, January 1989.

7. **Tangler, J., Smith, B., and Jager, D.,** SERI advanced wind turbine blades, presented at ISES Solar World Congress, Denver, CO, August 1991.

8. **Butterfield, C. P., et al.,** Dynamic stall on wind turbine blades, presented at Windpower '91 Conf., Palm Springs, CA, September 1991, NREL/TP-257-4510.

9. **Wright, A. D. and Thresher, R. W.,** Prediction of stochastic blade responses using measured wind-speed data as input to FLAP, SERI/TP-217-3394, November 1988.

10. **Musial, W. D., et al.,** Photoelastic stress analysis on a Phoenix 7.9-meter blade, NREL/TP-257-4512, October 1991.

11. **Vachon, W. A.,** The effect of controls on life and energy production of the 34-m test bed, presented at 8th ASME Wind Energy Symposium, January 1989.

Chapter 16

Assessing Renewable Energy Resources — Views Concerning the Federal Role

David S. Renné

CONTENTS

INTRODUCTION

Assessing the energy resource base for renewable energy technologies (such as wind, solar, biomass, hydro, and geothermal) often involves the use of geophysical data and collateral information developed through federal programs for unrelated purposes. For example, much of the information used in the early depiction of national wind energy resources was derived from hourly wind observations collected routinely at airports throughout the country by the National Weather Service (NWS). This approach to renewable energy resource assessment is in contrast to the more traditional "fixed stock" energy technologies (coal, oil, and nuclear). In these, measurement surveys are specifically directed toward assessing the total available and accessible resource base. Thus, data gathering programs within federal organizations such as the NWS, although well suited for their weather forecasting mission, were not originally designed to assess renewable energy resources specifically. Therefore, renewable resource assessments may contain uncertainties that are unacceptably high for deployment of an emerging renewable technology.

In this chapter, historical legislation defining a federal role in developing renewable energy resource assessments is presented, followed by a description of how some of these assessments have actually been conducted for two renewable technology areas: wind and solar. An examination is also made of how uncertainties in these assessments can influence renewable energy technology deployment decisions. This discussion helps suggest what the current federal and private roles should be in renewable energy resource assessments applicable to all technologies.

HISTORICAL PERSPECTIVE
ON RENEWABLE ENERGY ASSESSMENTS

The potential use of renewable energy technologies such as wind and solar emerged to the forefront of the American consciousness following the Arab oil embargo of 1973. However, relatively poor understanding of the spatial distribution of wind and solar energy resources was a significant source of uncertainty in evaluating their potential for meeting the nation's energy supply. During this time, the federal role in developing resource information was defined in part by legislation passed by the 93rd Congress as part of the Solar Energy Research, Development, and Demonstration Act of 1974 (P.L. 93-473). Section 5 of this Act, titled "Resource Demonstration and Assessment", states that a solar energy resource determination and assessment program shall be initiated "with the objective of making a regional and national appraisal of all solar energy resources..." The legislation goes on to say that "...the program shall emphasize identification of promising areas for commercial exploitation and development. The specific goals shall include:

*   Development of better methods for predicting the availability of all solar energy resources, over long time periods and by geographic location;
*   Development of advanced meteorological, oceanographic, and other instruments, methodologies and procedures necessary to measure the quality and quantity of all solar resources on a periodic basis;
*   Development of activities, arrangements, and procedures for the collection, evaluation, and dissemination of information and data relating to solar energy resource assessment."

Federal legislation such as this, and research programs initiated at that time, led to efforts for depicting the nation's spatial extent of wind and solar resources. However, the only significant nation-wide database to complete this work resides with the long-term measurements of wind and other meteorological parameters made at NWS stations, primarily airports. A small handful of these NWS stations have also collected solar irradiance data. The data collected at these locations are archived and stored at the National Climatic Data Center (NCDC) in Asheville, North Carolina. These are the primary data source for developing national assessments of wind and solar resources.

NATIONAL WIND ENERGY ASSESSMENT

The NCDC database has both significant value as well as serious shortcomings for assessing the nation's wind energy resources. For example, the NWS measurement network was designed to fulfill the primary meteorological mission of providing observations to allow for safe aircraft flight operations and to improve weather forecasts. Because of their general proximity to urban locations, weather observations at airport locations may not be representative of all conditions within the region. This was recognized early on when Pacific Northwest Laboratory developed the most comprehensive atlas of wind energy resources ever prepared for the U.S.[1] Virtually all the wind information archived at the NCDC was used in the development of this atlas. However, in certain regions of the country, particularly in coastal and mountainous regions, this database was inadequate for depicting some of the most important wind resource areas. Where few (if any) direct measurements of wind conditions were available in the NCDC archives, researchers had to make use of knowledge of wind flow characteristics in complex terrain and coastal regions where significant accelerations and decelerations of wind speed are possible. In this way, a detailed gridded wind resource atlas was developed for the U.S. despite the absence of wind data at or near many of the grid points. Since wind resources were estimated at many grid points, a "certainty rating" was assigned to each grid. The degree or rating of certainty tends to be quite low in regions of complex terrain such as the Rocky Mountains and the Appalachians. The rating is relatively high in regions where NWS measurements would be representative of large areas, like the Great Plains. An important point here is that many of the techniques used to "fill in" estimates of wind resources in areas with little or no data made use of a comprehensive knowledge of mesoscale and microscale meteorology that, in part, results from years of federally sponsored research programs.

NATIONAL SOLAR ENERGY ASSESSMENT

A similar, if not more constraining, situation exists for depicting the nation's solar resources. Major efforts to produce a comprehensive National Solar Radiation Data Base (NSRDB) were recently completed by the National Renewable Energy Laboratory (NREL), in conjunction with the NCDC.[2] The purpose of this database is to provide a more current and accurate national atlas of solar resources in the U.S. An earlier database, SOLMET/ ERSATZ, was completed in the late 1970s and described in SOLMET, Vol. 2.[3] Both databases used actual measurements of solar radiation made intermittently at about 40 NWS stations since 1952 (known as the NWS SOLRAD network). As with wind resources, mesoscale and microscale meteorological processes can result in far different solar resource values than that observable even at a nearby station. Consequently, an accurate depiction of the national resource is not possible with a spatial coverage of 40 stations or less, many of which did not operate concurrently.

NREL, therefore, developed a model that makes use of meteorological observations at 239 NWS stations to provide a more comprehensive database. This resulted in a database that consists of over 90% modeled, rather than measured, information. Even so, many areas of the U.S., particularly those in regions of complex terrain such as the Rocky and Sierra Nevada Mountains, do not have even modeled data; again, because of the lack of NWS station observations. As NREL produces solar atlases from the NSRDB, the certainty degree or rating associated with estimating resource levels at regions where neither measured nor modeled data exist will have to be specified in a way similar to that for the wind atlases.

Figure 1 shows the location of the 239 stations used in the NSRDB. Virtually all these stations are NWS stations.* The uncertainties in developing a comprehensive national atlas from such a station density, particularly in the complex terrain regions of the western U.S. and many coastal areas, is obvious.

Many data collection programs have been conducted in the U.S. to address wind and solar energy resources, but for one reason or another these are not available to the national databases (i.e., the proprietary nature of some data, cost of data assimilation into a national database, and uncertainties in data quality). Thus, although additional efforts can improve the spatial density of coverage shown in Figure 1 to some extent, other techniques (e.g., data interpolation methods, models, or surface radiation estimates derived from satellite data) are necessary to reduce the uncertainties in a gridded national atlas of renewable energy resources.

The future coverage of solar resource data using the federal network is unclear. The NWS is undergoing a modernization program to improve short-term weather forecasts. This program will pay dividends in improved public health and safety, safer aircraft operations, and decreased crop losses. However, a more cost-effective observational

---

* For both the wind and solar assessments, data collected by non-federal programs has been made available to improve the spatial coverage. However, this represents a small percentage of the total database.

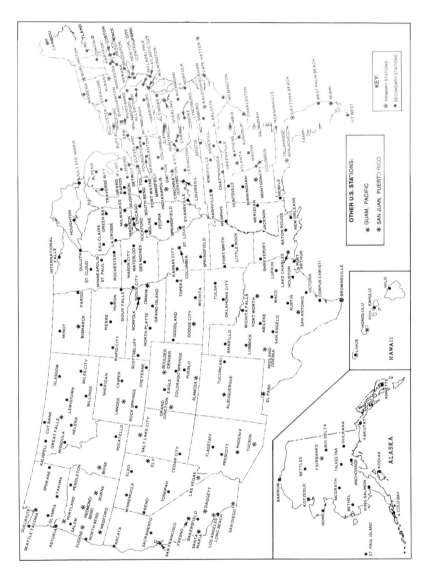

Figure 1 Location of sites used in the National Solar Radiation Data Base. The majority of these sites are at airport locations and are operated by the NWS.

strategy is also being implemented by NWS, which means the way in which weather observations are currently made will be changed or, in some cases, terminated. For example, the existing SOLRAD network is being curtailed, which will impede future efforts to upgrade the NSRDB. Upgrades of all databases are essential, since the industry will continually need an evaluation of resource levels relevant to the time period for which the projects will occur. Furthermore, using only outdated historical databases can result in important uncertainties regarding nonrepresentative periods associated with such phenomena as volcanic eruptions, climatic variability, and trends in human-induced pollutant loadings or microclimatic change. Examples of this are given in the next section. Unfortunately, these actions provide an example of a federally-supported initiative that has an important benefit to society (health and safety through improved short-term weather forecasts), but will result in a loss of information beneficial to another important and growing segment of society (energy independence through the use of renewable technologies).

HOW RESOURCE UNCERTAINTY CAN AFFECT DEPLOYMENT

There are many sources of uncertainty that occur when determining the renewable energy resource at a given location. This section presents an example of how understanding and reducing the solar energy resource uncertainty is important to renewable energy deployment.

SOURCES OF UNCERTAINTY IN RESOURCE ASSESSMENT

One source of uncertainty associated with resource assessment is the inherent error in the measurement methods and systems. This includes errors associated with instrument performance and calibration, and the way the instrument output is converted to engineering units. Reducing these uncertainties is possible by applying greater care in conducting a measurement program and assuring that the instrument calibrations are traceable to a known standard.

A second source of uncertainty is due to the spatial and temporal nature of the resource. Only a finite number of measurement stations are feasible and cost effective. Thus, sampling aliasing, both in space and time, can result in both spatial and temporal uncertainties in the measured resource. One way in which this source of uncertainties can be reduced for regional resource assessments is to increase the geographic distribution of measurements and to conduct the program over a long period of time (decades).

A third source of uncertainty in resource assessment is in the representativeness of the sites chosen to be included in the assessment. A measurement station that is representative of only the specific local microclimate in which it is installed will represent a much greater source of uncertainty when used to extrapolate to a regional depiction of resources than a station that is representative of a large geographic area. Thus, resource assessments in the Rocky Mountain region typically have much larger uncertainties than in the Great Plains because of a combination of conditions where individual measurement locations in the Rocky Mountains are not as representative of large areas as stations in the Great Plains, and where the spatial coverage of measurements in the Rockies tends to be of much lower density than in the Great Plains.

A fourth source of uncertainty in resource assessment is in the method used to either extrapolate or interpolate resource data to location where there are no data. Interpolation procedures can range from simple contouring routines to complex models; however, any procedure will have inherent uncertainties. Whenever resource information is interpolated to a site where there is no actual data, the uncertainty in the method used to perform the interpolation must be added to the uncertainty of the overall assessment.

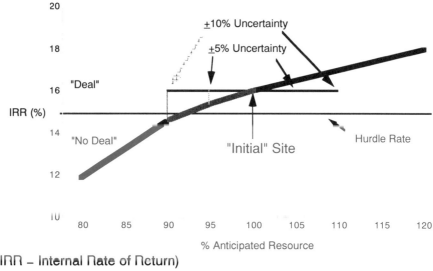

(IRR – Internal Rate of Return)

Figure 2 Effect of resource uncertainty on investor return.

A fifth major source of uncertainty in resource assessment is the natural climatic variability that occurs. This variability cannot be predicted for future projects and is not captured in short-term measurement programs. However, if a measurement program has been conducted for a long time and thus captures climatic variability of the type that would be expected to occur over the life of a project, this long-term information is useful for project planning. An example of the usefulness in knowing long-term resource variability is shown in a later section.

IMPORTANCE OF REDUCING RESOURCE UNCERTAINTY: A HYPOTHETICAL CASE STUDY

The following examines a hypothetical case study of a renewable energy project, where the economics or internal rate of return (IRR) is related to the anticipated resource at the proposed site. The IRR is a common measure of the financial viability of a proposed project. The IRR is defined as the discount rate that makes the net present value equal to zero over the lifetime of the project. Since a higher IRR means a better return on the investment, one would expect that the IRR would be a function of the resource available at the site. Typically, the higher the resource level, then the higher the IRR. Many other factors besides the resource level go into estimating the IRR of a project, such as the initial investment, depreciation, tax incentives, interest rates on loans, inflation, and operation and maintenance costs. For any given project, the IRR must typically exceed some minimum value, commonly referred to as the "hurdle rate", in order to attract investors and make the project viable. Although evaluation of IRR can be a complex process, our purpose here is to show a simple example to indicate the value in reducing resource uncertainty (or, for instance, finding a better wind or solar site than originally proposed) in the context of project deployment.

Figure 2 shows a hypothetical relationship between a project IRR and the percentage of anticipated resource at a site. The hurdle rate for this particular example is at an IRR of 15%. For this example, the anticipated resource at the proposed site actually achieves a hurdle rate of 16%, meaning there should be a "deal" to deploy the project based on the anticipated resource at the site. However, there may be considerable uncertainty associ-

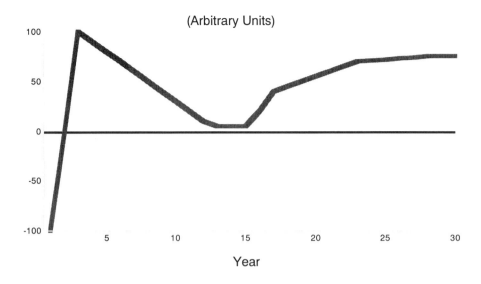

Figure 3 Hypothetical 30-year project cash-flow profile.

ated with the resource value (based on the discussion in the previous section) that should also be accounted for in the decision to deploy the project. If there is a 10% uncertainty in the resource, shown by accounting for the "worst case", the project IRR actually drops below the hurdle rate and will likely not be deployed. However, if that uncertainty can be reduced to 5%, even using a worst case assumption, the project would still be viable. Conversely, prospecting for a higher-than-anticipated resource site (further to the right along the curve) and remaining far enough above the hurdle rate[*] means that large uncertainty will not be a factor in canceling the project. Thus, this simple example demonstrates how reducing uncertainty in resource information reduces the risk in project decision-making, indicating the value to the renewable energy industry of high-quality, representative data in a resource assessment.

THE IMPORTANCE OF INTERANNUAL RESOURCE VARIABILITY

Figure 3 shows a hypothetical cash flow profile over the 30-year life expectancy of a renewable energy project, indicating the year-to-year variability of a project's economic viability. The hypothetical curve indicates a large negative cash flow at the initial investment of the project, shifting quickly to a positive cash flow for reasons such as tax incentives and delayed loan paybacks. Over time, the tax incentives may decrease and loans must be paid off, causing the cash flow to decrease. In later years of the project, cash flow can again increase once all the debt is paid off and revenues from energy sales remain high. However, from a profile such as this, it can be seen that if two or three low-resource years occur during periods when cash flow is expected to be low, the project could run into serious financial difficulties. For this reason, it is essential that long-term resource information be available for evaluating the performance of a project over its lifetime.

This profile, as well as the example in the previous section, shows how important it is to evaluate the representativeness of short-term resource data to long-term climatic conditions, and to assure that natural climatic variability is incorporated into the analysis. In the case of solar resource information, natural events such as volcanic eruptions can

---

[*] This additional site prospecting will likely add to the initial cost of the project.

252

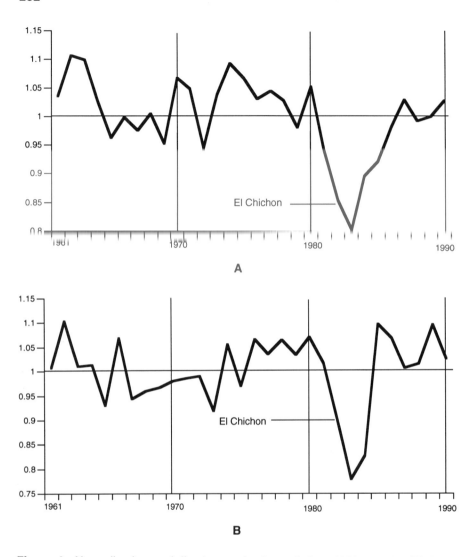

Figure 4 Normalized annual direct normal solar radiation, 1961 to 1990. (A) Boulder, Colorado; (B) Ely, Nevada.

influence the year-to-year variability of the resource. In order to demonstrate real examples of this variability, Figures 4A and 4B show the 30-year variability of the normalized direct normal solar radiation at two sites: Boulder, Colorado and Ely, Nevada. During the first 22 years of record, the direct normal data at both sites varies by no more than ±10% of the long-term mean. However, 1982 witnessed the massive eruption of El Chichon in Mexico. The large amount of volcanic ash injected into the stratosphere from this event, as witnessed by the large increase in aerosol optical depth (as described in Reference 2) contributed to a substantially reduced direct normal solar resource for several years. Clearly, if resource data collected at the proposed site in Figure 2 did not reflect this event, or if this event occurred during a time when a project's cash flow was already low (Figure 3), a renewable energy project could be at high risk. Although volcanic eruptions are not readily subject to long-term predictions as yet, having an historical record that shows the type of interannual variability that can be expected at a site may be helpful when evaluating proposed renewable energy projects.

THE FEDERAL AND THE PRIVATE ROLES
IN RESOURCE ASSESSMENT

From the discussion in the previous sections, one might ask the question: What is the appropriate federal role for reducing the uncertainties in depicting the extent of our national renewable energy resources? Since the federal government's current focus in the development and deployment of renewable energy technologies includes "market conditioning" (i.e., reducing barriers in implementing renewable energy projects), one can argue that a much more aggressive federal program to improve the spatial and temporal quality of renewable resource data is appropriate. Although the existing NWS database serves an important national need (improved weather forecasting, safe aircraft operations), important uncertainties exist when applying it to other national needs, i.e., accurate depiction of current renewable energy resources. However, these shortcomings of NWS data are offset, in part, through other aspects of federally sponsored research that allow for the interpolation or extrapolation of data to regions where no data exist. Furthermore, a national measurement strategy to accurately depict renewable resources at *all* locations would obviously be cost prohibitive. Since reliable data have *value* to the renewable energy industry, there is a point at which on-site measurements to reduce the uncertainty in resource characterization within a grid point of a national atlas are more appropriately a private activity. Given the trade-offs between national priorities and the needs of the renewable energy industry, the following scenario is suggested for the federal and the private roles in renewable energy resource assessments.

THE FEDERAL ROLE

The federal government should support the collection of *high-quality*, *calibrated* data at a number of representative locations throughout the U.S. to serve as long-term *baseline* stations for purposes of assessing and characterizing renewable energy resources and reducing uncertainties. What this means, of course, is that the current high-quality measurement of wind resources at NWS sites throughout the U.S. should continue. For solar resources, a multiagency effort is needed to reinstate a high-quality solar radiation monitoring network, improve its performance, and expand its coverage — perhaps at sites other than NWS sites. Other renewable technologies require a similar strategy of high-quality baseline measurements. This measurement strategy would serve multiple purposes; not only would it provide critical assistance to the renewable energy industry, it would assure that a measurement program was in place which would help in the monitoring of the cause and effects of global climate changes induced by human activities. These synergistic benefits would help justify the cost of such an expanded measurement program.

Another federal role is to develop and provide to the renewable energy industry the tools necessary to characterize a specific site through models and interpolation techniques that make use of such data. For example, if appropriate techniques are available to interpolate or extrapolate long-term baseline data to a given site at an acceptable level of accuracy, then the time to properly evaluate the proposed site can be shortened considerably, thereby reducing a major risk to renewable energy deployment. These tools can include ways to incorporate collateral observations, such as satellite data or data from other agencies and non-federal data sources, to augment the database. This implies a multiagency effort that involves not only DOE, but also NOAA, the National Aeronautics and Space Administration, the National Science Foundation, the Department of Defense, etc.

THE PRIVATE ROLE

On-site characterization of renewable energy resources is a necessary role for private developers in order to maximize positive cash flow and minimize risk in a project. The developer should have access to the federal database and tools with which to make these characterizations and to the state-of-the-art measurement strategies researched by the federal government. Then the developer should implement these tools with a measurement strategy and any other modeling or interpolation techniques appropriate to validate the site characterization efforts for long-term deployment of renewable energy technologies such as wind, solar, biomass, geothermal, hydro, etc.

ACKNOWLEDGMENTS

The author would like to thank Mr. Brian Parsons of the National Renewable Energy Laboratory for his very helpful and insightful discussions regarding the role of resource information in making renewable characterizations and decisions. The author also greatly appreciates the constructive comments provided by the reviewers of this chapter.

REFERENCES

1. **Elliott, D. L., et al.,** Wind Energy Resource Atlas of the United States, DOE/CH10094-4. Solar Technical Information Program, National Renewable Energy Laboratory, Golden, CO. 1987.
2. National Renewable Energy Laboratory, User's Manual: National Solar Radiation Data Base (1961–1990), Version 1.0, NSRDB-Volume 1, Distributed by National Climatic Data Center, Federal Building, Asheville, NC, 1992.
3. SOLMET, Vol. 2, Final Report — Hourly Solar Radiation-Surface Meteorological Observations, TD-9724 National Climatic Data Center, Asheville, NC, 1979.

Chapter 17

Areal Wind Resource Assessment of the United States

M. N. Schwartz and D. L. Elliott

CONTENTS

INTRODUCTION

Estimates of the electricity that could potentially be generated by wind power and of the land area available for wind energy development have been calculated for the contiguous United States, in support of the U.S. Department of Energy's National Energy Strategy. These estimates were based on the wind resource data published in a national resource atlas.[1] Estimates of the wind resource in this atlas are expressed in wind power classes ranging from class 1 to class 7, with each class representing a range of mean wind power density or equivalent mean speed at specified heights above the ground (Table 1). Wind turbine hub heights are generally 30 to 50 m above the ground. Areas designated class 4 or greater are suitable for most wind turbine applications. Power class 3 areas may be suitable for wind energy development using turbines on tall (hub height of 50 m or higher) towers. Class 2 areas are marginal, and class 1 areas are unsuitable for wind energy development. A map of the areal (percentage of land area) distribution of the wind resource digitized in grid cells (1/4° latitude by 1/3° longitude; Figure 1) shows that exposed areas with moderate to high wind resource (class 3 and greater) are dispersed throughout much of the contiguous U.S.

METHOD OF CALCULATING THE WIND ELECTRIC POTENTIAL

Several factors determine the amount of land area suitable for wind energy development within a particular grid cell in a region of high wind energy potential. The important factors include the percentage of land exposed to the wind resource and land-use and environmental restrictions. The land area exposed to the wind for each grid cell was estimated based on a landform classification and ranged from 90% for relatively flat terrain down to 5% for mountainous terrain. Estimates of land area excluded from wind energy development, in percent per grid cell, were made for various types of land-use (e.g., forest, agricultural, range, and urban lands). Environmental exclusion areas were defined as federal and state lands where wind energy development would be prohibited or severely restricted (including parks, monuments, wilderness areas, wildlife refuges, and other protected areas). Finally, additional land exclusions were estimated to account for transportation right-of-ways, locally administered park land, privately administered areas, and proposed environmental lands.

Table 1 Classes of wind power density

Wind Power Class	10 m (33 ft)[a]		30 m (98 ft)[a]		50 m (164 ft)[a]	
	Wind Power Density, W/m²	Speed[b] m/s (mph)	Wind Power Density, W/m²	Speed[b] m/s (mph)	Wind Power Density, W/m²	Speed[b] m/s (mph)
	0	0	0	0	0	0
1	100	4.4 (9.8)	160	5.1 (11.4)	200	5.6 (12.5)
2	150	5.1 (11.5)	240	5.9 (13.2)	300	6.4 (14.3)
3	200	5.6 (12.5)	320	6.5 (14.6)	400	7.0 (15.7)
4	250	6.0 (13.4)	400	7.0 (15.7)	500	7.5 (16.8)
5	300	6.4 (14.3)	480	7.4 (16.6)	600	8.0 (17.9)
6	400	7.0 (15.7)	640	8.2 (18.3)	800	8.8 (19.7)
7	1000	9.4 (21.1)	1600	11.0 (24.7)	2000	11.9 (26.6)

[a] Vertical extrapolation of wind power density and wind speed are based on the 1/7 power law.

[b] Mean wind speed is estimated assuming a Rayleigh distribution of wind speeds and standard sea-level air density. The actual mean wind speed may differ from these estimated values by as much as 20%, depending on the actual wind speed distribution and elevation above sea level.

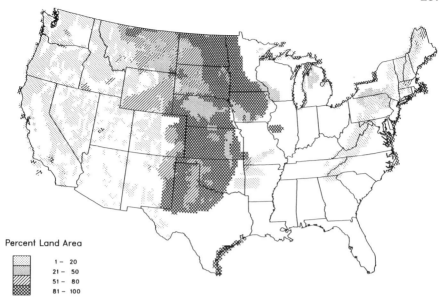

Percent Land Area

	1 – 20
	21 – 50
	51 – 80
	81 – 100

Figure 1 Percentage of land area with class 3 or higher wind resource.

The wind electric potential per grid cell was calculated from the available windy land area and the wind power classification assigned to each cell. The amount of potential electricity that can be generated is dependent on several factors, including the spacing between wind turbines, the assumed efficiency of the machines, the turbine hub height, and the estimated energy losses (caused by wind turbine wakes, blade soiling, etc.).

A wind turbine spacing of 10D (row spacing) by 5D (lateral spacing within a row), where D is the rotor diameter, has been assumed for the calculation of wind electric potential. This spacing is generally applicable to areas such as the Great Plains, where the power-producing winds blow from a variety of directions. Closer spacing of wind turbines occur in the mountain passes of California where the winds blow from one predominant direction.

The assumed turbine efficiency (25%) and the estimated energy losses (25%) are based on the characteristics of wind turbines installed during the last decade. Wind turbines with advanced technology features such as improved blade design are currently under development. These machines will be more efficient and have lower energy losses than turbines presently in operation.

RESULTS

A study in 1991[2] used a variety of land exclusion scenarios to estimate the available windy land and resultant wind electric potential for several levels of wind resource, both for the contiguous U.S. as a whole and for each of the 48 contiguous states. Figure 2 shows the overall wind energy potential based on terrain exposure and several land exclusion scenarios. Even under the most restrictive (severe) land exclusion scenario, the wind energy potential for areas with power class 3 and above is greater than total U.S. electricity consumption was in 1990. Figure 3 shows the contribution that the wind energy of each state could make to meet the total electrical needs of the nation, assuming a moderate land exclusion scenario. North Dakota alone has enough potential energy from windy areas of class 4 and higher to supply 36% of the total 1990 electricity consumption of the 48 contiguous states.

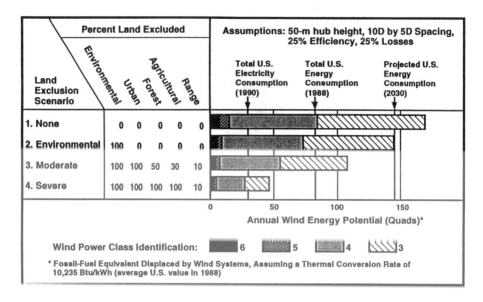

Figure 2 Wind energy potential of the contiguous United States.

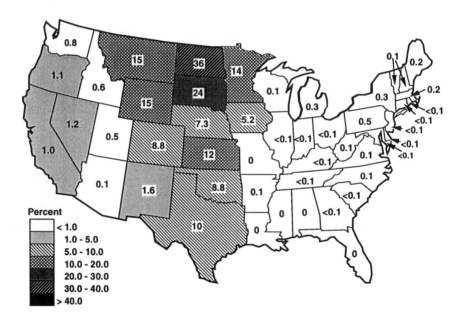

Figure 3 Percentage of the total 1990 electricity consumption of the contiguous United States that each state could produce (assumptions: areas designated class 4 and above, hub height of 30 m, moderate land exclusion scenario).

A study in 1992[3] recalculated estimates of windy land area and wind electric potential based on a more accurate mapping of environmental exclusion areas and a moderate land exclusion scenario. These new estimates were about 1 to 2% higher than the preliminary estimates made in 1991 based on the same land exclusion scenario. Overall, even with land being excluded from wind energy development for environmental and land-use considerations, the amount of windy land available and potential electric power from wind energy is surprisingly large. The amount of windy land available for power class 4 and above is approximately 460,000 km², or about 6.0% of the total land area in the contiguous U.S. The potential average power from areas with class 4 and higher is estimated at 500,000 MW. If advanced wind turbine technology is utilized to take advantage of areas with wind resource class 3 and higher, then the amount of windy land available is over 1,000,000 km², or almost 14% of the land area in the contiguous U.S. Moreover, the estimates show that a group of 12 states in the midsection of the country have enough wind energy potential to produce nearly four times the amount of electricity consumed by the nation in 1990.

CONCLUSION

The considerable wind electric potential has not been tapped before because the wind turbine technology was not able to utilize this resource. However, during the past decade, increased knowledge of wind turbine behavior has led to more cost-effective wind turbines that are more efficient in producing electricity. The price of the electricity produced from wind by these advanced turbines is estimated to be competitive with conventional sources of power, including fossil fuels. Because of the increasing competitiveness of wind energy, wind resource assessment will become essential in incorporating wind energy into the nation's energy mix. For example, the 1992 study[3] also produced gridded maps of wind electric potential and windy land area for the 48 contiguous states that show the distribution of these features within individual states, thus enabling utilities and wind energy developers to identify promising areas for wind energy. The importance of accurate wind resource assessment is also recognized in other parts of the world.

Detailed wind resource assessments have been proposed or are being considered as part of a plan to increase the use of wind energy in Europe, Asia, Latin America, and other regions. The decreasing cost of wind power and the growing interest in renewable energy sources should ensure that wind power will become a viable energy source in the U.S. and worldwide.

ACKNOWLEDGMENTS

We wish to thank Alan Peoples for his efforts on the development of the updated database for exclusion areas and Gene Gower for the development of the computer software for the project. This work was supported by the U.S. Department of Energy (DOE) under Contract DE-AC06-76RLO 1830 at the Pacific Northwest Laboratory, which is operated for the U.S. Department of Energy by Battelle Memorial Institute.

REFERENCES

1. **Elliott, D. L., Holladay, C. G., Barchet, W. R., Foote, H. P., and Sandusky, W. F.,** Wind Energy Resource Atlas of the United States, DOE/CH 10093-4, Solar Energy Research Institute, Golden, CO, 1987.
2. **Elliott, D. L., Wendell, L. L., and Gower, G. L.,** An Assessment of the Available Windy Land Area and Wind Energy Potential in the Contiguous United States, PNL-7789, Pacific Northwest Laboratory, Richland, WA, 1991.
3. **Schwartz, M. N., Elliott, D. L., and Gower, G. L.,** Gridded state maps of wind electric potential, in *Proceedings Windpower'92*, American Wind Energy Association, Washington, D.C., 1993, 50.

Chapter 18

The Biomass Resource Potential of Energy Crops

L. L. Wright, J. H. Cushman, and S. A. Martin

Biomass refers to green plants or almost any organic products derived from plants. Green plants capture solar energy and store it as chemical energy suitable for conversion to more convenient types of energy forms. Trees and grasses, starch and oil seeds, sawdust, wood wastes, agricultural residues, food-processing wastes, and paper are all biomass materials. Biomass can be used directly as a solid fuel to produce heat, or it can be converted to other energy carriers such as liquid and gaseous fuels. Biomass is the only renewable energy resource that can be converted to liquid fuels.

Using biomass for energy has a long history. Wood chunks and cow dung have been used for heating and cooking since humans discovered fire. Wood wastes and garbage are burned in many locations to generate heat and electricity. Half the transportation fuel used in Brazil is ethanol derived from fermentation and distillation of sugar from sugarcane. China and India convert animal and plant wastes into methane for heating, cooking, and electricity using bacteria to decompose biomass in biogas digesters. Biomass currently accounts for about 14% of the world's total energy use and 35% of the energy use in developing countries.[1]

Given the many potential uses of biomass and the large projected future needs for energy, there is considerable interest in quantifying the total amount of biomass resources that are available on a renewable basis in the U.S. Wood wastes, municipal solid wastes, and other biomass wastes supplied about 3.54 exajoules (EJ) or a little over 4% of the nations total energy requirement in 1990.[2] Estimates of additional resource availability from wastes, agricultural residues, and forest thinnings vary from 1 to 14 EJ; but when economic recoverability is considered, the estimates are closer to 1 EJ. Forests and grasslands could be harvested for energy, but management for other uses including forest products, recreation, and wildlife habitat is usually preferable. The millions of hectares of excess cropland which exists in the U.S. provide the potential for producing several exajoules of dedicated biomass energy feedstocks. Dedicated feedstock supply systems for energy can revitalize the agricultural sector of the U.S. economy by providing profitable uses for marginal cropland.

The dedicated feedstock supply systems (DFSS) being developed by the Department of Energy include several fast-growing trees, perennial grasses, and annual grasses (Table 1). Experimental results from field trials have generated optimism that selected and improved energy crops, established on cropland with moderate limitations for crop production, have the potential for producing high yields. Both trees and grasses, under very good growing conditions, have produced average annual yields of greater than 35 dry Mg/ha/year. Sorghum has shown especially high yields in the Midwest. Energy cane, a hybrid between sugarcane and its wild relatives, has yielded as much as 48 dry Mg/ha/year in Florida. These experimental results demonstrate that some species have the genetic potential for very rapid growth rates.

The new woody crops developed by the Department of Energy's Biofuels Feedstock Development Program (BFDP) offer, at a minimum, a 100% increase in biomass production rates over the 2 to 4 Mg/ha/year of dry leafless woody biomass produced by most natural forest systems. Short rotation woody crops (SRWC) are capable of producing biomass yields of 9 to 22 dry Mg/ha/year on typical cropland sites. These gains in yields

Table 1 **Average and highest yields of short rotation woody crops (SRWC) and herbaceous energy crops (HEC)**

	Highest Yield (Mg/ha)	**Average Yields (Mg/ha)**	**Production Area**	**Special Ecological Niche**
All yields are dry weight, Mg/ha/year				
SRWC species	43	9–??	Eastern U.S. and West Coast	Occasionally-flooded croplands
HEC species				
Thin-stemmed perennials	35	4–25	Eastern U.S.	Dry and erosion-sensitive croplands
Thick-stemmed perennials	48	12–48	Alabama, Florida, Hawaii, Georgia	South and subtropics
Annuals	37	1–37	Eastern U.S.	Crop rotation

over natural forest systems are a result of identifying superior varieties of fast-growing trees, collecting and propagating selected species and varieties, identifying the site requirements for those species and varieties, improving stand establishment success, and identifying appropriate weed control techniques. Yield data are from experimental plots established on cropland with moderate limitations.

Woody crop species with the most potential are poplar, black locust, sycamore, sweetgum, and silver maple.[3] Poplar species or hybrids have shown great potential for growth across many parts of the U.S. However, the susceptibility to fungal disease of many poplar clones must first be overcome to ensure dependable yields. Genetic improvements over wild types are either being initiated or have been achieved in black locust, silver maple, sweetgum, and sycamore. Eucalyptus has shown exceptional growth promise for Hawaii, Florida, southern Texas, and California. Some of the SRWC species may be a suitable crop for occasionally flooded croplands since, after the planting year, they would be less susceptible to crop loss. Intensive management is necessary only during the first 1 or 2 years or until the trees fully occupy the site. Fertilizer and pesticide applications are anticipated to be lower than levels used on most agricultural crops. Harvest would occur every 5 to 10 years. The ability of these trees to resprout after harvesting potentially means that establishment will occur only once every 15 to 30 years. However, if new improved plant materials are continuously produced, replanting after each harvest may be more common.

Several herbaceous energy crops (HEC) have been screened by BFDP on a range of sites. Grasses, including thin-stemmed perennials, thick-stemmed perennials, and annuals, have shown the best results.[3] Of these, the thin-stemmed perennials have promise for use across most of the present crop-growing regions of the U.S. These include cool-season grass species (e.g., wheatgrasses, smooth bromegrass, tall fescue) and warm-season grass species (e.g., big bluestem, bermudagrass, switchgrass). Thin-stemmed warm-season grasses native to the U.S. represent the best options among herbaceous species for energy crops for these reasons: compatibility with existing harvest and handling systems, erosion control capabilities, and relatively high productivity on a wide range of sites. Planting, fertilizing, harvesting, and storage operations are similar to those used for hay production. Once established, perennial grasses provide erosion control close to that achieved by undisturbed pasture. This is fortunate, as it means energy crops may be an acceptable alternative crop for erosion-prone croplands. Yields for various grass varieties have ranged from 4 to 25 Mg/ha/year once established in many test plots.

Experimental stands have been successfully established in the Southeast, the Midwest, and the Great Plains of the U.S. Very recent trials with switchgrass suggest that select varieties may yield as high as 35 Mg/ha/year in Alabama and Texas, and up to 20 Mg/ha/year in Virginia and Nebraska.

The remaining two herbaceous systems have different production niches. Thick-stemmed perennials such as energy cane, sugar cane, and napiergrass provide very high yields but are limited to the South and subtropics. Stands have been established in Alabama, Florida, Hawaii, and Georgia. Management handling as an energy feedstock will be different from current procedures for sugar production. Either drying or storage as a silage may be required. These types of grasses require high fertilizer levels and the heavy stalks can blow over, causing losses at harvest. Yields obtained have been very good, ranging from 12.2 to 47.7 Mg/ha.

The third group of energy crops, annuals, includes corn, sorghum, sudangrass, and their hybrids, with yields ranging as high as 37 Mg/ha. Annuals can be included in a crop rotation system, which may increase acceptance by the farmer. Annuals have a higher management requirement than perennials as they must be planted each year. They are also unsuited for environmentally sensitive areas unless used as part of a crop rotation that does not leave the soil uncovered and vulnerable to erosion between harvest and next planting.

SRWC and HEC systems offer several advantages over natural forests and long-rotation systems for supplying energy conversion facilities. Natural forests have recreational, wildlife, scenic value, soil stabilization, and other values that often are not compatible with intensive harvesting operations. In cases where natural forests or long-rotation systems are being harvested for timber or pulp, energy can provide an alternative market for residues, but the supply would likely be insufficient as a primary energy feedstock. Research indicates that it would not be economically or environmentally beneficial to replace natural forest systems with energy crops. DFSS provide an environmentally desirable alternative for providing dependable supplies of biomass within a reasonable haul distance of an energy facility.

The future DFSS resource in the U.S. is somewhere between 5 and 20 EJ (exajoules) raw biomass. Biomass wastes and residues already contribute about 3.6 EJ to the nation's energy supply. The total biomass energy contribution can be substantial relative to the 83 EJ fuels and electricity currently used in the U.S. Major physical factors determining the resource potential include the amount and type of land "available" and average biomass yields (Figure 1). Figure 1 expresses total mass of material produced converted to its energy equivalent based on an assumption of about 18.5 GJ/Mg of dry biomass; it does not consider energy used to produce the material or convert it to fuel. It is assumed that wood and grasses are produced in equal amounts. The net energy derived from biomass will depend greatly on whether the material is converted to liquid transportation fuels or electricity and on the technologies used for the production of the biomass.[4] Analyses suggest that net energy from DFSS converted to fuels or electricity will be 5 to 10 times greater than the energy inputs.

One reason for presenting the raw biomass potential rather than the amount of energy available once converted is that so many different options for conversion are available. Furthermore, ongoing research is rapidly changing the efficiency of those technologies. In direct combustion systems, new methods of using waste heat to lower the moisture level of the feedstocks can improve conversion efficiency. Jet turbines are being linked with small gasifiers to generate electricity with projected high efficiency. Gasifiers are being improved for the production of methanol from biomass using chemical catalysts, heat, and pressure. Pyrolysis, a process which applies heat to cause chemical changes in a substance, is being developed into a system to produce biocrude oils to substitute for petroleum products. Fuels cells that efficiently convert chemical energy from gases into

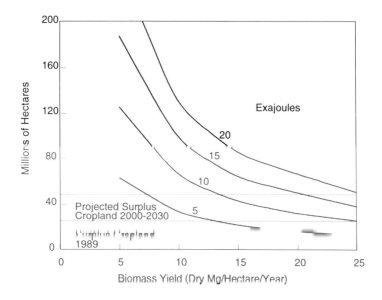

Figure 1   Relationship of yield and hectares to biomass resource potential in the United States.

electrical energy are available commercially for use in electric power plants and are being developed for use in homes and vehicles. Ethanol from sugarcane, sugarbeets, or corn is already being used as fuel, and this technology is being extended to convert biomass wastes to ethanol. Each system will vary in net energy produced.

Analysis has shown that about 160 million ha of the 170 million ha of cropland in the U.S. are capable of producing energy crops.[5] Much of the land capable of producing energy crops is currently being used for agricultural crops, pasture, or range. Yet, about 24 million ha cropland are currently being idled or removed from production as a result of government efforts to reduce surplus crop production and protect environmentally sensitive land.[6] USDA's 1990 Resource Conservation Act (RCA) analysis[7] projects that by 2030 as much as 52 million ha cropland may be idled and as much as 30 million ha land may be permanently removed from crop production (Table 2). The RCA analysis presumes that some of the 30 million ha permanently removed from agriculture will be converted to urban land, farmsteads, and other uses not suitable for agriculture or forestry purposes. However, about half of the 30 million ha is presumed to be Conservation Reserve Program (CRP) land that remains out of production.

Table 2 **Projected idle and harvested cropland in million hectares**

	Current (1990)	**Year 2000**	**Year 2030**
Total cropland	170	147	140
Cropped	138	88	88
Idle	32	59	52
Removed from cropland base	0	11	30

The figures in this table reflect an intermediate stress level that implies continuation of current trends in crop productivity and exports.[6,7]

Alternative uses for idled land are needed to maintain a healthy rural economy. The high-value specialty crops currently under development do not require large land areas to saturate their markets. Energy markets, in contrast, could potentially use all available land for production of energy feedstocks.

The amount of land that will be used in the future for production of energy crops is nearly impossible to predict. It will depend a great deal on energy prices, agriculture policy, export opportunities, and the success of energy and food crop research. The RCA analysis assumes significant increases in per-hectare yields of corn, soybeans, and wheat with about the same levels of inputs as currently used.[7] Although more than 50 million ha cropland is projected to be idled after the year 2000, about 75% of that land is located in the Great Plains[7] and requires irrigation for most crops. We presume that energy crop production will be limited to cropland not requiring irrigation because of the relatively high costs of irrigation and concern about depletion of aquifers. The amount of land projected to be idled in areas capable of supporting nonirrigated energy crop production is about 16 million ha. Additionally, much of the land in the CRP program today is also located in regions suitable for crop production though slope, wetness, thin soils, or other problems may limit productivity. Of the 14.3 million ha currently in CRP, about 10.5 million ha are in regions generally appropriate for energy crop production. Inasmuch as energy crops such as trees and perennial grasses have very low erosion rates compared to row crops, they could provide an environmentally sound alternative for using the land if soil quality is satisfactory or can be improved. Thus, the future cropland base potentially available and suitable for energy crops could include 16 million ha idled land and at least 10 million ha land projected to be permanently removed from the cropland base for a total of 26 million ha. The "available" land base is much larger and more of that land could become "suitable" for energy crop production with appropriate research advances, and policy or market changes.

The choices made by individual landowners will ultimately determine how much land is converted to energy crop production. Cropland is already being used for energy production. Substantial amounts of corn crops are being converted to ethanol fuels, and some soybean crops are being converted to biodiesel fuels. New energy crops that can be planted annually in rotation with food crops (such as sorghum or rapeseed) may be the easiest for landowners to adopt first. Perennial grasses, which are the same as or similar to forage grasses, may gain early acceptance since they will require relatively little change in cultural practices, attitudes, labor, and equipment. Wood energy crops have several attributes that could interest landowners if markets are available. These attributes include a redistribution of timing of labor demands, lower annual labor requirements after establishment, the ability to store the biomass on the stump to await better market conditions, and the ability to survive flooding conditions once established. However, woody crops also require new cultural practices, changes in attitudes about land use, and access to nontraditional farm equipment. Educational assistance to landowners will be critical for facilitating successful energy crop production on a large scale. Financial assistance and/or long-term contracts between energy developers and landowners may be necessary to increase supplies of energy crops.

Changes in both agricultural and environmental policies could greatly affect the profitability of producing energy crops. Reduction in price-support programs, international trade agreements for agricultural commodities, and carbon taxes on fossil fuel emissions are examples of policy changes that could favor energy crop profitability. Effecting a linkage between the energy and agricultural sectors of the U.S. economy by producing electricity and liquid fuels from energy crops has multiple benefits. All these benefits should provide a major stimulus for developing policies conducive to energy crop production and biomass energy commercialization.

There is a concern that biomass energy production could or does use land needed for food production and, therefore, that biomass energy should be ignored or abandoned as an energy supply solution. If the world population continues to rise and/or climate change occurs rapidly, concern over competition for land is legitimate. However, millions and millions of hectares of surplus good-quality cropland exist now in the U.S. and Europe. There are also millions of hectares of degraded land all over the world. If environmentally sensitive plant production technologies can be developed now for attaining both higher food crop yields and energy crop yields, this can only be a benefit to society now and in the future. Many agriculturists believe that large opportunities exist for improving production per unit area for both energy and food crops both through genetic selection and efficiency improvements.[8] The establishment of perennial crops (tree or grasses) on degraded lands could aid in improving the land for future production of food crops. Developing a greater understanding of plant physiology, genetics, and biotechnology for producing energy feedstocks can only have positive spin-offs for meeting human food needs. Research aimed at ensuring that energy crops are produced in a way that is environmentally sound will likely be applicable to food production techniques as well. The contribution of biomass energy systems will eventually be limited by land competition issues, but the positive benefits to be gained by developing biomass energy now should not be overlooked.

Biomass energy systems can have many positive environmental benefits. At a global level, substitution of biomass feedstocks for fossil fuel feedstocks can contribute to the reduction of anthropogenic carbon dioxide emissions. Wright et al.[4] present calculations showing that if 14 to 28 million ha land in the U.S. could produce energy crops, and if yields and conversion efficiencies were high, 6 to 20% of U.S. 1990 net fossil-fuel emissions could be avoided. Carbon dioxide released by the burning of cellulosic feedstocks is recycled back into sustainably produced energy crops. Additionally, perennial grasses and trees can provide soil stabilization, may require less use of chemicals, and will provide structural and compositional diversity in the agricultural landscape that could enhance wildlife habitat. Future energy crop practices are likely to incorporate mixtures of trees and grasses, use of leguminous trees and herbaceous crops to reduce the use of nitrogen fertilizers, and adoption of no-till establishment techniques to reduce erosion potential. Clearly, the degree of benefit obtained depends on the previous land use and the selection of appropriate crops and cropping techniques for the type of cropland available.

Economic competitiveness of energy crops is critical to their availability as an energy resource. Cost estimates for delivered woody feedstocks in the Midwest have changed from $4.13/GJ based on early 1980 assumptions[9] to $3.25/GJ in 1992.[10] The woody crop cost reductions are attributed to a 50% increase in yields, lower planting costs, lower site preparation costs, and lower maintenance costs. The cost estimates for 1992 are even lower if 1982 assumptions for land rental values are used. In the early 1980s, analysts assumed that woody crops could be grown on lower-value land not suitable for agricultural crop production. By 1992, research data had demonstrated that high woody crop yields could only be obtained on relatively good cropland. Although switchgrass appears to grow well on lower-quality cropland, cost estimates for switchgrass on Midwest sites are about $3.45/GJ due to the general high value of land, high harvesting costs, and storage losses. For both woody and herbaceous crops, the cost estimates assume a full return to the farmer for land, labor, and investment in equipment, as well as overhead costs for insurance, taxes, etc.

Currently, the price of delivered biomass feedstocks must be about $2.20/GJ or less to produce energy products (liquid fuels or electricity) that are competitive with fossil fuel energy systems. Feedstocks can be produced for that price today only in situations where the land value is compensated by federal subsidies, such as for land in the Conservation Reserve Program. The noncompetitive situation for biomass energy systems will change because of three separate trends. First, research is moving toward producing high-yielding crops on lower-quality cropland. Secondly, research is also leading to the development of high-efficiency conversion systems that can remain competitive at higher feedstock costs. Thirdly, pressure is mounting to ensure that the costs of fossil energy reflect environmental costs such as CO_2 emissions and acid rain damage. In the near future, it is likely that a combination of research advances and policy changes will place biomass energy systems in a very competitive position.

REFERENCES

1. **Scurlock, J. M. O. and Hall, D. O.,** The contribution of biomass to global energy use, *Biomass*, 21, 75–81, 1990.
2. **Overend, and Chum, ,**
3. **Wright, L. L.,** Production Technology Status of Woody and Herbaceous Crops, *Biomass and Bioenergy*, in press.
4. **Wright, L. L., Graham, R. L., Turhollow, A. F., and English, B.,** Growing short-rotation woody crops for energy production, in *Forests and Global Change Vol. 1: Opportunities for Increasing Forest Cover*, Sampson, N. and Hair, D., Eds., American Forests, Washington, D.C., 1992.
5. **Graham, R. L.,** An analysis of the potential land base for energy crops in the conterminous United States, *Biomass and Bioenergy*, in press.
6. U.S. Department of Agricultural, Agricultural Resources: Cropland, Water, and Conservation Situation and Outlook, AR-19, Resources and Technology Division, Economic Research Service, U.S. Department of Agriculture, Washington, D.C., September 1990a.
7. U.S. Department of Agriculture, The Second RCA Appraisal: Soil, water and related resources on nonfederal land in the United States, analysis of condition and trends, Miscellaneous Publication 1482, U.S. Department of Agriculture, Washington D.C., U.S. Government Printing Office: 1990-722-462:20154, 1990b.
8. **English, B. C., Maetzold, J. A., Holding, B. R., and Heady, E. O., Eds.,** Future Agricultural Technology and Resource Conservation, *Proc. RCA Symp.*, December 1982, Iowa State University Press, Ames, IA.
9. **Perlack, R. D. and Ranney, J. W.,** Economics of short-rotation intensive culture for the production of wood energy feedstocks, *Energy*, 12(12), 1217–1226, 1987.
10. **Turhollow, A. T.,** The economics of energy crop production, *Biomass and Bioenergy*, in press.
11. **English, B. C.,** 2000 RCA Baseline with moderate exports: Selected solution tables for the 10 farm production regions and a national summary, CARD Report SR-9, U.S. Department of Agriculture, Washington, D.C., March 1986.
12. **Osborn, C. T., Llacuna, F., and Linsenbigler, M.,** *The Conservation Reserve Program: Enrollment Statistics for Signup Periods 1–11 and Fiscal Years 1990–92*, Statistical Bulletin 843, U.S. Department of Agriculture, Washington, D.C., November 1992.
13. **Robertson, T., English, B. C., and Post, D. J.,** Assessment and Planning Staff Report: Documentation of the CARD/RCA Linear Programming Model Calibration Process, U.S. Department of Agriculture, Washington, D.C., July 1987.
14. **Wright, L. and Shaw, K.,** Bioenergy, in *The Encyclopedia of the Environment*, Boulanger, S., Ed., Houghton Mifflin, in press.

INDEX

A

Acetaldehyde, 76, 78–80, 93–94, 106
Acetone, 106
Acid Deposition and Oxidant Model (ADOM), 106, 107
Acid rain, 223
ADOM, see Acid Deposition and Oxidant Model
Advanced Oxidation Processes (AOP), 187, 188, 190–192
Advanced Wind Turbine Program, 242
Aeration, 229
Aerodynamics, 241
Air conditioning, 14
Air quality, see also Environment
 in Albuquerque, New Mexico, 65–66
 in Colorado, 76, 83
 geothermal energy and, 23–25
AIRSHED model, 77
Albuquerque, New Mexico study, 61–73
 formic acid in, 63, 65–69
 methods used in, 62–65
 peroxyacetylnitrate in, 61–65, 69–72
 peroxypropionylnitrate in, 61–65, 69–72
 results of, 65–72
 sample sites in, 62–65
 sampling protocols in, 62–63
Aldehydes, 76, 78, 79, 90, 92–99, see also specific types
 carbon monoxide and, 92–94
 descriptive statistics of, 93–94
 diurnal behavior and, 92
 vehicle emissions and, 94–99
Amorphous silicon, 145, 151–153
AOP, see Advanced Oxidation Processes
Artificial photosynthesis, 125–137
 electron transfer in, 128, 129
 molecular tetrads in, 129–131
 molecular triads in, 128–129
 reaction centers in, 126
  complex, 129–137
  pentad, 131–137
  simple, 127–128
Artificial ultraviolet light sources, 188–190
Auraria monitoring studies, 90–92, 96–97

B

Bacteriopheophytin, 126
Benzene, 79
Binary-plant electrical generation, 21
Biofuels, 214, see also specific types
Biomass, 214–215
Biomass-ethanol conversion process, 46
Biomass-ethanol fuel cycles, 30, 34, 37, 38–41
Biomass feedstock, 215
Biomass resource potential of energy crops, 261–267
1,3–Butadiene, 79
By-products of ocean thermal energy conversion, 14
CAAA, see Clean Air Act Amendments

C

Cadmium sulfide, 161
Cadmium telluride, 144–146, 153–164
 materials and devices for, 153–158
 recycling and, 159–160
 status of, 164, 165
 technical issues in, 158
Carbon cycle, 27
Carbon dioxide, 23, 31, 39, 52–53, 119, 120
Carbon monoxide
 in Albuquerque, New Mexico study, 61, 63, 65–66
 in Colorado study, 76–90
  data characteristics in, 80–81
  formaldehyde and, 92–94
  time series analyses of, 81–90
 formaldehyde and, 92–94
 in Ontario, Canada study, 107
 oxygenated fuel effects and
  in Albuquerque, New Mexico study, 61, 63, 65–66
  in Colorado study, 76–90
   data characteristics in, 80–81
   formaldehyde and, 92–94
   time series analyses of, 81–90
 TFCA and, 32, 46–47, 48
 time series analyses of, 81–90
Carotenoids, 126, 131

269